CLIMATE ENGINEERING AND THE LAW

Climate change is increasingly recognized as a global threat, and is already contributing to record-breaking hurricanes and heat waves. To prevent the worst impacts, attention is now turning to climate engineering – the intentional large-scale modification of the environment to reduce the impact of climate change. The two principal methods involve removing some carbon dioxide from the atmosphere (which could consume huge amounts of land and money, and take a long period of time), and reducing the amount of solar radiation reaching the Earth's surface, perhaps by spraying aerosols into the upper atmosphere from airplanes (which could be done quickly but is risky and highly controversial). This is the first book to focus on the legal aspects of these technologies: what government approvals would be needed; how liability would be assessed and compensation provided if something goes wrong; and how a governance system could be structured and agreed internationally.

Michael B. Gerrard is a professor and Director of the Sabin Center for Climate Change Law at Columbia Law School.

Tracy Hester is a lecturer at the University of Houston Law Center.

Climate Engineering and the Law

REGULATION AND LIABILITY FOR SOLAR RADIATION MANAGEMENT AND CARBON DIOXIDE REMOVAL

Edited by

MICHAEL B. GERRARD

Columbia Law School

TRACY HESTER

University of Houston Law Center

CAMBRIDGE
UNIVERSITY PRESS

CAMBRIDGE
UNIVERSITY PRESS

University Printing House, Cambridge CB2 8BS, United Kingdom

One Liberty Plaza, 20th Floor, New York, NY 10006, USA

477 Williamstown Road, Port Melbourne, VIC 3207, Australia

314–321, 3rd Floor, Plot 3, Splendor Forum, Jasola District Centre, New Delhi – 110025, India

79 Anson Road, #06-04/06, Singapore 079906

Cambridge University Press is part of the University of Cambridge.

It furthers the University's mission by disseminating knowledge in the pursuit of education, learning, and research at the highest international levels of excellence.

www.cambridge.org
Information on this title: www.cambridge.org/9781107157279
DOI: 10.1017/9781316661864

First published 2018

Printed in the United States of America by Sheridan Books, Inc.

A catalogue record for this publication is available from the British Library.

Library of Congress Cataloging-in-Publication Data
Names: Gerrard, B. Michael, 1951– editor. | Hester, Tracy, editor.
Title: Climate engineering and the law : regulation and liability for solar radiation management and carbon dioxide removal / edited by Michael B. Gerrard, Columbia University School of Law [and] Tracy Hester, University of Houston School of Law.
Description: Cambridge, United Kingdom; New York, NY, USA: Cambridge University Press, 2018. | Includes bibliographical references and index.
Identifiers: LCCN 2017057299 | ISBN 9781107157279 (hardback)
Subjects: LCSH: Climatic changes – Law and legislation. | Environmental management – Law and legislation. | Carbon dioxide mitigation – Law and legislation | Solar radiation – Management – Law and legislation. | Environmental management – Law and legislation – United States.
Classification: LCC K3585.5.C578 2018 | DDC 344.04/633–dc23
LC record available at https://lccn.loc.gov/2017057299

ISBN 978-1-107-15727-9 Hardback

Contents

Contributors

Michael B. Gerrard is Andrew Sabin Professor of Professional Practice and Director of the Sabin Center for Climate Change Law at Columbia Law School, and Chair of the Faculty of Columbia's Earth Institute. From 1979 through 2008 he practiced environmental law full-time in New York City, most recently as partner-in-charge of the New York office of Arnold & Porter (where he is now Senior Counsel). He formerly chaired the American Bar Association's Section of Environment, Energy, and Resources. He received his BA from Columbia and his JD from New York University School of Law.

Tracy Hester is a lecturer at the University of Houston Law Center, where he directs the speaker activities and academic events for its Energy, Environment and Natural Resources Center. Prior to joining academia, he practiced environmental law for 24 years, most recently as the partner-in-charge of Bracewell LLP's Houston environmental group (where he still acts as Counsel). He has also served on the governing Council of the American Bar Association's Section of Environment, Energy, and Resources, and chaired its enforcement, congressional relations, young lawyers, and law professor committees. He received his JD from the Columbia Law School and his BA from the University of Texas at Austin (Plan II).

Michael Burger is Executive Director of the Sabin Center for Climate Change Law and Research Scholar and Lecturer-in-Law at Columbia Law School. Mr. Burger graduated from Brown University and Columbia Law School. He also has an MFA in creative writing from New York University.

Justin Gundlach is a Fellow at the Sabin Center for Climate Change Law and an Associate Research Scholar at Columbia Law School. He holds a BA from Wesleyan University, an MSc from the London School of Economics, and a JD from New York University Law School. He previously worked as a clinical teaching fellow and staff attorney at Georgetown University Law Center's environmental litigation clinic.

Eli Kintisch is a contributing correspondent with *Science* magazine and an independent producer in Washington, DC. His book *Hack the Planet: Science's Best Hope – or Worst Nightmare – for Averting Climate Catastrophe* (2010) was given a starred review by *Publisher's Weekly*.

Albert C. Lin is a professor of law at the University of California, Davis School of Law. He previously was a trial attorney for the Environment and Natural Resources Division of the US Department of Justice and a law clerk to federal appellate judges James Browning and Merrick Garland. Professor Lin received his JD from the University of California, Berkeley School of Law and an MPP from the Kennedy School of Government.

Jesse L. Reynolds is a postdoctoral researcher at the Utrecht Centre for Water, Oceans and Sustainability Law of Utrecht University. He has degrees in international public law from Tilburg University; in environmental policy from the University of California, Berkeley; and in chemistry and environmental sciences from Hampshire College.

Preface and Acknowledgments

This book is born out of our shared experiences and convictions. We have stood on beaches of islands facing creeping extinction from rising oceans, or watched sky-blue glacial ice melt irrevocably during its retreat to a shrinking spawning ground. Each of us picked through wreckage in our home towns left by hurricanes that reached astonishing size and inflicted immense damage, perhaps fueled in part by ocean waters raised and warmed by climate change (Sandy in New York, Ike and Harvey in Houston). And both of us know that if predictions of climate change prove true – even the moderate forecasts – all of us could soon forever lose hosts of imperiled species, irreplaceable landscapes, and fragile ecosystems. More urgently, and tragically, untrammeled and accelerating climate change could mean unanswerable suffering for entire populations of people who live in the most vulnerable areas with the least resources to adapt, and many of these people may not survive these ordeals.

These facts and experiences emphasize that climate change is among the most fundamental and important challenges of our era, and certainly the greatest in the realm of environmental law and policy. Both of us also share the conviction that, based on past performance and likely future behavior, the world's nations and leaders have not taken action that is remotely close enough to answer the peril. Until we collectively find a way to reduce, then stop, and then remove our emissions of greenhouse gases into the atmosphere, the urgency of the challenge and the threat of losses will only grow. This situation demands that we explore all options and possible solutions.

The two of us also share a common history in another respect. We both had long careers practicing environmental law in the private sector (Michael in New York, Tracy in Texas) and then transitioned to academia. This has given us a sense of the power and limitations of law, and the importance of exploring new legal approaches to emerging problems, while keeping an eye on what is achievable in the real, messy world.

We both firmly believe that the best solution to the climate problem is a major reduction in greenhouse gas emissions, especially from the most important sector, energy use. We have both worked very hard in advancing legal approaches toward this end.[1] It is also clear, however, that despite the world's best efforts, this transformation will not happen fast enough to avoid many adverse impacts, and it will be necessary to adapt to the climate change that is certain to occur. We have written extensively on the legal issues associated with adaptation.[2]

But even adaptation will not be enough. Barring unanticipated and swift technological and political transformations, concerted efforts at both mitigation (reducing emissions) and adaptation (preparing for the impacts that cannot be prevented) will still not be enough to prevent disasters in many parts of the world – maybe not next year, but in the decades to come.

This leads us to exploration of something we have also been writing about for several years:[3] climate engineering – the intentional manipulation of the Earth's environment and atmosphere to offset the effects of man-made climate change. These technologies are still in their earliest stages, and unsurprisingly the very concept has proven highly controversial. But as attempts to reduce greenhouse gas emissions continue to fall short and the prospects for sufficient future reductions grow dimmer, the calls for research and limited testing have grown more persistent and persuasive.

No one prefers climate engineering as the first and primary response to climate change, but we will almost certainly need large-scale carbon dioxide removal from the ambient atmosphere in the future, and the prospects that solar radiation management will be pursued appear strong enough to justify mapping out a governance plan for it before anyone attempts to test or deploy that technology. And, most important, the challenges of all climate engineering approaches make it all the more pressing to continue our efforts for effective and real reductions in global greenhouse gas emissions.

This volume helps further that debate and exploration. It is the first book devoted exclusively to the legal challenges posed by climate engineering research, development, and deployment. We are fortunate to have several extraordinary academics and practitioners join us in analyzing the legal issues posed to climate engineering by international law, national domestic environmental laws, and general principles for liability. The front lines of climate engineering – research and testing – receive their own separate consideration. For the most significant legal hurdles, we have tried to provide practical and useful perspectives on approaches to resolve the concerns raised by the legal challenge (where appropriate).

Michael would like to thank his colleagues at Columbia Law School, the Sabin Center for Climate Change Law, and Columbia's Earth Institute for stimulating conversations and legal and scientific insights; and Cullen Howe for his help in putting the manuscript in final form. Most importantly, he wants to thank his wonderful family – Barbara, David, William, Grace, and Anya – for their love and support.

Tracy would like to thank the many helpful colleagues who made his contribution to this book possible, including research grant assistance from the University of Houston Law Center and insights and feedback from his invaluable research assistants (Spencer Taylor and Lauren Stelck). And most important, he wants to thank Linda, his ever-supportive, wonderfully intelligent, and always inspiring wife. He couldn't have done this book without her.

NOTES

1 E.g., *see* Michael B. Gerrard, ed., *The Law of Clean Energy: Efficiency and Renewables* (American Bar Association, Section of Environment, Energy, and Resources 2011); Michael B. Gerrard & John Dernbach, eds., *Legal Pathways to Deep Decarbonization in the United States* (forthcoming 2018).

2 E.g., *see* Michael B. Gerrard & Katrina Fischer Kuh, eds., *The Law of Adaptation to Climate Change: US and International Aspects* (American Bar Association 2012); Michael B. Gerrard & Gregory E. Wannier, eds., *Threatened Island Nations: Legal Implications of Rising Seas and a Changing Climate* (Cambridge University Press 2013).

3 See, e.g., T. Hester, A *Matter of Scale: Regional Climate Engineering and the Shortfalls of Multinational Governance,* 3 CARBON AND CLIMATE LAW REVIEW. 168 (2013); *Climate Change Geoengineering: Legal, Political and Philosophical Perspectives*, ed. W. Burns (Oxford University Press 2012) (chapter by T. Hester, Applying US Environmental Laws to Climate Engineering); T. Hester, *Remaking the World to Save It: Applying U.S. Environmental Law to Climate Engineering Projects*, 38 ECOLOGY LAW QUARTERLY. 851 (2011).

1

Introduction and Overview

Michael B. Gerrard

When the United Nations Framework Convention was negotiated in Rio de Janeiro in 1992, its goal of preventing "dangerous anthropogenic interference with the climate system" seemed well within reach. However, the chief instrument to achieve this goal – the Kyoto Protocol of 1997 – failed miserably. Nearly two decades of negotiations to develop a successor led to the Paris Agreement of 2015, but the voluntary pledges that almost all countries made under it do not take effect until 2020 and the current commitments add up to a world that is very much in the danger zone.[1] The countries promised to come back every five years and do better. Meanwhile, the election of Donald Trump set back US efforts considerably.

Thus a generation after the Rio conference, and after countless meetings and studies, global greenhouse gas (GHG) emissions are much higher, and we have very little time to get them down to safe levels. Keeping global average temperatures well below 2°C (3.6°F) above pre-industrial conditions – the international goal set in Paris – is still theoretically possible but increasingly unlikely. Some serious voices are now saying that, barring a technological revolution, a seismic shift in global politics, or a worldwide economic depression, the only way to avert global climate catastrophe may be climate engineering.

This outcome would not be so bad if we had confidence that climate engineering could be accomplished quickly and without disastrous side effects. However, all the major methods under discussion are either extremely slow or might well cause other terrible problems, and all are unproven at a large scale.

Nonetheless, there is a real chance that in the next few years, someone – perhaps a desperate nation, or a rogue group, or wealthy individual – will attempt large-scale climate engineering. That effort could threaten dire consequences in parts of the world, and we would have all the ingredients for a global conflict.

Hence the need for this book: to assemble in one place what is known about the existing law that is relevant to climate engineering, to identify the open issues, and to describe what a governance structure would need to cover. That way, if the time comes when the world is ready to tackle the problem of actually creating such a structure, or (in a much less pleasant scenario) we are facing immediate choices about deploying these technologies, lawyers and policymakers can be better equipped to make informed decisions.

1.1. WHAT IS CLIMATE ENGINEERING?

The most commonly used definition of "geoengineering" is probably that employed by the Royal Society: "the deliberate large-scale intervention in the Earth's climate system, in order to moderate global warming."[2] We use the term "climate engineering" in the book title simply to avoid confusion with large-scale earth-moving and other activities that do not involve the climate.

Whatever term is used, there are two essential elements: large scale (big enough to affect the climate of the planet or at least regions of it) and specific intent to alter the climate (as opposed to doing so as an unintended side effect, as our industrial civilization has been doing for many years). Virtually all agree that the best way to deal with the climate problem is to reduce GHG emissions; climate engineering arises because efforts to do so have largely failed so far, and there is concern that the planet may be crossing tipping points that will cause irreversible grave damage. Climate engineering might also be attempted in response to a calamitous event, such as a lethally unrelenting heat wave, or the drowning of a city under circumstances that suggested that others might be threatened.

As described in detail in Chapter 2, almost all climate engineering proposals can be divided into two very different categories: solar radiation management and carbon dioxide removal.

Solar radiation management (also called "albedo modification") attempts to decrease the amount of sunlight reaching or staying on the Earth's surface by scattering or reflecting some of it back into space. The most commonly discussed method involves spraying aerosols into the upper atmosphere (using airplanes, balloons, or other means) that would form highly reflective particulates. Another idea is to launch very large numbers of small mirrors or reflectors into a near-Earth orbit. Closer to ground, "cloud whitening" – spraying seawater into the air to make clouds more reflective – could also increase the Earth's albedo. Other variations have been suggested.

Solar radiation management appears to have three major virtues: it is fast, cheap, and reversible. For a cost of perhaps a few billion dollars a year, temperatures might be lowered a degree or two (or more) in less than a year. However, it has a long list of negatives. The principal ones include the following:

- It does not reduce GHG emissions or levels in the atmosphere, and thus does not address climate change's twin threat: ocean acidification, in which carbon dioxide lowers the pH of ocean waters and harms marine life.
- It might disrupt regional weather patterns, and cause droughts in some places and extreme precipitation in others.
- It would alter ecosystems in unpredictable ways by dimming the light, changing the availability of water, and allowing carbon dioxide levels to rise.
- Once begun, it might have to be continued for a long time or indefinitely, because stopping it could cause rapid and disastrous warming unless carbon dioxide levels in the atmosphere had dropped meanwhile.

The best evidence of what solar radiation management would accomplish (for good and for bad) comes from the eruption of Mount Pinatubo in the Philippines in 1991. It injected 20 million tons of sulfur dioxide into the atmosphere, which decreased the amount of sunlight absorbed and lowered global average temperatures about 0.3°C for a period of three years.[3] However, it caused an unprecedented increase in the size of the "ozone hole" and it was followed by floods along the Mississippi River and drought in the Sahel area of Africa that may be attributable to climatic shifts caused by the eruption.[4]

A few ideas have been proposed that could have regional rather than global effects, such as increasing the albedo over the Arctic to reduce melting, or over cities that are suffering from extreme heat. It is not known if these would really work or would have only regional impacts.

The other category of climate engineering is *carbon dioxide removal.* It removes carbon dioxide from the ambient air. (In contrast, carbon capture and sequestration involves removing the gas from an emissions stream, such as that of a power plant, before it reaches the ambient air; since it operates locally, we do not consider it climate engineering.) A wide variety of technologies have been proposed to accomplish this: for example, machines that directly remove the carbon dioxide from the air; fertilizing the oceans with iron filings or other substances that spur algal growth, which absorbs carbon dioxide before sinking to the bottom; creating charcoal that holds the carbon; growing and burning large quantities of biomass for energy, and capturing and sequestering the carbon dioxide; and exposing certain carbon-absorbing rocks to the air in a process called "enhanced weatherization."

A massive program of tree planting might also fall under this category, but it is inherently temporary – trees release their carbon into the air when they die, burn, or are cut down – and it is the subject of separate bodies of law that are beyond the scope of this book.

Carbon dioxide removal is more benign and less risky than solar radiation management. It mitigates the problem of ocean acidification, and it has much less potential for unanticipated side effects. However, at least with today's technology, it appears to be slow and expensive, and may require a great deal of land, and possibly water and fertilizer. It seems that many years of carbon dioxide removal would be required to make much of an impact on global temperatures. Moreover, it is not clear where all the captured carbon dioxide would be held (perhaps in depleted oil and gas reservoirs, coal beds, saline aquifers, or certain kinds of rock formations), whether anyone would want to live on top of these massive gas reservoirs, and how long the gas would stay there. Research is underway to convert the gas to useful or at least easier-to-manage substances. Various technologies are being tested, and it is always possible that the years to come will see a technological breakthrough that enables rapid and inexpensive removal of carbon dioxide with minimal useless residue; however, we cannot count on this happening.

1.2. HISTORY OF CLIMATE ENGINEERING DISCUSSIONS

Humans have imagined altering the weather, and then the climate, for many years. In the 1840s the first meteorologist employed by the US government, James Espy, proposed lighting large fires to make rain through convective updrafts.[5] In 1908, three years after winning the Nobel Prize in chemistry, Svante Arrhenius, the first scientist to calculate how carbon dioxide affected surface temperatures, suggested that burning fossil fuels could help prevent another ice age.[6] In the ensuing decades, there were many attempts to alter local or regional weather patterns, mostly to alleviate droughts. In 1966 the US National Science Foundation produced a report, *Weather and Climate Modification*, followed in 1973 by an update from the National Science Foundation, *Weather and Climate Modification: Problems and Progress*.

An Italian physicist, Cesare Marchetti, coined the term "geoengineering" in the early 1970s to describe the idea of "disposing" of atmospheric carbon dioxide in the deep oceans.[7] Around that time a Russian scientist, Mikhail Budyko, became the first to propose that the Earth's climate could be cooled with the intentional release of aerosols into the upper atmosphere.[8]

Concern about global climate change escalated in 1988 when the United Nations Environment Programme and the World Meteorological

Organization established the Intergovernmental Panel on Climate Change (IPCC). That same year a NASA scientist, James Hansen, famously testified before Congress about the danger that uncontrolled GHG emissions could cause dangerous global warming.

The topic of geoengineering as a solution to climate change remained somewhat taboo, however, until 2006 when Paul Crutzen, who a decade earlier had won the Nobel Prize in chemistry for his work on chemical threats to the ozone layer, published an article entitled "Albedo Enhancement for Stratospheric Sulphur Injections: A Contribution to Resolve a Policy Dilemma?"[9] Important studies followed from the Royal Society in 2009 and 2011,[10] the US Government Accountability Office in 2010,[11] the National Academies of Sciences, Engineering, and Medicine in 2010,[12] and the National Research Council in 2015.[13] A group of more than 100 leading researchers and thinkers met at the Asilomar Conference Center in California in 2010 and prepared a set of recommended principles for research into climate engineering techniques.[14] Several other conferences have followed.

Beginning in 1996 each of the major assessment reports of the IPCC has discussed (though not endorsed) climate engineering and emphasized the risks and uncertainties involved with solar radiation management. However, the IPCC's Fifth Assessment Report (2014) included several scenarios that relied on one form of geoengineering (though not using that term) – bioenergy production with carbon dioxide capture and storage – while stating that the availability and scale of such technologies were uncertain.[15]

The US House of Representatives' science committee held three hearings on geoengineering in 2009 and 2010, and another in 2017. As this is written at the end of 2017, legislation is under discussion but has not yet advanced.

1.3. CLIMATE ENGINEERING IN THE CONTEXT OF CLIMATE POLICY

Climate policy can be seen as having four possible components: mitigation, adaptation, carbon dioxide removal, and solar radiation management. This section describes the four and how they fit together.

Mitigation – As noted above, the principal objective of global climate policy is avoiding "dangerous anthropogenic interference with the climate system." The universally preferred method to do this is reducing GHG emissions (termed "mitigation"). However, though there has been a great deal of activity aimed at reducing emissions, on a global scale it is difficult to point to much measurable success in achieving the desired result. In 1992, the year of the Rio conference, global carbon dioxide emissions were 22.7 gigatons; in 2015 they were 36.2 gigatons – an increase of 59 percent.[16] Global concentrations of greenhouse gases increased from 356 parts per million in 1992 to around 410 parts per million of carbon dioxide in 2017.[17]

There are several reasons why it is extraordinarily difficult to bring down GHG levels in the atmosphere. First, GHGs spread globally very quickly and some (in particular CO_2) persist for hundreds of years, and thus the present ambient GHG levels result from the cumulative emissions of countless sources over the past two centuries. There is a lag between the emissions occurring and the response of the climate system. As Juan B. Moreno-Cruz and David W. Keith have written, "[t]he inertia of the carbon-climate system makes it impossible to quickly reduce climate risk by reducing emissions, as it is expected that 40% of the peak concentration of CO_2 will remain in the atmosphere for 1000 years after the peak is reached."[18] These factors also reduce the incentive for individual actors to lower their emissions, because they would incur the costs immediately and others would reap the benefits in mostly remote times and places.

Second, global GHG emissions (and many other forms of pollution) can be described as the product of three factors: technology, population size, and affluence.[19] Climate policy aims only to affect technology (such as energy source). Many things influence population size, and government policy plays only a limited direct role. Difficult issues of religion, morality, and culture arise whenever government does attempt to influence population size. Affluence is almost universally desired, and it is difficult to imagine a government policy that overtly attempted to reduce it (though of course some policies in some places attempt to redistribute it). Many developing countries have regarded climate policies as efforts to restrain their growth into the affluence enjoyed by the developed world. Some encouraging figures emerged in 2015, suggesting that GHG emissions and economic growth were beginning to lose their tight relationship,[20] but it is too early to tell whether that development will be a long-term trend.

Third, the needed change in technology would entail a massive transformation of the global energy system away from fossil fuels and toward cleaner sources – renewables (such as wind, solar, hydropower, and maybe bioenergy), and possibly nuclear, together with major improvements in efficiency. Though expanding these technologies would yield major benefits in the form of fewer air pollution deaths and reduced environmental degradation from fuel extraction and processing,[21] these advantages for the most part do not appear in national and corporate finances, and energy transformation would cause a great deal of economic dislocation. This dynamic naturally leads to powerful opposition by the interests that would be harmed. Such opposition has played a major role in the paralysis that the United States Congress has experienced in the climate policy area since 1990, and in the policies of the Trump administration.

Fourth, while energy use is responsible for about three-quarters of global GHG emissions, deforestation and agricultural emissions from livestock, soil, and nutrient management cause about one-quarter.[22] The United Nations' REDD program (Reducing Emissions from Deforestation and Forest Degradation) aims to reduce deforestation. As the countries with the largest amounts of deforestation, such as Brazil and Indonesia, have learned, it is very challenging for authorities to control activities in remote areas. Moreover, remaining forest land will face increasing pressures from the rising food demand of growing populations; the sprawl of cities as they expand to accommodate migration from rural areas as well as internal growth; the desire of affluent societies to eat meat, which consumes much more land per unit of nutrition than non-meat products; the use of biofuels as a low-carbon substitute for fossil fuels; and the loss of arable land due to sea level rise and drought.

Adaptation – All the foregoing means that even if the world made its best efforts to reduce GHGs (which it does not), atmospheric GHG levels will continue to rise for at least decades, and climate change will get worse. Thus, in addition to the imperative of mitigation, we also have the necessity of adaptation – efforts to moderate, cope with, and prepare for the current and anticipated impacts of climate change on human and natural systems.[23] For years some policy circles frowned on any discussion of adaptation because it could be seen as an admission of defeat regarding mitigation efforts; but it has become apparent that, under even the most optimistic scenarios, a great deal of adaptation will be needed. In contrast to mitigation, efforts at adaptation can yield short-term and local benefits. However, the places in the greatest need of adaptation to sea level rise, drought, and glacial melt (such as large parts of Africa, Asia, and South America, as well as the small island states) also tend to be the places with the fewest resources to adapt, and the level of international assistance is far below what is needed.[24]

There are severe limitations on what adaptation can accomplish. Houses can go up on stilts, water can be redirected, and crop planting can be adjusted. But while sea walls can hold back ocean waters over very limited areas, protecting entire islands (except for the tiniest) or coastlines of longer than a few miles seems to be prohibitively expensive, and this approach has adverse environmental impacts of its own. Tens or hundreds of millions of people live in the Ganges Brahmaputra Delta, the Mekong Delta, the Nile Delta, the Mississippi Delta, the low-lying islands of the Pacific and Indian Oceans and the Caribbean Sea, and other places that cannot be protected by any sea walls that we can today imagine building. Protracted drought, loss of glacial meltwater, and episodes of unbearable heat will likely make substantial areas unsuitable for much agriculture. Fish stocks may become severely depleted as a result of ocean warming and acidification, as well as other forms of pollution and overfishing.

It is all but certain that much suffering will result from climate change despite efforts to adapt. This dark future in turn has led to demands by the most vulnerable nations for an international mechanism to compensate them for "loss and damage" – what they will endure after adaptation has been insufficient. The topic remains on the United Nations agenda after the Paris conference, but no one can have confidence that any meaningful compensation will be provided, as the wealthy countries cannot be forced to pay.

Carbon dioxide removal (CDR) – This brings us to the third component of climate policy – CDR. Though this fact is perhaps not yet widely recognized, current global climate policy already assumes a great deal of CDR, for without it, achievement of the goal of keeping the increase in global average temperatures well below 2°C (3.6°F) seems to be impossible. (The Paris Agreement also declared an aspiration of keeping the increase to 1.5°C (2.7°F), which would require an even greater amount of carbon dioxide mitigation and removal.) This reality is illustrated by the representative concentration pathways (RCPs) used by the IPCC in making projections under various future scenarios. (The IPCC has no RCP under which the world is likely to stay at 1.5°C.) Table 1.1 shows the four RCPs used by the IPCC, in somewhat simplified form.[25]

In other words, the only scenario under which we keep global surface temperature increases under 2°C relies on extensive use of one form of CDR (bioenergy plus carbon capture and storage, or BECCS). Even keeping to 3°C – a scenario that entails grave consequences – requires large-scale afforestation (i.e., establishing forests or stands of trees where there have previously been none).

Other scenarios have been assessed. The database for the IPCC's Fifth Assessment Report includes 400 scenarios that have a 50 percent or better chance of no more than 2°C warming (with three scenarios removed due to incomplete data). Of these, 344 assume the successful and large-scale uptake of "negative emissions technologies," such as BECCS, direct air capture, enhanced weatherization, or others, and the other 56 assume that global emissions peak around 2010, which did not happen.[26]

TABLE 1.1 *IPCC Representative Concentration Pathways*

Name	CO_2e (ppm)	Likely temperature anomaly over 1850 to 1900	Description	Assumed carbon dioxide removal
RCP 8.5	>1000	> 4°C	Business-as-usual growth	None
RCP 6.0	850	4°C	Stabilization without overshoot	None
RCP 4.5	650	3°C	Stabilization without overshoot	Large-scale afforestation
RCP 2.6	450	2°C	Peak before 2100 and then decline	Bioenergy with carbon capture and storage (BECCS), leading to negative emissions

Source: Intergovernmental Panel on Climate Change, Climate Change 2014: Synthesis Report Summary for Policymakers (2014), Table SPM.1, www.ipcc.ch/pdf/assessment-report/ar5/syr/AR5_SYR_FINAL_SPM.pdf.

The need for negative emissions is implicitly embedded in the Paris Agreement, which declares that:

> [i]n order to achieve the long-term temperature goal [of less than 2°C with an attempt to keep to 1.5°C], Parties aim to reach global peaking of greenhouse gas emissions as soon as possible … and to undertake rapid reductions thereafter in accordance with best available science, so as to achieve a balance between anthropogenic emissions by sources and removals by sinks of greenhouse gases in the second half of this century.[27]

The reality is that, if the temperature goal is to be met, any substantial continued use of fossil fuels in the second half of this century will require a great deal of CDR. Fossil fuel use emits about 32 gigatons of carbon dioxide per year. Other sources, such as methane leakage, cement manufacture, and other industrial processes, add another 5–7 gigatons carbon dioxide equivalent. Deforestation and agriculture, forestry, and other land use changes (but subtracting emissions sequestered by forest growth) add yet another 10–12 gigatons a year. This all adds up to about 49 gigatons. However, global carbon sinks remove only about 18 gigatons per year (8.8 to the oceans, 9.2 to land, not including land use changes).[28]

Thus the sinks take up about the equivalent of the non-fossil sources. To achieve a "balance" between emissions and sinks, we need to just about end the release of GHGs from fossil fuels unless there is a radical increase in sinks or reduction of non-fossil fuel emissions. This shift would have to be achieved well before the end of the century. One study suggests that carbon dioxide from electricity generation would have to be brought close to zero by 2050, and by then around 25 percent of energy required for transportation would also need to come from electricity (up from less than 1 percent now).[29]

There seem to be only three ways to continue to use fossil fuels for electricity in the second half of the twenty-first century (and for transport by the end of the century) and still meet the temperature goal, and it appears we would need to implement all three on a large scale:

- Capture the carbon dioxide from power plants and other large industrial sources before it escapes into the air, and utilize or sequester it
- Initiate a massive program of CDR from the ambient atmosphere
- Create new sinks, such as through the immediate halt to deforestation and a worldwide program of tree planting.

All three of these methods raise a question of how long the carbon will be stored; we do not know how long carbon will stay in reservoirs. We do know that trees do not live forever, and when they burn or die, they release their carbon. Moreover, most methods of CDR, when deployed at a scale necessary to stay within 2°C, would require extremely large amounts of one or more scarce resources – land, energy, water, nutrients, and investment capital. In view of these resource demands, serious questions have been raised about the feasibility of removing sufficient carbon dioxide from the atmosphere to avoid dire consequences.[30] The technology that is receiving perhaps the greatest attention, BECCS, which involves producing large amounts of biomass from fast-growing trees, switchgrass, agricultural waste, or other sources, burning it in power plants, and capturing and storing the carbon, is still in the development stage, and there is not yet the assurance that this approach (or any of the other negative emissions technologies) can be performed economically at scale.[31]

Solar radiation management (SRM) – It is still possible, at least theoretically, that in the next few decades the world will engage in sufficient programs of mitigation, adaptation, and CDR to prevent the widespread damage that can be caused by severe climate change. Some technological breakthrough may be our best hope of this. After all, the technology of hydraulic fracturing, which had been under quiet development for decades, burst forth in the late

2000s and revolutionized (for better or worse) oil and natural gas production in just a few years. Not long before, both cellular telephone technology and the Internet, after years in obscure laboratories, exploded globally and changed the world. But it is difficult in 2017 to be confident that anything of this speed and scale will happen with CDR or with clean energy production, especially in the absence of a carbon tax or some other strong economic incentive, and it is irresponsible to count on this happening. (The prices of solar and wind technology are plummeting, but not enough to induce most of the world's energy users with give up their current fossil fuel-fired equipment and vehicles.) This reality brings us to the fourth prong of climate policy, the one that almost everyone fervently hopes to avoid: SRM.

SRM has the potential for all the horrible side effects noted above, and maybe more. However, in a truly grim situation, it is imaginable that someone will attempt it. In this way it resembles chemotherapy – it will make the cancer patient very sick, will cause her hair to fall out, and might kill her, but if it's the only known option that may keep her alive, she may try it. SRM should be thought of as chemotherapy for the planet: a desperate last chance of avoiding the worst. Moreover, the possibility that it could be done unilaterally by one country at a manageable cost in a short period of time makes its use all the more foreseeable.

The possibility of SRM poses the risk of a moral hazard: that the belief it could work could induce some people to be less diligent in pursuing the far superior (but more expensive) pathway of mitigation. Already we have heard talk of SRM as a tolerable fallback in case the temperature goals are not met. At a minimum, it could be seen as a way of buying time until new technologies for mitigation and CDR are developed and deployed on a mass scale. This additional time might be needed if it appeared that some irreversible environmental tipping point might soon be crossed, or if forecasts showed an unacceptable near-term probability of a catastrophic event affecting a country with the capability of launching a SRM program.

The uncertainties involved in all these projections are extremely high, and many judgment calls involving politics, economics, and morality will be required in deciding whether and how to deploy SRM. Who has the power to make and enforce these judgment calls, through what process, and based on what considerations? That is where the law comes in.

1.4. CURRENT STATE OF INTERNATIONAL LAW

As discussed in detail in Chapter 3, no legally binding international treaties or other instruments are presently in force that explicitly apply to climate engineering. However, several binding instruments are relevant, and a number of nonbinding international documents do speak directly about climate engineering.

1.4.1. *Binding Multilateral Agreements*

Binding multilateral environmental agreements are the most important category of international instruments. Several of them focus on human impacts on the atmosphere; others concern the marine environment; and still others relate to procedural obligations. However, it must be noted that being "binding" as a matter of international law does not necessarily mean that there is an enforcement mechanism under which a court or other tribunal can impose remedies such as injunctions or monetary damages or penalties. Implementation of most treaties depends on the continued good faith of the countries that have adopted them, and on their domestic implementing laws.

Atmospheric agreements – The United Nations Framework Convention on Climate Change of 1992 (UNFCCC) establishes the global governance system for climate change. As shown above, its objective is to avoid "dangerous anthropogenic interference with the climate system." The key agreements under the UNFCCC aimed at achieving that objective are the Kyoto Protocol of 1997 and the Paris Agreement of 2015. Under the Kyoto Protocol, the world's developed countries (except for the United States, which never joined) committed to emissions reductions against a 1990 baseline, as specified for each country. Kyoto did not impose emission reduction obligations on developing countries, including those that have been rapidly developing, such as China and India. After Canada dropped out in 2011, Kyoto applied to only about 19 percent of the world's GHG emissions,[32] and the obligation to reduce even those was met in part by exporting the manufacturing to developing countries. Kyoto's principal successor, the Paris Agreement, is more comprehensive but less binding. Almost every country (representing 95 percent of the world's GHG emissions)[33] made a nonbinding pledge to control its emissions, though in June 2017 President Trump announced that the USA (which emits about 12 percent of global emissions)[34] would withdraw from the Paris Agreement and not agree to meet its pledge.

Of most specific relevance to climate engineering, the UNFCCC requires all parties to "promote and cooperate in the conservation and enhancement of sinks and reservoirs." As noted previously, the Paris Agreement implicitly relies on CDR methods that would fit within the meaning of "sinks" or otherwise utilize "reservoirs" to store the captured carbon.

SRM could (at least for a time) reduce some of the dangers of climate change, but due to the possibility of negative side effects it could also itself be seen as "dangerous anthropogenic interference with the climate system."

While several international conventions address atmospheric obligations of signatory nations, none of them sets out a clear framework to govern SRM. For example, the UNFCCC speaks generally of the development and sharing of technology that could apply to climate engineering research. The Kyoto Protocol continues the call for sinks and reservoirs, and for technology development and sharing, but is not otherwise particularly relevant. As noted above, the Paris Agreement implicitly calls for CDR. The Vienna Convention for the Protection of the Ozone Layer of 1985 and its Montreal Protocol of 1987 seek to protect the stratospheric ozone layer. This goal would be assisted by CDR. However, some methods of SRM could degrade the ozone layer; further scientific research is needed on this point.

The Convention on Long-Range Transboundary Air Pollution of 1979 (CLRTAP) is a regional agreement with 51 parties in North America, Europe, and Central Asia. Like the UNFCCC and the Vienna Convention, it is a framework convention that establishes general goals and procedures but that leaves the substantive decisions to protocols. None of CLRTAP's eight protocols explicitly concerns climate change. Three of the protocols could be implicated by a SRM program that injected sulfates into the atmosphere; this outcome could require certain notices and consultations but not impose substantive restrictions, as discussed in Chapter 3.

Marine agreements – The 1982 United Nations Convention on the Law of the Sea (UNCLOS), among many other things, seeks to prevent, reduce, and control pollution of the marine environment. Its coverage includes airborne pollution that enters oceanic waters. Existing GHG emissions that contribute to ocean acidification arguably violate UNCLOS, though there is no clear enforcement mechanism. The objectives of UNCLOS would be advanced by CDR that reduces acidification. However, methods that involve dumping materials into the oceans (such as iron filings), putting materials under them (such as carbon dioxide injected below the seabed), or otherwise trying to manipulate the oceans may raise UNCLOS issues.

The London Convention on the Prevention of Marine Pollution by Dumping of Wastes and Other Matter of 1972, and the 1996 London Protocol to it, aim to reduce marine pollution from dumping. The governing body that administers both has adopted a nonbinding resolution that sets forth procedures for conducting ocean fertilization experiments. A proposed amendment to the London Protocol, not yet in force, has binding rules regarding ocean fertilization and various other activities in or affecting the oceans; it would be the first binding instrument of international law that is specifically concerned with climate engineering.

Procedural obligations – The 1991 Convention on Environmental Impact Assessment in a Transboundary Context (the Espoo Convention) obligates its 45 parties to "take all appropriate and effective measures to prevent, reduce and control significant adverse transboundary environmental impact from proposed activities." It also contains various requirements for assessment, disclosure, and consultation. In addition, the 1998 Convention on Access to Information, Public Participation in Decision-Making and Access to Justice in Environmental Matters (the Aarhus Convention) requires its 47 parties to pursue the three objectives given in the Convention's title. Unlike the Espoo Convention, the Aarhus Convention is not limited to actions with transboundary impacts.

Other multilateral environmental agreements – The 1992 Convention on Biological Diversity now includes almost all of the world's countries except the United States. To the extent that climate engineering activities either protect or impair biological diversity, this Convention could be relevant. It adopts as guiding principles that states have a sovereign right to exploit their own natural resources, and that they have a concomitant responsibility to prevent transboundary harm. Like several other agreements, it calls for impact assessment, notice, and consultation. Its Conference of the Parties has adopted several decisions that explicitly concern climate engineering. They call for a precautionary approach that would avoid large-scale climate engineering unless there is an adequate scientific basis for concluding it would not impair biodiversity. However, the language is nonbinding and somewhat vague.

The Antarctica Treaty System broadly protects freedom of scientific investigation in the region, presumably including climate engineering research, and encourages international cooperation. Parties are to limit their adverse environmental impacts, including avoiding adverse effects on climate or weather patterns.

The 1977 Convention on the Prohibition of Military or Any Other Hostile Use of Environmental Modification Techniques (ENMOD) has a definition of "environmental modification" that encompasses almost all climate engineering proposals. However, it applies only to "military or any other hostile use of environmental modification techniques having widespread, long-lasting or severe effects as a means of destruction, damage or injury to any other State Party." Moreover, the agreement "shall not hinder the use of environmental modification techniques for peaceful purposes," and it recognizes that such techniques may "contribute to the preservation and improvement of the environment for the benefit of present and future generations." Thus, if done for benign purposes and with suitable precautions, climate engineering would be entirely consistent with ENMOD.

1.4.2. *Nonbinding Multilateral Environmental Agreements*

The United Nations has convened several major international summits that enjoyed widespread participation and resulted in statements that are influential though not binding.

The 1972 UN Conference on the Human Environment in Stockholm generally supports interventions into the natural world to improve humanity's well-being. (There is less attention here than in subsequent agreements to preserving nature for its own sake.)

The Conference on Environment and Development in Rio de Janeiro in 1992 gives greater emphasis to the protection of the natural world and to the importance of economic development of poorer countries. It calls on states to avoid transferring sources of environmental harm to other states.

Further summits on sustainable development were held in Johannesburg in 2002 and again in Rio de Janeiro in 2012. For the most part they reaffirmed the principles adopted in Stockholm and Rio 1992. However, the report of the 2012 Summit expresses concern about the potential negative environmental impacts of ocean fertilization.

Two other nonbinding international statements, not resulting from UN summits, are of note.

In 1980 the UN Environment Programme (UNEP), in cooperation with the World Meteorological Organization, approved its Provisions for Cooperation Between States in Weather Modification. These Provisions define "weather modification" broadly enough to encompass most forms of climate engineering, and they encourage such activities "for the benefit of mankind and the environment," but only after conducting an environmental assessment. This statement was approved by the UNEP Governing Council.

In 2007 the United Nations General Assembly approved a resolution on oceans that urged states to exercise caution when considering proposals for ocean fertilization, but that resolution also encouraged research in that area.

1.4.3. *Customary International Law*

A number of binding rules do not come from treaties or other such instruments, but rather derive from the repeated behavior of states undertaken out of a sense of legal obligation (*opinio juris*).

Two relevant elements of customary international law appear in most multilateral environmental agreements, and are linked together: (1) a state's sovereign right to exploit its natural resources, and (2) its obligation to avoid transboundary harm. This pairing implies that states may conduct climate engineering activities within their own territory, but only if there is not a significant risk of causing harm outside that state's territory. This limitation does not mean that zero risk is required; the probability and magnitude of the harm are relevant. Large-scale SRM activities certainly pose a significant risk of causing transboundary harm. Most forms of CDR (except perhaps ocean fertilization) do not. Where there is such a significant risk, states must assess, disclose, and consult; but the precise scope of these obligations, and what happens if there are objections, remains unclear.

If a state's actions in violation of international law do cause transboundary harm, the customary international law doctrine of state responsibility requires the state to pay compensation. However, as explored in Chapter 5, there are no mechanisms in place to determine whether and how much compensation would be required in the case of climate engineering, or to whom it would be paid. There would also be very difficult issues of causation and attribution. The necessity defense could also come into play if a state acted to prevent grave and imminent peril.

1.4.4. *Principles of International Environmental Law*

Several principles of international environmental law are not themselves binding, but are reflected in many treaties and other international instruments. They can be looked to for guidance, but their precise meaning and legal significance are unclear. Those principles that seem most relevant to climate engineering are:

- The environment as a common concern of humankind: All states have individual and collective legitimate interests in and responsibilities for the maintenance of the climate, biological diversity, stratospheric ozone, Antarctica, and the marine environment.
- Common but differentiated responsibilities: All states have some responsibilities toward such common objectives as preventing climate change, restoring stratospheric ozone, and conserving biological diversity, but the wealthier countries must shoulder their fair (and typically larger) share of the cost.
- Precaution: States should take precautionary measures to anticipate, prevent, or minimize adverse impacts on the global commons; where

there are threats of serious or irreversible damage, lack of full scientific certainty should not be used as a reason for inaction.
- Polluter pays: The polluter, not the victim, should pay for environmental damage.
- Cooperation: States should cooperate with each other in good faith in addressing environmental problems.
- Equity: Current populations affected by environmental problems should have a voice in the relevant decisions, and the interests of future generations should also be considered.

All of the above principles appear in one form or another in UNFCCC. They have bearing not only on decisions that determine the levels of GHG emissions, but also on the duties of states with respect to research about and implementation of SRM and CDR, the way decisions on these matters are made, who should pay to carry them out, and who should bear responsibility if something goes wrong. However, the principles are all so broad that there is ample ground for disagreement about how any of them applies in climate engineering contexts.

1.4.5. *Other Domains of International Law*

Outside of the realm of environmental law, several other areas of law are pertinent to certain kinds of climate engineering.

Civil aviation – Some forms of SRM involve the release of various substances from aircraft. The principal international agreement on aircraft flights is the 2006 Convention on International Civil Aviation, also known as the Chicago Convention. The Convention is implemented by the International Civil Aviation Organization.

Space law – The foundational treaty is the 1967 Treaty on Principles Governing the Activities of States in the Exploration and Use of Outer Space, including the Moon and Other Celestial Bodies, also known as the Outer Space Treaty. All states with space programs are parties. It establishes a "freedom of scientific investigation in outer space," and calls for space activities to be conducted "for the benefit and in the interests of all countries … in accordance with international law … in the interest of maintaining international peace and security and promoting international cooperation and understanding" and "with due regard to the corresponding interests of all other States Parties." Under the 1972 Convention on International Liability for Damage Caused by Space Objects, the launching party has absolute liability for damage from their space objects.

Human rights – Numerous human rights theories arguably apply to climate change.[35] Some of them can be read as meaning that allowing excessive levels of GHG emissions would violate human rights. Some of them could also be construed to impose an obligation to perform research on and implement CDR if that action would reduce the adverse impacts of GHGs. The situation with SRM would be more complex, because it could help some regions of the world and perhaps harm others. Except for the European Court of Human Rights, there are few tribunals with the power to take enforcement actions against violators of human rights obligations or to impose sanctions.[36]

Intellectual property – The 1994 Agreement on Trade-Related Aspects of Intellectual Property Rights (TRIPS) of the World Trade Organization harmonizes the national laws on intellectual property. TRIPS provides some limited circumstances under which states may require a patent holder to license an invention on particular terms if necessary for the protection of humans and the environment. If climate engineering technologies prove important to protect human health and welfare or critical aspects of the environment, these compulsory licensing provisions of TRIPS might arguably apply.

International disputes – The United Nations General Assembly, the United Nations Security Council, and the International Court of Justice all have the ability to resolve certain kinds of disputes, though many sorts of decisions from these bodies are nonbinding.

1.5. CURRENT STATE OF US LAW

Were the United States to engage in climate engineering activities, several federal and state laws would be relevant. Some of them apply to all different kinds of activities; some are more narrowly applicable. Chapter 4 discusses all of these in detail.

1.5.1. *Laws of General Applicability*

The National Environmental Policy Act of 1970 applies to a wide variety of actions that are directly undertaken, financed, or approved by the federal government. It requires the preparation of an environmental impact statement for "major federal actions significantly affecting the quality of the human environment." It is a procedural and disclosure law; it does not impose substantive requirements. Fifteen states, the District of Columbia, and Puerto Rico have enacted analogous laws for state and local actions.

The Weather Modification Reporting Act of 1972 applies to federal and nonfederal actions, whether public or private. Any person engaging in any weather modification activity must submit a report to the National Oceanic and Atmospheric Administration (NOAA). It is primarily a disclosure law, but NOAA may make recommendations if it concludes that the proposed project "may significantly depart from the practices or procedures generally employed in similar circumstances to avoid danger to persons, property, or the environment."

Many states have their own laws on weather modification. These laws were mostly adopted to address cloud-seeding operations and to license cloud seeders.

The federal Endangered Species Act protects federally listed species. Actions that threaten to harm individual members of a protected species or significantly modify their habitat may require an "incidental take" permit. If there is federal involvement, an interagency consultation process is triggered with strong substantive requirements. Many states have their own laws on the protection of wildlife. There are also federal laws designed to protect migratory birds and marine mammals.

Several laws grant extraordinary powers to federal, state, and local governments to act in emergency situations. These laws could allow an agency to halt a climate engineering project, and could conceivably allow an agency to start one. The Robert T. Stafford Disaster Relief and Emergency Assistance Act is the principal statute authorizing the Federal Emergency Management Agency to act when the president has declared an emergency or a disaster at the request of the governor of an affected state. Many environmental laws also provide exemptions (usually temporary and limited) in case of emergency. Moreover, the Commander-in-Chief Clause and the Executive Powers Clause of the Constitution, and arguably the constitutionally mandated oath to "preserve, protect and defend the Constitution," give the president inherent powers to act against threats to the nation, though the scope of these powers is much contested.

1.5.2. *Laws Applicable to Specific Techniques*

The preceding section discussed US laws that could apply to any kind of climate engineering. Other laws would apply only in more limited circumstances.

Projects that involve the dispersal of substances into the atmosphere – as would most types of SRM – could invoke the Clean Air Act. This is a long and complex statute that creates several different permitting programs. Some involve stationary sources (such as power plants); some involve mobile sources (such as automobiles and airplanes); one concerns acid deposition, and another regulates ozone-depleting substances.

The Clean Water Act regulates the discharge of water pollutants. Depending on where it occurred and whether ocean iron fertilizer is deemed to be a "pollutant," the release of fertilizer into the ocean could require a Clean Water Act permit, but more likely such actions would be governed (if at all) by the federal Ocean Dumping Act.

Several federal laws govern the manufacture, use, disposal, and clean-up of various chemical substances. These laws could be invoked if any of these substances are involved in a climate engineering project.

The Federal Aviation Administration would become involved in any delivery of aerosols into the stratosphere by use of airplanes, rockets, artillery shells, or balloons. Unmanned rockets over a certain size would require permits under the Commercial Space Launch Act.

The genetic modification of crops to increase their albedo has been proposed. No federal statute specifically aims to govern genetically modified organisms (GMOs); rather a patchwork of laws apply in particular circumstances. Some localities have adopted restrictions on the use of GMOs, and some states have enacted legislation to block such local restrictions. A few states have adopted their own restrictions on GMOs.

Requirements to use white roofs or other albedo modification techniques would involve local building and land use rules.

Air capture devices for CDR would need to comply with any local zoning and building requirements.

1.6. LIABILITY AND COMPENSATION

SRM appears to have considerable potential for adverse side effects. Though it is designed to lower global average temperatures and thus prevent the worst effects of climate change, computer modeling plus the experience after the Mount Pinatubo eruption suggest that it may cause severe adverse effects in some parts of the world, including contributing to drought and to extreme precipitation. It may also have other adverse effects that cannot now be anticipated. Some forms of CDR (especially ocean iron fertilization) have their own potential for negative impacts. The planned operation of still other methods may consume large amounts of land, water, and energy, creating different kinds of losses.

Those suffering losses will want compensation. As discussed in Chapter 5, establishing a compensation system will require addressing the following questions:

1. Which tribunals will hear these claims? No existing court has clear jurisdiction over such claims. Will new tribunals be established, or will existing ones be utilized? Will they be international, regional, or national courts? What happens if different tribunals render inconsistent findings?
2. What degree of proof of causation will a plaintiff need to establish, and through what method? Adverse weather events have always occurred, and distinguishing the effects of climate engineering from what would occur under natural viability will be a major challenge.
3. What legal theories will be applied? International laws and doctrines such as those against transboundary harm or guaranteeing human rights? Conventional tort theories at the state or provincial level such as nuisance, negligence, and trespass? Or another set, possibly custom designed for this purpose?
4. Will liability be imposed only on those who violated rules that will be established for the conduct of climate engineering projects, or will there be liability for all consequences, regardless of whether misconduct occurred?
5. Who will pay? The government that undertook the climate engineering, the particular contractors, the beneficiaries of the climate engineering, the whole world community if the actions were authorized by an international body?
6. Who will receive compensation? Will states take in the money and distribute it as they see fit? Will individuals, corporations, and entities need to file their own claims? What sorts of losses will be compensable – personal injury, property damage, economic losses? If localized or regional starvation results, or population displacement occurs, how will losses be measured and compensation provided?
7. Will some sort of trust fund or insurance mechanism be established in advance, so that the funds to pay any judgment are readily available?

The international community has established special liability systems for particular kinds of risks, such as those involving nuclear accidents, oil spills, rockets and satellites, and biological agents. Some of these systems contain features that could be utilized in creating a liability and compensation system for climate engineering.

1.7. RESEARCH

No artificial techniques of either CDR or SRM have ever been tested at larger than a laboratory scale, with the exception of a few poorly controlled tests (some of uncertain legality) of ocean iron fertilization and aerosol sulfate injection, and one small commercial plant for CDR opened in Switzerland in 2017. Much of what we think we know about how CDR and SRM would work derives from modeling calculations and from natural occurrences such as the eruption of Mount Pinatubo. Considerable research is needed in order for society to even begin to understand what techniques would work, how long they would take, what side effects they would have, and how much they would cost. Without this information, it would be difficult for anyone to reach rational decisions (either positive or negative) about deployment.

As discussed in detail in Chapter 6 of this book, several factors have inhibited climate engineering research:

1. The absence of governmental mandates and of a widespread price on carbon means few private entrepreneurs have an incentive to develop these technologies, making most research dependent on either governmental financial assistance or philanthropy.
2. The absence of a clear set of rules for this research makes its conduct legally risky, as experimenters cannot know what restrictions and penalties might be imposed retroactively, or at least after significant investments have been made. Even the identity of the institutions that would establish and enforce the rules has not been determined.
3. Many people are opposed to the very concept of climate engineering out of concerns about its negative side effects and about its possible use as an excuse not to engage in GHG reduction. The absence of a clear line between research and deployment makes some who would accept research fear a slippery slope to full-scale deployment and to the lock-in of particular technologies.
4. Some types of geoengineering may have transboundary or even global impacts; indeed, they would need to if they are to meaningfully address climate change. Thus some sort of international governance regime is needed even for large-scale research, as no state (no matter how large) can set rules with confidence that they will be embraced internationally. Most states are reluctant to be branded as rogue nations, and to subject their individual and corporate citizens to undue legal risks abroad.

At the Asilomar International Conference on Climate Intervention Technologies in 2010, the following five principles of climate engineering governance were adopted, and variations on them have been widely accepted:

1. Climate engineering research should be aimed at promoting the collective benefit of humankind and the environment;
2. Governments must clarify responsibilities for, and, when necessary, create new mechanisms for, the governance and oversight of large-scale climate engineering research activities;
3. Climate engineering research should be conducted openly and cooperatively, preferably within a framework that has broad international support;
4. Iterative, independent technical assessments of research progress will be required to inform the public and policymakers; and
5. Public participation and consultation in research planning and oversight, assessments, and development of decision-making mechanisms and processes must be provided.

One of the issues that must be decided in selecting a governance regime for climate engineering research is when various kinds of rules should apply. Most attention here goes to SRM, since it appears to have much greater potential for negative unintended consequences than CDR. SRM technology is likely to proceed through these phases: laboratory, technology development, process study, scaling test, and climate response test. The first of these phases entails no environmental impacts, and the second and third have little risk of significant impacts. Thus only when a scaling test is undertaken (a test at a large enough scale to validate the models and smaller tests) is there such a risk that the international community needs to have a say. However, concern about the slippery slope and technological lock-in has convinced many that international governance is needed at earlier stages, at least if the experiments exceed certain size thresholds (which still need to be set).

Another issue that must be decided is which international institutions will be in charge. A few institutions currently exist with authority over comparable subject matter (particularly the United Nations Framework Convention on Climate Change and the Montreal Protocol on Ozone Depleting Substances), but their mandates would need to be expanded. Alternatively, a new global agreement on climate engineering could establish a wholly new institutional structure.

1.8. GOVERNANCE STRUCTURES

No regulatory structure is in place or under official negotiation to cover either SRM or CDR. Since most forms of CDR have only local impacts, international governance may not be essential for its conduct, though an international system to give each country incentives to conduct CDR activities (such as a globally harmonized carbon tax, or emission reduction quotas for each country) would be very helpful.

However, SRM is designed to have positive global impacts and could have locally or regionally negative impacts in unknown locations. As discussed in Chapter 7, it is important that a governance system be established. The chances of staying within the Paris temperature goals are not very promising, and the odds of catastrophic climate conditions are high enough that the world community may want to seriously consider deploying SRM. Some state (or even very wealthy individual) may well attempt to unilaterally deploy SRM.

Rules are needed to govern such decisions. In particular, institutions and processes will be needed to decide whether to undertake SRM; the appropriate manner and scale of action; when SRM can be discontinued; who will pay for SRM; what actions to take if SRM is undertaken without required authorization; what liability and compensation rules to apply; and adjudication of liability.

We identify several core principles that should be followed in governing SRM:

- Tailor the governance to the technology.
- Distinguish governance of research and small-scale field tests from management of larger deployments.
- Craft governance that can also respond on regional and local scales as well as globally.
- Governance should assure that climate engineering remains reversible and additive.

The best initial strategy is probably an evolutionary approach keyed to existing domestic and international authorities, rather than an attempt to formulate a comprehensive agreement all at once. Such an approach should integrate existing international and domestic legal authorities into a consistent and logical governance framework for SRM. It should assure that each nation's governance framework approach coordinates in a consistent and coherent fashion with each other. It should also encourage the development of parallel technologies needed to assure compliance, and remain open to bottom-up, emergent governance of SRM.

The UNFCCC could serve as a useful existing platform to conduct discussions and hopefully reach international agreement. It is an established body with participation by essentially every country in the world. Because its deliberations have proven themselves to be very time-consuming, initial proposals might be put forward by unofficial groups of experts and stakeholders, some of which have already begun meeting.

The governance regime should have at least these elements:

- Definition of large-scale projects
- Authorization required for large-scale projects
- Authorization process
- Authorization principles
- Identification of eligible applicants
- Monitoring, reporting, and verification
- Transparency
- Public deliberation
- Impact assessment
- Liability rules
- Compensation rules
- Establishment of a climate engineering fund
- Penalties
- Dispute resolution.

A research program should begin fairly soon, and simultaneously discussions should take place to devise a governance system with rules that address the issues identified above, and which are discussed in greater detail in Chapter 7.

NOTES

1 Climate Action Tracker, http://climateactiontracker.org/global.html.

2 Royal Society, *Geoengineering the Climate: Science, Governance and Uncertainty*, London: The Royal Society, 2009, p. ix. Several other definitions, but all largely amounting to the same thing, are set forth in Rob Bellamy et al., *A Review of Climate Geoengineering Appraisals*, 3 *WIREs Climate Change* 597, 599–600 (2012), doi: 10.1002/wcc.197.

3 National Research Council, *Climate Intervention: Reflecting Sunlight to Cool Earth* (2015) at 7.

4 Stephen Self et al., The Atmospheric Impact of the 1991 Mount Pinatubo Eruption, in Christopher G. Newhall & Raymundo S. Punongbayan, eds., *Fire and Mud: Eruptions and Lahars of Mount Pinatubo, Philippines* (US Geological Survey 1996), *available at* http://pubs.usgs.gov/pinatubo/self/; Kevin E. Trenberth & Aiguo Dai, Effects of Mount Pinatubo Volcanic Eruption on the Hydrological Cycle as an Analog of Geoengineering, 34 *Geophysical Research Letters* L15702 (2007).

5 James Rodger Fleming, *Fixing the Sky: The Checkered History of Weather and Climate Control*, New York: Columbia University Press, 2010 at 4.
6 *Id.* at 5.
7 Rob Bellamy et al., *supra* note 2, at 599.
8 National Research Council, *supra* note 3, at 28.
9 Paul J. Crutzen, Albedo Enhancement by Stratospheric Sulfur Injections: A Contribution to Resolve a Policy Dilemma?, 77 *Climatic Change* 211 (2006).
10 Royal Society, *supra* note 2; Royal Society, *Solar Radiation Management: The Governance of Research*, London: The Royal Society, 2011.
11 US Government Accountability Office, A *Coordinated Strategy Could Focus Federal Geoengineering Research and Inform Governance Efforts*, GAO-10–903 (2010).
12 National Research Council, *Advancing the Science of Climate Change*, Washington DC: National Academies Press, 2010, ch. 15.
13 National Research Council, *Climate Intervention: Carbon Dioxide Removal and Reliable Sequestration*, Washington DC: National Academies Press, 2015; National Research Council, *supra* note 3.
14 Asilomar Scientific Organizing Committee, *The Asilomar Conference Recommendations on Principles for Research into Climate Engineering Techniques* (Climate Institute, November 2010).
15 Arthur Peterson, "The Emergence of the Geoengineering Debate Within the IPCC," Case Study, Geoengineering Our Planet Working Paper and Opinion Article Series, September 9, 2014. This technique aims to remove carbon dioxide from the ambient air. The term "carbon capture and sequestration" is more commonly used in the context of the removal of carbon dioxide from the waste stream of a power plant.
16 PBL Netherlands Environmental Assessment Agency and European Union Joint Research Centre, Trends in Global CO_2 Emissions: 2016 Report (2016), Table 2.7, pp. 42–43. It appears that there was a slight decline from 2014 to 2015, due largely to reduced coal use in China, but there are questions about the data from China, and the likely growth in emissions from India and other rapidly developing countries may continue to drive global emissions upward. *See* Robert B. Jackson et al., "Reaching Peak Emissions," 6 *Nature Climate Change* 7 (2016).
17 National Oceanic & Atmospheric Administration, Global Monitoring Division, Trends in Atmospheric Carbon Dioxide, www.esrl.noaa.gov/gmd/ccgg/trends/graph.html.
18 Juan B. Moreno-Cruz & David W. Keith, Climate Policy under Uncertainty: A Case for Solar Geoengineering, 121 *Climatic Change* 431, 432 (2013).
19 Paul R. Ehrlich & John P. Holdren, Impact of Population Growth, 171 *Science* 1212 (1971).
20 Robert B. Jackson et al., *supra* note 16.
21 The nuclear fuel cycle is one exception.
22 Intergovernmental Panel on Climate Change, *Climate Change 2014: Mitigation of Climate Change*, New York: Cambridge University Press, 2014, p. 816
23 *See generally* Michael B. Gerrard & Katrina Fischer Kuh, *The Law of Adaptation to Climate Change: U.S. and International Aspects*, American Bar Association, 2012.

24 United Nations Environment Programme, The Adaptation Gap Report 2014 (2014).
25 *See also* Detlef P. van Vuuren et al., The Representative Concentration Pathways: An Overview, 109 *Climatic Change* 5 (2011); Richard H. Moss, The Next Generation of Scenarios for Climate Change Research and Assessment, 463 *Nature* 747 (2010).
26 Kevin Anderson, Duality in Climate Science, 8 *Nature Geoscience* 898 (December 2015); *see also* Pete Smith et al., Biophysical and Economic Limits to Negative CO_2 Emissions, 6 *Nature Climate Change* 42 (January 2016).
27 Paris Agreement, Art. 4 Par. 1.
28 Gabriel Blanco et al., Drivers, Trends and Mitigation, Chapter 5 in Intergovernmental Panel on Climate Change, *supra* note 22, at 354, 357–360.
29 Joeri Rogelj et al., Energy System Transformations for Limiting End-of-century Warming to Below 1.5 C, 5 *Nature Climate Change* 519 (2015), doi: 10.1038/nclimate2572; *see also* Kelly Levin et al., "Understanding the Paris Agreement's Long-term Goal to Limit Global Warming," World Resources Institute, December 15, 2015, www.wri.org/blog/2015/12/insider-understanding-paris-agreement%E2%80%99s-long-term-goal-limit-global-warming.
30 Pete Smith et al., *supra* note 26.; Sabine Fuss et al., Betting on Negative Emissions, 4 *Nature Climate Change* 850 (October 2014).
31 Richard Martin, "The Dubious Promise of Bioenergy Plus Carbon Capture," *MIT Technology Review* (January 8, 2016).
32 Calculated from World Bank, *Total Greenhouse Gas Emissions (kt of* CO_2 *equivalent)*, *available at* http://data.worldbank.org/indicator/EN.ATM.GHGT.KT.CE. Does not include emissions from land use change or from short-cycle biomass burning (such as agricultural waste burning and Savannah burning); figures not available for these sources.
33 Climate Policy Observer, In Depth: Assessing the INDCs: a comparison of different approaches, January 6, 2016, *available at* http://climateobserver.org/in-depth-assessing-the-indcs-a-comparison-of-different-approaches/.
34 World Bank, *supra* note 32.
35 *See* United Nations Environment Programme and Sabin Center for Climate Change Law, *Climate Change and Human Rights* (2015); William C. G. Burns, Human Rights Dimensions of Bioenergy with Carbon Capture and Storage: A Framework for Climate Justice in the Realm of Climate Geoengineering, in Randall Abate, ed., *Climate Justice: Case Studies in Global and Regional Governance Challenges*, Washington DC: Environmental Law Institute, 2016, 149–70; William C. G. Burns, *The Paris Agreement and Climate Geoengineering Governance: The Need for a Human Rights-Based Component*, Waterloo, Ontario: Center for International Governance Innovation 2016.
36 Domestic courts in some nations would have the ability to enforce international human rights obligations.

2

Technologies

Eli Kintisch

This chapter examines the two main methods of deliberately tinkering with the atmosphere to cool the planet: solar radiation management and carbon dioxide removal.

2.1. SOLAR RADIATION MANAGEMENT

Solar radiation management (SRM), also known as albedo modification, seeks to address the energy imbalance that is at the heart of the global warming challenge (the word albedo comes from *albus*, the Latin for white). Two hundred thirty-five watts of energy from the Sun pour onto every square meter of the planet, getting absorbed by the climate system. Before the buildup of anthropogenic greenhouse gases, a roughly equal flux of energy, in the form of heat, flowed out of the planet. But scientists estimate the Earth system is now absorbing a small amount of that outgoing energy – one study estimated the amount was a little more than half a watt per square meter – because of the buildup of greenhouse gases in the atmosphere[1] (a doubling of the concentration of CO_2 in the atmosphere would cause a total of about 4 watts per square meter of additionally absorbed energy[2]). That small imbalance, over the entire planet's surface, is the fundamental reason that the Earth's temperature has risen by roughly one degree Celsius since preindustrial times.

One of the key findings regarding SRM is that small perturbations in the Earth's atmosphere can have immediate and profound effects on the globe. That gives SRM the potential to be used in a hurry: According to a 2015 report of the National Research Council, SRM technologies are the "only ways that have been suggested by which humans could potentially cool Earth within years after deployment."[3] It also explains why estimates of their costs are surprisingly low given their potential to affect the global climate. "Some SRM approaches are thought to be low in cost, so the scale of SRM deployment will likely depend primarily on considerations of risk."[4]

Albedo modification techniques could theoretically be implemented at several different levels of the atmosphere: in the upper atmosphere (Section 2.1.1), in space (Section 2.1.2), in low-lying clouds at the sea surface (Section 2.1.3), in cirrus clouds at the lower atmosphere (Section 2.1.4), or at the Earth's surface (Section 2.1.5).

2.1.1. *SRM in the Upper Atmosphere: Stratospheric Aerosol Albedo Modification (SAAM)*

The most commonly proposed technique for managing solar radiation is to mimic the natural cooling effect of volcanoes by spreading sulfurous particles in the atmosphere. Russian scientist Mikhail Budyko was the first to propose this technique in 1974, but Nobel laureate Paul Crutzen gave the idea a measure of legitimacy in an editorial essay proposing research into SAAM in 2006.[5]

Harvard physicist David Keith calls injection of aerosols into the stratosphere as a means of geoengineering "fast, cheap, imperfect and uncertain."[6] Natural aerosols, including dust or chemicals emitted by marine or terrestrial plants, not only transport chemicals and microbes around the planet but also affect its climate. The particles emitted by volcanoes are particularly potent, as seen in studies of the 1991 eruption of Mount Pinatubo, which sprayed 14 to 26 megatons of aerosol into the stratosphere as sulfur dioxide.[7] Those particles created a haze in the sky, scattering light away from the Earth and cooling the planet by 0.3°C over three years.[8] Previous volcanoes during the Earth's geologic history have been even more powerful in terms of their cooling effect.

Modeling studies suggest that if a similar quantity of sulfur aerosol were artificially injected into the stratosphere, the cooling would begin within months – which on geologic timescales is essentially instantaneous.[9] To compensate for the warming that a doubling of CO_2 would cause, an albedo modification scheme would have to block between 1.7 percent and 2.5 percent of incoming solar radiation.[10]

In the lower portion of the atmosphere, the troposphere, the weather washes aerosols out of the atmosphere in days. But models show that in the stratosphere, a layer that begins roughly 8 kilometers above the ground and extends to about 50 kilometers, aerosols mix readily, can last for years before removal, and spread about the planet in a matter of weeks. For that reason SAAM involves releasing aerosols into the stratosphere, where they can modify the Earth's albedo. "[I]f one can inject aerosol a bit above the bottom of the stratosphere in the tropics, it would be carried upwards and then towards the poles, resulting in a long lifetime and a relatively even distribution of aerosol throughout the stratosphere," wrote one proponent of the idea, based on modeling.[11]

Modeling has shown some of the opportunities and risks of the global application of the technique. Global mean temperatures may be lowered using albedo modification, but the effect would likely be patchy. In a widely-cited comparison of 12 major climate models from around the world, researchers found that applying SAAM alone to attempt to restore a preindustrial climate would overcool the tropics and insufficiently cool the Arctic.[12] Several modeling studies have suggested that SAAM could be deployed to protect and regrow Arctic ice, including the Greenland ice sheet.[13]

Different studies have made varying conclusions when it comes to the impact of SAAM on precipitation. In a simulation of SAAM sufficiently deployed to counteract a quadrupling of preindustrial atmospheric CO_2, the multimodel study found that precipitation was affected, but not much: Over 92 percent of the globe, the difference in either direction was less than 0.2 mm of rain per day.[14] But in a study in which one model was used to simulate five different albedo modification techniques, "[p]otentially damaging changes in regional precipitation were a feature of all [the methods]." The reason for the difference may be that in the multimodel study the differences between the models can cancel one another as their results are averaged.[15]

2.1.1.1. Techniques for Implementing SAAM

The technique proposed for albedo modification that has received the most attention is release of aerosols by planes from within the stratosphere.

Rockets and naval guns have also been proposed as vehicles for the delivery of aerosols to altitude but have received little serious attention. An even simpler approach has also been proposed: using a high-altitude balloon and a tube. Intellectual Ventures, a Seattle-based firm run by US inventor Nathan Myrvold, has promoted the Stratoshield, a scheme for delivering aerosols to high altitude that allows the production of the sulfates to be done on the ground, after which they would be delivered to altitude and released there. As discussed in detail in Chapter 6, in 2010, British academics began an experiment that would use the same basic approach. Funded by the UK government's Engineering and Physical Sciences Research Council, the Stratospheric Particle Injection for Climate Engineering (SPICE) experiment, sought to study how the release of harmless water particles affected the stratosphere. But it was halted before any particles could be released.

Only a handful of close analyses of costs of SAAM have been conducted, but the initial findings are that the cost of deploying various approaches would be on the order of billions of dollars per year. In 2012 Aurora Flight Science Corporation found that the technology to deliver sufficient aerosols to altitude in order to counteract the warming caused by a doubling of preindustrial CO_2 "exists today."[16] The study looked at a number of techniques to deliver aerosols to high altitude. While dirigibles were seen as having potential for geoengineering operations, given their potential to carry large payloads and stay aloft for extended periods, Aurora found that their use at high altitude would be more risky than using airplanes known to be able to operate there. Use of rockets and naval guns had estimated annual costs of $390 billion and $137 billion, respectively – one or two orders of magnitude more expensive than using airplanes.

Aurora Flight Sciences offered this analysis of the blimp-and-tube approach: "Pipes suspended by floating platforms provide low recurring costs to pump a liquid or gas to altitudes as high as 20 km, but the research, development, testing and evaluation costs of these systems are high and carry a large uncertainty; the pipe system's high operating pressures and tensile strength requirements bring the feasibility of this system into question."[17] The analysis concluded that delivery of material required for stratospheric solar radiation management appears to be feasible in terms of engineering considerations.[18]

For airplanes, delivering this quantity – on the order of 1 million to 5 million metric tons of aerosol – would require an initial investment of $1 billion to $11 billion, with yearly operating costs between $0.6 billion and $2.6 billion (the costs on the high end of the estimate are linked to building and operating new aircraft with specialized capabilities for cargo delivery and dispersal in the stratosphere). The size of that operation would be roughly equivalent to flying 25 fully loaded 747 jets each day.[19]

Operating a global SAAM system would presumably require a robust observational system to monitor its operation and impacts. Such a system would likely have costs on the order of those for standard Earth-observing satellites, with a cost to build and operate in the billions of dollars per year.[20]

2.1.1.2. Environmental Impacts of SAAM

2.1.1.2.1. PHOTOSYNTHESIS An important difference between using a sunlight-reflecting approach in space and spraying aerosols in the upper atmosphere is that, by deploying aerosols, a large amount of diffuse solar radiation reaches the ground (that is why some scientists believe the spread of these particles in the upper atmosphere could make the sky look hazy). Several studies have suggested that this increase in diffuse light could potentially increase photosynthesis in plants, since such photons can reach plants in the canopy that cannot receive direct light from the Sun. In a 2009 study, researchers used a global climate model to study the last four decades of the twentieth century, when a "global dimming" was caused by aerosol pollution in the lower atmosphere, including sulfates emitted by power plants. From 1960 to 1980, for example, they found that the increased diffuse light caused plants to take up an additional 4.4 million tons of carbon each year via enhanced photosynthesis.[21] But this effect has not been directly considered in modeling studies of stratospheric aerosol geoengineering.

2.1.1.2.2. STRATOSPHERIC OZONE SAAM can disrupt the chemical cycle that maintains the stratospheric ozone layer, a chemical layer that protects biological life from ultraviolet sunlight. Theoretical and modeling studies suggest that sulfates in the lower portion of the stratosphere can provide surfaces for chlorine to change into forms that deplete ozone. After the 1991 Pinatubo eruption, total ozone measured in tropical and temperate latitudes declined by 4 percent, though scientists say that anthropogenic emissions of chlorine and bromine aerosols may have played a role. The chlorine in the stratosphere comes from long-lived chlorofluorocarbon (CFC) pollutants, whose levels peaked in the 1980s and continue to drop. So the impact of SAAM on the ozone layer in part depends on when it is attempted and the CFC levels in the atmosphere at the time.

Most scientists who have examined the matter believe that there is significant potential for sulfate aerosols to damage the ozone layer, but that is only one aspect of the risk to biological life. Stratospheric aerosols may scatter ultraviolet rays in the upper atmosphere before they can pass to the lower atmosphere. So even if the ozone layer is weakened, the effect on human health and ecosystems may be attenuated if fewer harmful rays pass through the stratosphere to the planet.

2.1.1.2.3. COLOR OF THE SKY In the same way that city residents sometimes experience a brightening of their sky due to pollution, SAAM would tend to whiten the sky. A 2012 study found that blocking just 2 percent of the sun's light would cause the skies to appear three to five times brighter. "Although our study did not address the potential psychological impact of these changes to the sky, they are important to consider," a coauthor of the report said in a press release.[22]

2.1.1.3. Halting Albedo Modification

Once a global albedo modification scheme begins, suddenly halting it could have profound effects. First, such a halt could cause warming to resume at a rate many times faster than the current warming rate. "Whereas the non-geoengineered world warms relatively slowly with relatively slow increases in atmospheric CO_2, in the case of a catastrophic failure of a solar geoengineering system, Earth would experience a large climate forcing at the time of system failure and would warm rapidly for several decades."[23] Second, as shown in a 2007 modeling study,[24] such a sudden warming would trigger the release of stored carbon in the ocean and land, compounding the total amount of greenhouse warming.

2.1.2. *In Space: Solar Sunshades*

A number of proposals to cool the Earth envision deploying solar sunshades, launched into orbit around the Earth or the Sun. Soviet scientists in 1960, for example, proposed creating rings of dust in the upper atmosphere that would reflect additional sunlight onto the planet to warm certain areas, a proposal in line with real attempts to warm Siberia by rerouting Russian rivers.[25] Subsequent published proposals involve orbiting clouds of dust particles placed in an equatorial plane like the rings of Saturn, or random clouds of mirrors flying in low Earth orbit. A key challenge for sunshades launched into Earth orbit, however, is to balance weight and costs. Heavier ones incur greater costs to launch, but lighter ones are vulnerable to being "blown out of orbit by the light pressure force exerted by the sunlight they are designed to scatter."[26]

Such a tradeoff would be less of a problem if a ring of sunshades were launched into orbit near the so-called LaGrange 1 point. There, in a position between Earth and the Sun, the gravitational forces of each body pulling on the ring, along with the force of the Sun's rays pushing on the scattering ring, could counteract one another. In a 2006 paper[27] astronomer Roger Angel elaborated on a previous proposal[28] to install a "space sunshade" at such a position with the aim to reduce the solar energy striking the planet by 1.8 percent to counteract the warming effects of a doubling of CO_2.[29]

Angel's envisioned project would involve a swarm of roughly ten trillion disks, each roughly 0.1 mm thick and about 60 cm in diameter. Angel proposed using magnetic energy to launch a stack of one million such disks from the Earth's surface every second for 30 years.[30] That rate has led some experts to conclude that deploying such a space-based approach would be infeasible for decades.[31] Moreover, if coal plants were used to generate the needed electricity, the sunshade would require 30 kilograms of coal for each kilogram of sunshade launched. But each kilogram of sunshade launched would mitigate the direct warming effect of 1,000 kilograms of coal.

But other problems loom. As with using stratospheric aerosols to cool the planet, reducing solar radiation with solar shields does not precisely counteract the additional warming caused by greenhouse gases. It leads to cooler tropics – since the Sun's heating is more dominant there than greenhouse gas heating – and warmer poles, which are warmed more by the greenhouse effect than by the Sun. Turning down the Sun in this way would reduce the gradient between the tropics and poles, a key driver of air in global circulation patterns. The effects could be profound, including a reduced "amplitude of the seasonal cycle … giving warmer winters and cooler summers." Plus, as with proposals to cool the planet by injecting stratospheric aerosols, modeling studies have shown that by preferentially cooling the tropics, this method could reduce precipitation worldwide, which is driven by sunlight striking the tropics.[32] Among scientists considering various geoengineering proposals, solar sunshades have received far less attention than other methods. But the Royal Society says they should not be discounted out of hand.[33]

If in the future it became probable that some form of geoengineering would be needed for a period approaching a century or longer, on such a timescale (and with the continual advance of technical capabilities) it is quite possible that the best examples of this type may offer a cheaper and less risky approach to SRM than approaches that seek to alter either the higher or lower layers of the atmosphere.[34]

2.1.3. *Lower Atmosphere: Marine Cloud Brightening*

Low-lying, layered clouds, known as stratocumulus clouds, provide a reflective white cover on top of the dark ocean, to an altitude of up to 1,500 meters.[35] Several scientists have proposed increasing the albedo of such clouds, which cover 20 to 40 percent of the world's ocean,[36] to enhance their ability to reflect sunlight. This geoengineering technique is known as marine cloud brightening, or MCB.

The idea is to spray salt particles into these low clouds where they would serve as additional nuclei for cloud droplets.[37] Those particles would attract existing water droplets in the clouds to condense around them, creating clouds with smaller drops but more of them. This rearrangement of water in the cloud would mean fluffier clouds with more surface area. Clouds with more surface area are more reflective.[38]

A real-life example of how this would roughly work comes from ocean-going ships emitting aerosol pollution from their smokestacks. These particles catalyze the formation of white cloudy "shiptracks" in their wake, visible by satellite. The carbon emissions from these ships contribute to greenhouse warming. But during the moments when they happen to be under clouds, under the correct conditions to allow brightening, unwitting geoengineers in cargo ships may cause twice as much cooling as warming over a century timescale, one study estimated.[39]

That study involved running smoke generators on a small vessel operating off the California coast, to measure its ability to whiten clouds. The project found an even better ratio of 50 times more cooling (due to whitening clouds) than warming (due to carbon emissions). The experiment also found, however, that MCB worked only a fraction of the time, even when conducted beneath cloud conditions that were theoretically conducive.[40] One reason may be that as more nuclei particles are added to clouds, the effectiveness of each additional particle at growing the cloud is reduced.[41] A 2013 modeling study found another problem: Once sprayed out, aerosol particles can coagulate, slashing MCB's efficacy in half.[42]

Climate modeling of the technique to understand its potential usefulness and risks presents a big challenge for atmospheric scientists. The problem is that representing the interplay of clouds and aerosols, and their impact on global patterns, is "a huge challenge" for models, in fact "one of the largest sources of uncertainty" in Earth system modeling.[43] The cloud features in the models are crude versions of the real world, and field studies have shown that today's models do a mediocre job at predicting how local perturbations can affect cloud structure.

That said, such model studies have suggested that the technique could be used to mitigate regional warming patterns, which could be useful in the mitigation of specific impacts of climate change. Simulations also suggest that MCB could be effective at restoring the global climate in a world in which emissions of greenhouse gases continue to rise, though as with other varieties of solar geoengineering, the cooling is not uniform and the technique cannot restore both temperature and precipitation to preindustrial conditions.

"Models consistently indicate that MCB can reduce temperatures," wrote the National Research Council in its 2015 geoengineering report. "Model simulations show that MCB targeted at susceptible marine stratocumulus will cool preferentially the eastern North/South Pacific and eastern South Atlantic, and will also cool globally and reduce Arctic warming."[44]

Most model studies show that MCB reduces evaporation and precipitation over tropical oceans. On impacts on precipitation more broadly, models show a variety of results, but that range could be as a result of using different models. A 2013 study used three different state-of-the-art Earth system models to obtain more robust results about the effectiveness of MCB. The team used the same experimental design with each model to see how MCB would perform under a scenario of relatively moderate greenhouse gas emissions in the twenty-first century. The amount of MCB deployed in the experiment was meant to counteract the total greenhouse warming at a constant level beginning in 2020. As with previous modeling studies, the effort showed that MCB can cancel warming for the century. It also found, however, that each of the three models showed that the intervention would make tropical and mid-latitude continents cloudier, wetter, and less prone to drought.[45]

Given that MCB operates over maritime waters, scientists have explored how the technique would affect the oceans in two specific ways that are pertinent to a warming world: the melting of the Arctic and the global loss of reefs. A study published in 2012, using a climate model, suggests that in a future scenario in which CO_2 concentrations have risen to twice the preindustrial levels, MCB could roughly restore global temperatures and ice cover to levels predicted in 2020.[46] MCB may be particularly effective in targeting polar regions because it can reduce a feedback loop in which a warming Arctic leads to more water vapor, a greenhouse gas, and less reflective ice cover, which amplifies the warming still further.[47]

In 2013 a study used a climate model to see how MCB could be used to lower sea surface temperatures in three regions suffering from coral bleaching: the Caribbean, French Polynesia, and the Great Barrier Reef. In the simulation, the scientists more than quintupled the number of cloud droplets in low clouds, from 60 per square centimeter to 375 in the three regions to make them more reflective. In a scenario in which global CO_2 was doubled, the model showed that cloud brightening successfully lowered sea temperatures, almost entirely eliminating coral bleaching events, which generally happen when sea temperatures rise by more than 1 degree Celsius over a period of three months or more.[48]

Side effects of the technique include the fact that blocking sunlight from striking reefs could deprive photosynthetic organisms of the key energy they need; one paper estimated that the technique would reduce the annual mean sunlight striking the ocean beneath brightened clouds by 20 percent, though this effect has not been measured directly.[49]

Additionally, if the technique were to be used as a palliative as global greenhouse gas levels continued to rise, "as with any SRM geoengineering technique, the MCB process would have to be continued indefinitely. Cessation of MCB ... could result in very rapid warming, which for coral reefs is the most common condition leading to coral bleaching."[50] (The authors do note that MCB, like all other methods of albedo modification, does nothing to stop ocean acidification, which along with pollution and fishing harms reefs and has many other adverse impacts on the marine environment.)

Using the MCB approach on a global scale would require an armada of boats. To counteract the global warming caused by current anthropogenic carbon emissions, researchers have calculated that 45,000 kilograms of seawater per second would be needed to be lofted into the sky in areas where marine clouds were found. This translates, roughly, into a fleet of 1,500 vehicles, each spraying 10^17 nuclei per second in the form of salt particles.[51]

To actually maintain this level of spraying would require having ships ready to move when nearby clouds are experiencing conditions making them conducive to being brightened. "The largest cooling effects could be achieved by staging several fleets around the world that are available for deployment on a daily basis and that can be scaled back to reduce energy and emission expenditures when suitable track-forming conditions are not available."[52]

In terms of hardware, conducting spraying at sea may be a technical challenge itself, though not impossible to achieve. A 2012 study looked at various spraying techniques and determined that long, thin capillaries made of silicon or short polymer capillaries could both work.[53]

2.1.4. *SRM in the Lower Atmosphere: Cirrus Cloud Seeding*

One proposed technique for solar radiation management that has received relatively little attention is seeding cirrus clouds. The goal is to dissipate these high-altitude clouds, which form in the troposphere above 6,000 meters and contribute to the greenhouse effect by trapping heat reflected up from the ground, warming the atmosphere. Most solar radiation management techniques seek to diminish the amount of the Sun's energy that strikes the Earth. By contrast, the goal of seeding cirrus clouds is to reduce the amount of heat that gets trapped in the atmosphere after infrared radiation is reflected up from the planet's surface.

Scientists have explored cirrus cloud seeding because it lacks some of the flaws that could plague aerosol injections, including damage to stratospheric ozone or alteration of the global hydrological cycle. The approach is based on the widely-used weather modification technique known as cloud seeding: spraying particles into clouds in an attempt to cause rain in one place or deter it in another. Despite little scientific proof that it works, farmers and land managers spend billions on the technique globally, and it is usually deployed to affect low clouds. But as a geoengineering tool this technique would focus on clouds that sit high in the troposphere.

A 2014 modeling study suggested that the technique could cool the planet by 1.4 degrees Celsius.[54] Furthermore, utilizing the technique at high latitudes could be more efficient than at middle latitudes or in the tropics. This is because in addition to trapping reflected heat via the greenhouse effect, cirrus clouds block a portion of the Sun's rays that strike the atmosphere in the first place. The latter effect, however, is more acute at lower latitudes when sunlight strikes the Earth more directly.

For this technique, scientists have proposed using aerosols that would serve as nuclei for ice crystals that would otherwise form normal cirrus clouds. These could include bismuth tri-iodide, a chemical related to the iodide-containing species that has been used for decades for low-altitude cloud seeding. The seeding particles could be released via tanks carried by high-altitude planes, a technique "safer and more viable than adding material to the fuel or exhaust of commercial aircraft, as has previously been suggested," asserted the authors of the 2014 study.[55] "Cirrus seeding appears to represent a powerful [SRM technique] with reduced side effects." Scientists have estimated that 140 tons of bistmuth tri-iodide, for example, could effectively counteract the current level of anthropomorphic greenhouse gas warming, and is easily deliverable by a small fleet of planes.[56]

2.1.5. *Albedo Modification at the Earth's Surface*

2.1.5.1. Crop Albedo Bioengineering

The mixture of plants growing on the Earth's surface can have a profound impact on the regional and global climate both in terms of their effect on the carbon cycle and their reflectivity. A 2007 study, for example, found that the deforestation in agricultural regions of the northern hemisphere have cooled those areas by 2 degrees Celsius since 1750 because of albedo: Croplands were more reflective than the forests they had replaced.[57]

A few studies have subsequently explored how deliberate selection and planting of more reflective plants could increase the planet's albedo and cool the climate. A 2009 study proposed "a 'bio-geoengineering' approach to mitigate surface warming, in which crop varieties having specific leaf glossiness and/or canopy morphological traits are specifically chosen to maximize solar reflectivity."[58] Some plant varieties offer a range of reflectivities, the authors report: Waxy types of barley are 16 percent more reflective than non-waxy varieties, and waxier varieties of sorghum, a tall grain, are 19 percent more reflective than less waxy types. In several cases, the genes that are expressed in relevant cultivars have been identified in previous studies.[59] A subsequent modeling study looked at several factors that crop geneticists successfully select for – leaf area, vertical profile, and reflectivity – and found that the scientists could increase the albedo of the crop by 34 percent without a change in water use.[60]

A computer model used in the 2009 study suggested that raising global crop albedos from .20 to .28 would lower global surface air temperatures by roughly 0.2 degrees C.[61] That is a relatively inefficient cooling scheme compared to other geoengineering techniques, but the experiment suggested strong regional effects: Converting plants to more reflective varieties could cool summer regional surface temperatures in mid-latitude North America and Eurasia by 1 degree Celsius.[62] Cooling also extended into the Barents Sea and North Atlantic, the modeling showed. "An unexpected benefit of cropland albedo change could thus be a small delay in Arctic sea-ice retreat."[63] But a 2015 study that compared the technique to other albedo modification techniques found a minimal global impact. "This scheme could at best be used to reduce warming locally and deliver local scale benefits to soil moisture and primary productivity," the authors wrote.[64]

Compared with other albedo modification schemes, bioengineering plants to make them more reflective could offer certain logistic advantages. While injecting particles into the stratosphere or launching sunshades into orbit would require new infrastructure, "the infrastructure required to create and propagate specific physiological leaf and canopy traits to large-scale cultivation is already in place. In addition, because arable crops are primarily grown for food, the annual replanting of modified varieties that is needed in order to retain continued climatic benefits is automatically achieved."[65]

2.1.5.2. Other Surface Albedo Modification Technologies

According to one overview of geoengineering methods:

> [a]chieving substantial global-mean temperature reductions through altering land-surface albedo represents a daunting challenge … Because land represents somewhat less than one-third of the planetary surface and approximately half of the land surface is cloud covered, ~10 percent of radiation incident on the global land surface would need to be reflected to offset the radiative forcing from a doubling of atmospheric CO_2 content.[66]

A group of Californian academics and private engineers calling themselves ICE911 have proposed adding materials to Arctic ice and snow in order to increase the albedo of those materials.[67] Another proposal is to create artificial whitewater – known as microbubbles – to enhance the albedo of those materials.[68]

In 2015 a study used a climate model to see the regional and global effects of raising surface albedo in the Arctic, concluding that "although ocean albedo alteration could lead to some sea ice recovery, it does not appear to be an effective way of offsetting the overall effects of CO_2 induced global warming."[69]

2.1.6. *Comparative Efficacy and Costs of Various Albedo Modification Technologies*

Another study used a single model to analyze the effect of various albedo modification schemes on temperature, precipitation, and other factors. It found that stratospheric SO_2 injection, marine cloud brightening, seeding cirrus clouds, and brightening the ocean surface all are "potentially able to return surface air temperature to 1986–2005 climatology under future greenhouse gas warming," though cirrus cloud thinning may not be able to cool the planet more than 1 degree Celsius.[70]

The study found each method has drastic impacts on the hydrological cycle. "Widespread regional scale changes in precipitation over land are significantly different from the 1986–2005 climatology and would likely necessitate significant adaptation despite geoengineering." The study's findings on land albedo modification suggested that the technique would cause "severe shifts in tropical precipitation." Other than the cirrus approach, it said, among the other schemes, "none has significantly less severe precipitation side effects than other schemes."[71]

In its review of albedo modification technologies, the National Research Council focused on SAAM and marine cloud brightening, the two most studied approaches. In summarizing its findings, the authors stressed the importance of further scientific and technical research on both methods. It may be possible to change the Earth's energy balance significantly:

> via either of these technologies without the need for major technological innovations. However, albedo modification strategies may introduce major and rapid perturbations to the planet with secondary and tertiary effects on environmental, social, political, and economic systems that are very difficult to predict currently and with effects that could be severely negative. Without further information on these risks, the low initiation costs of albedo modification cannot be balanced against other potential costs and risks of not deploying albedo modification methods.[72]

2.2. CARBON DIOXIDE REMOVAL (CDR)

Annual emissions of CO_2 globally are about 30 billion metric tons, and some 600 billion tons of anthropogenic CO_2 have accumulated in the atmosphere since the preindustrial era.[73] As discussed in Chapter 1, it will be essential to remove a great deal of CO_2 from the atmosphere, even if global emissions peak and begin to decline, in order to meet the temperature goals of the Paris Agreement.

In contrast to albedo modification strategies, which involve myriad moral, geopolitical, and economic challenges, CDR methods are mostly uncontroversial. According to the National Research Council, CDR strategies:

> are generally of lower risk and of almost certain benefit given what is currently known of likely global emissions trajectories and the climate change future. Currently, cost and lack of technical maturity are factors limiting the deployment of carbon dioxide removal strategies for helping to reduce atmospheric CO_2 levels. In the future, such strategies could, however, contribute as part of a portfolio of responses for mitigating climate warming and ocean acidification.[74]

This technique may require long timescales to be useful. A 2013 study estimated that "large scale application of [CDR] approaches could remove up to ~150ppm of CO_2 from the atmosphere," but that the methods "have only marginal potential to affect atmospheric CO_2 this century."[75] A 2015 modeling study found that to maintain global warming below 2 degrees Celsius above preindustrial values, negative emissions between 0.5 gigatons of CO_2 per year and 11 gigatons per year would be needed between now and 2100, depending on how optimistic one is about future energy and emissions trends.[76] The authors concluded that because technologies that could produce such negative emissions "have not been shown to be feasible … development of negative emission technologies should be accelerated, but also that conventional mitigation must remain a substantial part of any climate policy aiming at the 2 degree C target."[77]

Scientists routinely emphasize that such approaches should not be viewed as an alternative to cutting carbon pollution. "There is no substitute for dramatic reductions in the emissions of CO_2 and other greenhouse gases to mitigate the negative consequences of climate change, and concurrently to reduce ocean acidification," the National Research Council (NRC) wrote in its assessment of CDR technologies.[78]

CDR has come into focus in recent years as countries have concentrated on limiting global warming to 2 degrees Celsius or less by 2100, roughly equivalent to CO_2 concentration in the atmosphere of 450ppm. To accomplish that goal, models evaluated by the Intergovernmental Panel on Climate Change in 2014 show that negative emissions would play a central role to reach major goals of limiting emissions and climate change. The panel said in its summary for policymakers that most scenarios that achieve the 2-degree Celsius target "rely on the availability and wide-spread deployment of" carbon-negative technologies in the second half of the century.[79] The two technologies that the panel highlighted in this regard were bioenergy with carbon capture and storage and afforestation.

2.2.1. *Direct Air Capture and Storage*

Relying heavily on CDR as part of a climate response strategy to remove CO_2 emissions from the sky could mean creating a massive industry – perhaps the biggest engineering project in human history – to steadily remove this mass of gas from the atmosphere one molecule at a time.

Direct air capture is the technique of scrubbing CO_2 directly from the sky through large facilities. The technique can be used to obtain carbon for making a byproduct or fuel. Direct air capture and storage (DACS) means storing that CO_2 in a long-term reservoir. Multiple studies have concluded that this approach "could theoretically remove total annual global anthropogenic CO_2 emissions."[80]

The technology to remove CO_2 from the atmosphere economically is in its early stages of development. "Direct air capture of CO_2 is probably decades away from commercialization, even though its fundamental chemistry and processes are well understood."[81] A study by the American Physical Society (APS) in 2011 concluded that collecting CO_2 directly from the atmosphere "is not currently" economically viable despite "optimistic" technical assumptions.[82] It estimated that the basic cost of a system that could be built today would be about $600/ton, an order of magnitude more than the estimate for low carbon energy sources, though, as set forth below, experts in the field have challenged that number.

The two main approaches involve either (1) liquids or surfaces that chemically interact with CO_2, or (2) membranes that physically trap the CO_2 gas on solid surfaces. A handful of startup companies are trying to commercialize DACS. They are performing carbon capture from air in an open cycle in which CO_2 collected from the sky is used to make products. Some of these products will eventually release their carbon back into the atmosphere, like fuels.

The largest commercial demand in the US for CO_2 is for enhanced oil recovery, which uses the gas to push oil or gas from partly depleted reservoirs.

There is an ongoing debate whether this really advances the cause of addressing the climate problem. Some argue that it merely releases more fossil fuels, whose use negates the benefit of capturing CO_2; others argue that fossil fuels will be extracted somewhere in order to satisfy the demand for them, and therefore enhanced oil recovery just moves the location of the extraction while storing CO_2.[83]

As is apparent from the sales materials of companies that produce and sell gases to industry,[84] there are several current uses of CO_2, including beverage carbonation; flour and dough cooling; freezing and chilling; greenhouse growing; meat mixing; respiratory therapy; controlling pH in water and wastewater treatment plants; welding and metal fabrication; and various uses in the pulp and paper industry. However, the current demand for CO_2 is not nearly enough to utilize the quantity that would be captured if CO_2 removal were conducted on a massive scale. Several efforts are underway to develop new uses for CO_2 to help make chemicals, construction materials, plastics, and other products. This would be done through a variety of electrochemical, petrochemical, biochemical, and thermochemical processes. Almost all of these require the use of large quantities of energy; the plummeting cost of renewable energy (especially solar power) means that it may be possible to conduct these processes without creating large new CO_2 emission sources. One nonprofit organization has formed with the objective of finding ways to capture 10 percent of annual global CO_2 emissions and transform them into valuable products.[85] If these efforts are successful, they could create carbon-negative processes that would result in a reduction of CO_2 in the atmosphere.

2.2.1.1. Methods of DACS

The basic technologies required to do DACS with chemicals build off commercial absorption techniques that work in submarines and space shuttles to clean air of CO_2 gas. In its 2009 report on geoengineering, the UK Royal Society said that, given the commercial operation of these techniques, there is no "obvious limit" to the amount of CO_2 that could be collected from the atmosphere using chemical means.[86]

The first commercial-scale direct air carbon capture plant, by Swiss firm Climeworks, near Zurich, began operations in May 2017. It uses excess heat from a solid waste incineration plant to produce a concentrated stream of CO_2 for greenhouses nearby that are growing food. The company is also partnering with Danish Union Engineering to produce carbonated beverages, but the eventual goal includes producing fuel with carbon taken from the air, and eventually creating carbon-negative technologies. Advocates of carbon-negative approaches say this is important news. "The fact they're getting to commercial-scale prototypes is incredibly encouraging," Noah Deich, executive director of the Center for Carbon Removal in Berkeley, California told *Nature* magazine.[87]

In late 2015 Climate Engineering in Squamish, British Columbia partially opened a pilot demonstration plant that captures one metric ton of CO_2 per day. Right now it is releasing that gas into the atmosphere, but there are plans to install a small fuel synthesis plant to combine that CO_2 with hydrogen to make synthetic gasoline; the facility has plans to provide fuel for buses. The plant works by reacting air with a strongly basic solution, creating a chemical called potassium carbonate. Through a regeneration step that occurs at 900 degrees Celsius, the chemical is stripped of its carbon to create pure CO_2, and the basic solution is regenerated. Natural gas is burned to produce the heat for the reaction, but the CO_2 produced from that combustion is collected and processed, so the process produces no new carbon emissions.

Global Thermostat is a company that is using plastic surfaces coated with chemicals called amines to bond with CO_2 from the air. The firm has won investment from major energy firms, including energy giant NRG for its CO_2-scrubbing process, which occurs at under 100 degrees Celsius. Its 40-foot-high test facility is in Menlo Park, California.[88]

While carbon-scrubbing technologies are finding initial promise in niche markets, the cost of scaling up such technologies to affect the global concentration of CO_2 in the sky will be high. The precise amount of that cost is a matter of much contention among experts. The APS's $600/ton estimate is roughly six times the estimate of the cost to capture and store a ton of CO_2 from a power plant, which itself is much higher than the price of CO_2 in any of the world's carbon markets. The figure was based on an optimized case study in which the cost of capturing the CO_2 with a reagent and regenerating that reagent were both considered. A well-regarded critique of that report in 2012 estimated the cost of just the capture phase at $60/ton.[89] The two estimates used different assumptions about a number of factors, making it "difficult to directly compare these estimates."[90]

A number of approaches to DACS using membranes have been proposed, including materials called zeolites which have micropores that have been used to trap gases in industrial applications. The technology remains promising but has yet to exit the laboratory stage.

Driving the early development of direct air capture is demand for CO_2 for enhanced oil recovery and carbon-based fuels, sometimes aided by government policies promoting fuels with a low carbon footprint. In order to significantly reduce the amount of CO_2 that has built up in the atmosphere would require that air capture be scaled up into a massive global project. To reduce the atmospheric concentration of CO_2 by 50ppm, for example, requires sequestering 400 billion tons of CO_2,[91] presumably underground in saline aquifers and other formations. The authors of a recent study on carbon capture and storage concluded that an "effective" global geological storage capacity exists for 13,500 billion tons of CO_2, and a "practical" capacity of 3,900 billion tons of CO_2.[92]

2.2.2. *Bioenergy with carbon capture and storage*

Policymakers have focused on bioenergy with carbon capture and storage (BECCS) as one of the most important potential technologies for creating negative emissions. As biomass in plants grows, it removes carbon from the atmosphere via photosynthesis; BECCS works by burning or processing that carbonaceous material to create energy or chemicals, including fuels, and then storing the CO_2 that is formed in that process in a long-term storage facility such as an underground reservoir. The storage step is similar to the approach in a carbon capture and storage facility connected to a fossil fuel plant.

According to the IPCC in its 2014 assessment, "[t]here is uncertainty about the potential for large-scale deployment of BECCS."[93] Other scientists wrote "credibility as a climate change mitigation option is unproven and its widespread deployment in climate stabilization scenarios might become a dangerous distraction."[94]

First, there are questions about obtaining sufficient biomass. A major review of the potential of the technology for CDR found that the technique could likely be scaled to produce 100 exojoules of energy a year,[95] about half the current energy produced by the fossil fuel industry.[96] In its 2015 report, the NRC estimates that up to 500 million hectares of dedicated cropland would be needed for that scale – a whopping 10 percent of existing crop and grazing land worldwide. One study estimated that sequestering 1 gigaton of CO_2 from the atmosphere with this technique would require more than 52 million hectares of land for feedstock plants.[97] Probably the only way to free up sufficient land for BECCS to have a global impact would be to shift human diets away from meat, since direct food crops require less land per calorie provided. Otherwise, using massive new tracts of land to produce biomass for energy could create new competition with farms that produce feed for animals.

The 2015 NRC report argues that there is no reason to deploy carbon capture and storage with bioenergy until fossil fuel use has been reduced significantly. "Prior to that point, there is no difference in net carbon emissions to the atmosphere whether the CCS is tied to bioenergy or fossil fuel use."[98]

Converting forests to cropland to make BECCS fuel stocks creates a short-term spike in greenhouse emissions, as a result of the trees removed; to recover that lost carbon and get into "the red" may require "decades or more."[99] Depending on the current use of land that is converted to biomass production, biomass production at massive scale would likely also divert water supplies, increase use of fertilizer, alter wild ecosystems, and affect global food markets.

BECCS capture technology builds on existing technologies currently deployed at a few coal and gas facilities nationwide to capture carbon emissions. A handful of industrial plants planning to demonstrate the technique at full scale have yet to come online, including facilities in Decatur, Illinois and North Yorkshire, England.

Burning biofuels, like burning coal, creates a concentrated stream of CO_2, making its cost to capture and store the gas lower than that of DACS, which uses a diffuse stream. Available technologies include using biomass for combined heat and power. This process involves gasifying biomass fuel stocks to produce power and fuel simultaneously. Estimates range from $60 to $250 per ton of CO_2 sequestered, along the lines of carbon capture and storage from fossil fuel plants.[100]

2.2.3. *Storage or Utilization of CO_2*

As with carbon capture and storage from power plants, DACS and BECCS both rely on large-scale storage facilities or utilization to provide a final destination for the carbon they seek to sequester. Among technologies related to CDR, geologic storage has been a major focus of the scientific community. The US National Energy Technology Laboratory believes "the capacity for sequestering CO_2 in deep underground saline formations is vast enough to store essentially all CO_2 emissions from coal-fired power plants within the United States."[101] Formations of similar size exist on other continents, and scientists have proposed storing CO_2 in other environments, including in the deep ocean.[102] For decades the oil industry has stored CO_2 underground as part of efforts to obtain oil from old fields, so the technology for injection is mature. An unknown issue is the readiness of the technology to scale this storage up to a far larger size.

A study published in 2016 reported a successful effort in Iceland to inject CO_2 into a basalt formation, where it quickly mineralized into a harmless solid.[103] Basalt is the most common igneous rock in the world. If this technique is replicated in other types of locations, including possibly under the ocean, and if it could be employed economically, it could provide a satisfactory way to store massive quantities of CO_2.

Injection of gas into underground formations raises public concerns. Fracking to obtain natural gas has been shown capable of causing small- to medium-sized earthquakes and has led to allegations of water contamination, fueling public opposition to the practice and intense study of the phenomenon.[104] Large-scale storage of CO_2 underground could face similar political hurdles (there is a complex nexus between fracking and CO_2 storage; one factor is that shale sites viable for CO_2 storage could also be targeted for fracking projects, which would involve fracturing the caprock needed to secure the CO_2 long term). As discussed above, utilizing the CO_2 would be better than storing it, if it becomes embedded in products that have a long life.

2.2.4. *Ocean Fertilization*

Ocean fertilization seeks to take advantage of the ocean's natural carbon pump, which uses CO_2 at the sea surface and incorporates the carbon, via photosynthesis, into biological tissues which can fall or be transported to the deep ocean. Over long time scales, models show, the process described below can change atmospheric levels of CO_2 by more than 100ppm.[105]

Certain areas of the ocean, including the Southern Ocean, have plentiful nutrients but lack iron, a key trace micronutrient that sea plants known as phytoplankton need to grow. So fertilization with iron has been proposed as a means of accelerating the carbon pump and increasing the size of the ocean carbon sink. (Other fertilization techniques that have received less attention involve the addition of nutrients like nitrogen, in the form of urea, or phosphate.)

Small-scale experiments designed to study how phytoplankton blooms grow and die have been executed by marine biologists since 1993. As part of the 2004 European Iron Fertilization Experiment, German oceanographers and colleagues released some 13 metric tons of iron sulfate fertilizer into a rotating eddy in the Southern Ocean, catalyzing the creation of an algae bloom. In 2012 they reported in *Nature* that at least half the biomass of the bloom sank "below a depth of 1,000 metres and that a substantial portion is likely to have reached the sea floor. Thus, they argued, iron-fertilized diatom blooms may sequester carbon for timescales of centuries in ocean bottom water and for longer in the sediments."[106] Some colleagues agreed that the experiment had successfully sequestered carbon. Writing in an accompanying piece in *Nature*, geochemist Ken Buesseler said that the experiment shows "how the addition of iron to the ocean not only enhances ocean productivity, but also sequesters carbon" and that its success means that "larger and longer [ocean iron fertilization] experiments should be performed to help us to decide which, if any, of the many geoengineering options at hand should be deployed."[107]

Others have questioned Smetacek's conclusions regarding the fate of the bloom. In 2014 a study conducted a simulation of geoengineered carbon particles at depth. It found that two-thirds of the carbon sunk during the fertilization exercise was re-exposed to the atmosphere, taking an average of 38 years in their simulation.[108]

Commercial interests have looked into the approach as a means of CDR. In the summer of 2012 the Haida Salmon Restoration Corporation conducted privately funded commercial iron fertilization with the goal of increasing salmon returns.[109] The project, which involved the addition of more than 100 tons of iron into the Gulf of Alaska, drew criticisms from scientists and policymakers because it appeared to contravene an international treaty on dumping and another on biodiversity, as described in Chapter 6. Due to the bloom's small size and duration, the amount of annual carbon drawdown by the operation was roughly one-tenth the size of natural carbon fertilization events, including one catalyzed by iron-laden dust spewed by a volcano in the same area of the North Pacific in 2008.[110]

One technique of enhancing ocean uptake of atmospheric carbon is to use pipes that stimulate upwelling via the motion of waves. Theoretically, the tubes could pull nutrient-rich waters from the deep to the surface, where they could feed algae blooms, enhancing carbon uptake. But several studies have raised concerns that the technique would create waters with enhanced CO_2 levels in them "that would effectively cancel most, if not all, of the benefit of biological carbon drawdown."[111]

Ocean fertilization has also been informally proposed by commercial firms as a means of mitigating acidification.[112] But a climate modeling exercise in 2010 showed this approach is probably counterproductive in this regard. It explored a scenario in which greenhouse gas emissions continue to grow and the average global surface pH level had dropped by .44 units relative to preindustrial levels by 2100. An aggressive iron fertilization effort could only reduce total ocean pH change by .06 units, and this marginal improvement comes at the cost of acidifying the deep ocean, where organisms are highly sensitive to small changes in pH.[113]

2.2.5. *Biochar*

Biochar is a material created when organic material like agricultural waste is burned in the absence of oxygen. That process, called pyrolysis, creates a carbon-rich product that is stable or "biologically recalcitrant." Some call biochar "green charcoal" because it is produced in the same way as the popular fuel, and contains essentially the same ingredients. But biochar is used not for combustion, but as a soil additive. Its role as a means of CDR is that, by transforming biomass into charcoal, the carbon in the plant material is locked up instead of being released into the atmosphere when the biomass is burned or biodegraded in soil (biochar has also been used on a small scale for wastewater remediation).

If biochar production and use were to be scaled globally, some scientists believe the technique has a large potential to mitigate global greenhouse gas emissions. By allowing farmers to use less fertilizer, the technique reduces emissions from the production of those chemicals. Biochar also alters the way microorganisms process nitrogen, possibly reducing emissions of N_2O from the soil. Writing in *Nature Communications* in 2010, researchers estimated that this technique could draw down atmospheric greenhouse gases while reducing emissions of CO_2, methane, and nitrous oxide – greenhouse gases produced by agriculture – with a net reduction of 1.8 gigatons of CO_2-equivalent per year.[114]

But that depends on how the biochar was obtained; researchers warn that obtaining biochar at mass scale from forested land cleared to produce feedstock would be "disastrous" for global carbon emissions, for example. Harvesting lands improperly would also cause soil erosion and biodiversity loss, while also reducing the total carbon sequestered in the technique.[115]

A 2015 meta-analysis looked at 24 studies that estimated CO_2 emissions from biochar in soil. The scientists found that "biochar can persist in soils on a centennial scale," during which time 97 percent of its carbon is sequestered; during that time, additionally, the substance stimulates microbial activity, a boon to agriculture in poor soils.[116] Other studies have pointed to other advantages for agriculture, including slowing carbon return to the atmosphere during photosynthesis, improving soil productivity in degraded soils, the retention of nutrients, water and agricultural chemicals by runoff, and the use of biomass waste as a feedstock.[117] The authors of the 2010 *Nature Communications* paper wrote that, in terms of its side benefits, compared to "the possible strategies to remove CO_2 from the atmosphere, biochar is notable, if not unique, in this regard."[118]

But the technology is in its early stages of development. Globally, the biochar market is focused on niche applications for farmers, and was valued at $4.27 million in 2015, a minuscule size relative to global agriculture and carbon emissions.[119] As such, in a 2011 technology assessment of the method as a tool for global CDR, the US Government Accountability Office rated the maturity of biochar, on a scale of 1 to 9, at 2, citing "uncertainties in experimental data demonstrating the efficacy of biochar as a carbon sink."[120]

2.2.6. *Enhanced Weathering*

CO_2 emitted into the atmosphere reacts with water to form first carbonic acid, in the sky, and then, via a process of reaction with rocks, known as weathering, bicarbonate ions. (This generally occurs because the carbonic acid is weakly acidic, while the rocks are weakly basic.) Over a long time scale, via flow to the ocean and other chemical reactions, these ions form carbonate sediments on the floor of the ocean. One carbon dioxide removal technique is to accelerate this process to store massive quantities of carbon as either dissolved bicarbonate at sea or carbonate compounds on land, enhancing a natural carbon sink. "Enhanced weathering employs naturally occurring minerals and reactions, and therefore falls in the category of 'soft geoengineering' along with reforestation and agricultural techniques increasing soil carbon storage."[121]

Some proposed strategies would operate on land, others in the ocean. On land, engineers could distribute ground-up silicate rocks across terrestrial landscapes, where it would react with rainwater via the weathering process to lock up CO_2. A modeling of this method in 2015 found that an application of 1 kilogram of the pulverized rock on every square meter in an area comprising roughly one-third of the Earth's tropics could lower atmospheric CO_2 by 30ppm by 2100. Adding 5 kilograms of silicates on every square meter of the same area, paired with aggressive mitigation of carbon emissions by the end of the century, would cut atmospheric CO_2 by 300ppm and reverse ocean acidification.[122] (This is a separate technique from the idea of adding alkaline powder, like olivine minerals, to seawater to directly buffer the acidifying ocean, though the concept is similar.) Potential impacts include disruption of land from where the rocks would be taken and alteration of soil, stream, river, and ocean chemistry, with subsequent effects on plants and aquatic and marine ecosystems.

Ocean-based methods involve deriving alkaline materials from silicate rocks and dissolving them in the ocean, which would cause the sea to take in more CO_2. One source of that alkalinity could be lime, produced by the heating of limestone. Another is to grind up carbonate rocks and react them with CO_2 produced in a concentrated stream from a power plant.[123] Side effects could include alterations of ocean chemistry and a reduction of ocean acidification.

The cost of either category of weathering geoengineering is considerable because of the scale and complexity of the operation, which would require mining the minerals, processing them, grinding them up, and then transporting them to sites for distribution. In an analysis of several studies, the National Research Council concluded that weathering approaches could remove 1 gigaton of CO_2 per year between now and 2100, at a cost of between $50 and $100 per ton of CO_2. The Council suggested that the approach deserves further study, including kinetic studies to see how to more efficiently dissolve carbonates and silicates, and research into mineral extraction and seawater pumping. Thus far, only laboratory-scale experiments focused on the ocean-based approach have been done.[124]

NOTES

1 J. Hansen et al., Earth's Energy Imbalance and Implications, 11 *Atmospheric Chemistry and Physics* 24 13421–49 (2011).
2 *See* Realclimate, *The CO_2 Problem in 6 Easy Steps* (August 6, 2007), www.realclimate.org/index.php/archives/2007/08/the-co2-problem-in-6-easy-steps/.
3 Marcia McNutt et al., *Climate Intervention: Reflecting Sunlight to Cool Earth* (National Academies Press 2015).
4 Ken Caldeira et al., The Science of Geoengineering, 41 *Annual Review of Earth and Planetary Sciences* 231–56 (2013), www.annualreviews.org/doi/abs/10.1146/annurev-earth-042711-105548.
5 Paul J. Crutzen, Albedo Enhancement by Stratospheric Sulfur Injections: A Contribution to Resolve a Policy Dilemma? 77 *Climatic Change* 3–4 211–20 (2006).
6 D. Keith, Towards a Federal Research Program on Solar Radiation Management (Presentation, Washington DC, April 12, 2011).
7 McNutt et al., *supra* note 3.
8 B. Kravitz et al., The Geoengineering Model Intercomparison Project (GeoMIP), 12 *Atmospheric Science Letters* 162–67 (2011).
9 K. Caldeira and H. D. Matthews, Transient Climate-Carbon Simulations of Plantetary Geoengineering, 104 *Proceedings of the Natl Academy of Sciences* 9949–54 (2007).
10 K. Caldeira and L. Wood, Global and Arctic Climate Engineering: Numerical Model Studies, 366 *Philosophical Transactions of the Royal Society A: Mathematical Physical and Engineering Sciences* 4039–56 (2008), doi: 10.1098/rsta.2008.0132.
11 David Keith, *A Case for Climate Engineering* (MIT Press 2013).
12 B. Kravitz et al., Climate Model Response from the Geoengineering Model Intercomparison Project (GeoMIP), 118 *J. Geophysical Research Atmospheres* 8320–32 (2013), doi: 10.1002/jgrd.50646.
13 S. Tilmes et al., Can Regional Climate Engineering Save the Summer Arctic Sea Ice? 41 *Geophysical Research Letters* 3 880–85 (2014).
14 B. Kravitz, et al., *supra* note 12.
15 J. A. Crook et al., A Comparison of Temperature and Precipitation Responses to Different Earth Radiation Management Geoengineering Schemes, 120 *Journal of Geophysical Research: Atmospheres* 18 9352–73 (2015), 2015JD023269.
16 Justin McClellan, David W. Keith, and Jay Apt, Cost Analysis of Stratospheric Albedo Modification Delivery Systems, 3 *Environmental Research Letters* 34019 (2012), doi: 10.1088/1748–9326/7/3/034019.
17 *Id.*
18 *Id.*
19 *Id.*

20 McNutt et al., *supra* note 3.
21 Lina M. Mercado et al., Impact of Changes in Diffuse Radiation on the Global Land Carbon Sink, 458 *Nature* 1014–17 (2009), doi: 10.1038/nature07949.
22 Ben Kravitz, Douglas G. MacMartin, and Ken Caldeira, Geoengineering: Whiter Skies?, 39 *Geophysical Research Letters* L11801 (2012), doi: 10.1029/2012GL051652.
23 Ken Caldeira et al. The Science of Geoengineering, 41 *Annual Review of Earth and Planetary Sciences* 231–56 (May 2013), doi: 0.1146/annurev-earth-042711-105548.
24 Caldeira and Matthews, *supra* note 9.
25 E. Kintisch, *Hack the Planet: Science's Best Hope – or Worst Nightmare – for Averting Climate Catastrophe* (Wiley 2010).
26 The Royal Society, *Geoengineering the Climate: Science, Governance and Uncertainty* (September 1, 2009), https://royalsociety.org/topics-policy/publications/2009/geoengineering-climate/.
27 Roger Angel, Feasibility of Cooling the Earth with a Cloud of Small Spacecraft Near the Inner Lagrange Point (L1), 103 *Proceedings of the National Academy of Sciences* 46 17184–89 (November 14, 2006), doi: 10.1073/pnas.0608163103.
28 James T. Early, Space-Based Solar Shield to Offset Greenhouse Effect, 42 *Journal of the British Interplanetary Society* 567–69 (December 1, 1989).
29 Bala Govindasamy and Ken Caldeira, Geoengineering Earth's Radiation Balance to Mitigate CO_2-Induced Climate Change, 27 *Geophysical Research Letters* 2141–44 (July 15, 2000), doi: 10.1029/1999GL006086.
30 *Id.*
31 C. R. McInnes, Space-Based Geoengineering: Challenges and Requirements, 224 *Proceedings of the Institution of Mechanical Engineers, Part C: Journal of Mechanical Engineering Science* 571–80 (March 1, 2010), doi: 10.1243/09544062JMES1439.
32 The Royal Society, *supra* note 26.
33 *Id.*
34 John G. Shepherd, *Geoengineering the Climate: Science, Governance and Uncertainty* (Royal Society 2009).
35 McNutt et al., *supra* note 3.
36 L. M. Russell et al., Eastern Pacific Emitted Aerosol Cloud Experiment (E-PEACE), Bulletin of the American Meteorological Society (2013).
37 John Latham, Control of Global Warming?, 347 *Nature* 339–40 (1990), doi: 10.1038/347339b0.
38 McNutt et al., *supra* note 3.
39 L. M. Russell et al., *supra* note 36.
40 *Id.*
41 McNutt et al., *supra* note 3.
42 G. S. Stuart et al., Reduced Efficacy of Marine Cloud Brightening Geoengineering Due to In-Plume Aerosol Coagulation: Parameterization and Global Implications, 13 *Atmos. Chem. Phys.* 10385–96 (2013).
43 McNutt et al., *supra* note 3.
44 *Id.*
45 Kari Alterskjær et al., Sea-Salt Injections into the Low-Latitude Marine Boundary Layer: The Transient Response in Three Earth System Model, 118 *Journal of Geophysical Research: Atmospheres* 21 12, 195–206 (November 16, 2013), doi: 10.1002/2013JD020432.

46 Ben Parkes et al., The Effects of Marine Cloud Brightening on Seasonal Polar Temperatures and the Meridional Heat Flux, *International Scholarly Research Notices* 1–7 (2012), doi: 10.5402/2012/142872.
47 *Id.*
48 John Latham et al., Can Marine Cloud Brightening Reduce Coral Bleaching?, 14 *Atmospheric Science Letters* 4 214–19 (2013), doi: 10.1002/asl2.442.
49 John Latham et al., Global Temperature Stabilization Via Controlled Albedo Enhancement of Low-Level Maritime Clouds, 366 *Philosophical Transactions of the Royal Society of London A: Mathematical, Physical and Engineering Sciences* 1882 3969–87 (November 13, 2008).
50 John Latham et al., *supra* note 48.
51 S. Salter, G. Sortino, and J. Latham, Sea-Going Hardware for the Cloud Albedo Method of Reversing Global Warming, 366 *Philosophical Transactions of the Royal Society A: Mathematical, Physical and Engineering Sciences* 1882 3989–4006 (2008), doi: 10.1098/rsta.2008.0136.
52 McNutt et al., *supra* note 3.
53 Gary Cooper et al., A Review of Some Experimental Spray Methods for Marine Cloud Brightening, 4 *International Journal of Geosciences* 1 78–97 (2013).
54 T. Storelvmo, W. R. Boos, and N. Herger, Cirrus Cloud Seeding: A Climate Engineering Mechanism with Reduced Side Effects?, 372 *Philosophical Transactions of the Royal Society A: Mathematical, Physical and Engineering Sciences* 2031 20140116 (December 28, 2014), doi: 10.1098/rsta.2014.0116.
55 *Id.*
56 Michael Marshall, Get Cirrus in the Fight Against Climate Change, 2901 *New Scientist Environment* 14 (2013).
57 Richard Betts, Biogeophysical Impacts of Land Use on Present-Day Climate: Near-Surface Temperature Change and Radiative Forcing, 2 *Atmospheric Science Letters* 1–4 39–51 (June 2001), doi: 10.1006/asle.2001.0023.
58 Andy Ridgwell et al., Tackling Regional Climate Change by Leaf Albedo Bio-Geoengineering, 19 *Current Biology* 2 146–50 (January 27, 2009), doi: 10.1016/j.cub.2008.12.025.
59 *Id.* Perhaps counterintuitively, studies suggest more reflective leaves conduct more photosynthesis, probably because they keep the plants cooler, allowing less water loss via evaporation under low water conditions.
60 Darren T. Drewry, Praveen Kumar, and Stephen P. Long, Simultaneous Improvement in Productivity, Water Use, and Albedo through Crop Structural Modification, 20 *Global Change Biology* 6 1955–67 (June 2014), doi: 10.1111/gcb.12567.
61 Ridgwell et al., *supra* note 58.
62 *Id.*
63 *Id.*
64 J. A. Crook et al., *supra* note 15.
65 Ridgwell et al., *supra* note 58.
66 Caldeira et al., *supra* note 23.
67 *See* www.ice911.org.
68 Russel Seitz, Bright Water: Hydrosols, Water Conservation and Climate Change, 105 *Climatic Change* 3–4 365–81 (December 17, 2010), doi:10.1007/s10584-010-9965-8.

69 Ivana Cvijanovic, Ken Caldeira, and Douglas G. MacMartin, Impacts of Ocean Albedo Alteration on Arctic Sea Ice Restoration and Northern Hemisphere Climate, 10 *Environmental Research Letters* 4 44020 (2015), doi:10.1088/1748–9326/10/4/044020.
70 J. A. Crook et al., *supra* note 15.
71 *Id.*
72 McNutt et al., *supra* note 3.
73 *See* trillionthtonne.org.
74 Marcia McNutt et al., *Climate Intervention: Carbon Dioxide Removal and Reliable Sequestration* (National Academies Press 2015).
75 Caldiera et al., *supra* note 23.
76 T. Gasser et al., Negative Emissions Physically Needed to Keep Global Warming below 2 °C, 6 *Nature Communications* 7958 (August 3, 2015), doi:10.1038/ncomms8958.
77 *Id.*
78 McNutt et al., *supra* note 74.
79 Climate Change 2014: Mitigation of Climate Change, Contribution of Working Group III to the Fifth Assessment Report of the Intergovernmental Panel on Climate Change, O. Edenhofer et al., eds. (Cambridge University Press 2014).
80 US Government Accountability Office, Technology Assessment: Climate Engineering (July 2011).
81 *Id.*
82 Robert Socolow et al., *Direct Air Capture of CO_2 with Chemicals: A Technology Assessment for the APS Panel on Public Affairs* (American Physical Society 2011).
83 Thomas Overton, Is EOR A Dead End for Carbon Capture and Storage?, *Power*, April 12, 2016, www.powermag.com/is-eor-a-dead-end-for-carbon-capture/.
84 *See* Praxair, *Uses of Carbon Dioxide (CO_2) Gas and Liquid Carbon Dioxide by Industry*, www.praxair.com/gases/buy-liquid-or-compressed-carbon-dioxide-gas?tab=industries.
85 *See* The Global CO_2 Initiative, www.globalco2initiative.org/.
86 John G. Shepherd, *supra* note 34.
87 Daniel Cressey, Commercial Boost for Firms that Suck Carbon from Air, 526 *Nature* 306–307 (October 10, 2015).
88 Eli Kintisch, Can Sucking CO_2 Out of the Atmosphere Really Work?, *MIT Technology Review* (October 7, 2014), www.technologyreview.com/s/531346/can-sucking-co2-out-of-the-atmosphere-really-work/.
89 G. Holmes and D. W. Keith, An Air-Liquid Contactor for Large-Scale Capture of CO2 from Air, 370 *Philosophical Transactions of the Royal Society A: Mathematical Physical and Engineering Sciences*, 4380–403 (2012), doi: 10.1098/rsta.2012.0137.
90 McNutt et al., *supra* note 74.
91 Socolow et al., *supra* note 82.
92 J. J. Dooley, Estimating the Supply and Demand for Deep Geologic CO_2 Storage Capacity Over the Course of the 21st Century: A Meta-analysis of the Literature, 37 *Energy Procedia* 5141–50 (2013).
93 Climate Change 2014: Mitigation of Climate Change. Contribution of Working Group III to the Fifth Assessment Report of the Intergovernmental Panel on Climate Change, *supra* note 79.

94 Sabine Fuss et al., Betting on Negative Emissions, 4 *Nature Climate Change* 850–53 (September 21, 2014).
95 Felix Creutzig et al., Bioenergy and Climate Change Mitigation: An Assessment, 7 *GCB Bioenergy* 916–44 (2015), doi: 10.1111/gcbb.12205.
96 McNutt et al., *supra* note 74.
97 Manya Ranjan and Howard J. Herzog, Feasibility of Air Capture, 4 *Energy Procedia* 2869–76 (2011).
98 McNutt et al., *supra* note 74.
99 *Id.*
100 *Id.*
101 US Government Accountability Office, *supra* note 80.
102 *See* Chapter 3.
103 Juerg M. Matter et al., Rapid Carbon Mineralization for Permanent Disposal of Anthropogenic Carbon Dioxide Emissions, 352 *Science* 1312–14 (June 10, 2016).
104 Hilary Boudet et al., "Fracking" Controversy and Communication: Using National Survey Data to Understand Public Perceptions of Hydraulic Fracturing, 65 *Energy Policy* 57–67 (2014).
105 McNutt et al., *supra* note 74.
106 Victor Smetacek et al., Deep Carbon Export from a Southern Ocean Iron-Fertilized Diatom Bloom, 487 *Nature* 313–19 (July 19, 2012).
107 Ken O. Buesseler, Biogeochemistry: The Great Iron Dump, 408 *Nature* 305–306 (July 18, 2012).
108 J. Robinson et al., How Deep Is Deep Enough? Ocean Iron Fertilization and Carbon Sequestration in the Southern Ocean, 41 *Geophysical Research Letters* 7 2489–95 (2014).
109 P. Xiu, A. C. Thomas, and F. Chai, Satellite Bio-optical and Altimeter Comparisons of Phytoplankton Blooms Induced by Natural and Artificial Iron Addition in the Gulf of Alaska, 145 *Remote Sensing of Environment* 38–46 (April 2014).
110 *Id.*
111 McNutt et al., *supra* note 74.
112 David Freestone and Rosemary Rayfuse, Ocean Iron Fertilization and International Law, 364 *Marine Ecology Progress Series* 227–33 (July 28, 2008).
113 Long Cao and Ken Caldeira, Can Ocean Iron Fertilization Mitigate Ocean Acidification?, 99 *Climate Change* 303–11 (October 30, 2009).
114 Dominic Woolf et al., Sustainable Biochar to Mitigate Global Climate Change, 1 *Nature Communications* 56 (August 10, 2010), doi:10.1038/ncomms1053.
115 Swansea University, "Sustainable Biochar Can Offset 7–12% of Anthropogenic Greenhouse-Gas Emissions," www.swansea.ac.uk/science/news/sustainablebiocharcanoffset7-12ofanthropogenicgreenhouse-gasemissions.php.
116 Jinyang Wang, Zhengqin Xiong, and Yakov Kuzyakov, Biochar Stability in Soil: Meta-Analysis of Decomposition and Priming Effects, 8 *GCB Bioenergy* 512–23 (May 2015).
117 Woolf et al., *supra* note 114.
118 *Id.*
119 P&S Market Research, Global Biochar Market Size, Share, Development, Growth and Demand Forecast to 2023 – Industry Insights by Application (Gardening, Agriculture, Household), Feedstock (Agriculture Waste, Forestry Waste, Animal

Manure, Biomass Plantation), Technology (Microwave Pyrolysis, Batch Pyrolysis Kiln, Gasifier and Cookstove, Others), Manufacturing Process (Gasification, Fast and Intermediate Process, Slow Pyrolysis, Others) (November 2015), www.psmarketresearch.com/market-analysis/biochar-market.

120 US Government Accountability Office, *supra* note 80.

121 L. L. Taylor et al., Enhanced Weathering Strategies for Stabilizing Climate and Averting Ocean Acidification, 6 *Nature Climate Change* 402–406 (December 14, 2015).

122 *Id.*

123 Ken Caldeira and Greg Rau, Accelerating Carbonate Dissolution to Sequester Carbon Dioxide in the Ocean: Geochemical Implications, 27 *Geophysical Research Letters* 225–28 (January 15, 2000), doi: 10.1029/1999GL002364.

124 McNutt et al., *supra* note 74.

3

International Law

Jesse L. Reynolds

3.1. INTRODUCTION

Once the risks posed by climate change became widely known in the late 1980s, the policies needed to reduce these risks were immediately recognized as a matter of international law.[1] Although activities that emit greenhouse gases (GHG) benefit the emitters themselves, the accumulation of atmospheric GHGs will harm people and the environment across borders and generations. Limiting these emissions (through abatement, also called mitigation) would cost the country or other actor that took such steps, yet the benefits would be shared by the entire world, accruing especially to the future. Those who are willing to limit their emissions need assurance that others will also do the same, and not merely "free ride" on others' efforts. Some mechanism, often a legal one, is needed for this assurance of commitments and minimization of free riding. In addition to emissions abatement, other international actions – such as the adaptation of societies and ecosystems to a changed climate, research coordination, information sharing, and knowledge transfer – are necessary to reduce climate change risks, and they also are furthered by international legal mechanisms to ensure their implementation. Moreover, there are other important international legal issues, such as interactions between emissions abatement and trade law. Despite this early recognition and subsequent efforts, emissions abatement and adaptation remain insufficient and disappointing more than a quarter century later. It is clear that reducing climate change risks is a very difficult problem, arguably the most difficult one that humanity currently faces.

Some scientists and others who are concerned about climate change are increasingly considering climate engineering – intentional, large-scale interventions in natural systems to prevent or counter climate change. As described in more detail in Chapter 2 of this volume, these proposals come in two primary categories: carbon dioxide removal (CDR) and solar radiation management (SRM).[2] Like climate change, climate engineering was quickly recognized by scholars as a matter of international relations and cooperation, and specifically one of international environmental law.[3] The most important aspect of climate engineering for the purpose of international law is that its large-scale outdoors research and implementation would affect the environment across borders and in areas beyond national jurisdictions. Other issues of international law and cooperation more generally arise as well. For example, the climate change regime gives wealthy countries "differentiated responsibilities," including an obligation to share knowledge and technology – perhaps including that of climate engineering – with poorer developing countries. Likewise, climate engineering research would presumably be more efficient and legitimate through international coordination.

Aspects of climate change mitigation and climate engineering present problem structures that are, in various ways, both similar and distinct. They are public goods – something whose enjoyment is nonexcludable and nonrivalrous – with large spatial and temporal dimensions.[4] More specifically, CDR climate engineering resembles emissions abatement, in which individual actors choose whether to take immediate and costly actions whose climatic benefits would be widely dispersed and delayed in time. These actions may offer net benefits for the world, but they usually pose a net loss for the provider.[5] The primary challenges with these "aggregate effort global public goods" are getting the actors to provide the public good, assuring potential providers that others will also fulfill their promises, and limiting free riding. Together, these challenges are called a collective action problem. In contrast, some SRM methods presently appear to be so inexpensive and rapidly acting that, even though their expected beneficial effects would still be widely dispersed, the benefits that would immediately accrue to their provider could outweigh their direct financial costs.[6] The primary challenges with such a "single best effort global public good" are preventing the actors who have the capacity to provide it from providing it excessively or prematurely, and coordinating their efforts. These distinct problem structures mean that the implications of existing international law and the probable development of new law will vary, mainly between emissions abatement and CDR in one column and SRM in the other.

This chapter provides an overview of the existing international law of climate engineering. Although there are presently no international instruments that are legally binding, in force, and specific to climate engineering, numerous components of international law are relevant and have implications of varying clarity for climate engineering. Furthermore, some international legal instruments are specific to climate engineering, but are nonbinding or are not yet in force. The scope of the discussion here is limited to the implementation or large-scale outdoor research of climate engineering in a manner where international law might apply. For the most part, this would arise when the climate engineering activity would pose a significant risk of environmental harm that would either be transboundary or would occur in an area beyond national jurisdiction. Other aspects of international law are pertinent to climate engineering even in the absence of transboundary environmental risk, such as the prevention of international conflicts; the creation of international institutions for deliberation; legally binding commitments or hortatory calls for states to cooperate in research, to share knowledge, and to transfer technology; the expression of states' priorities; impacts on human rights; and intellectual property standards.

Three caveats are in order. First, this exploration is inherently speculative. Climate engineering proposals remain at early stages. How and whether these proposals will develop into real, effective, and socially acceptable technological means to reduce climate risks are highly uncertain. The best that can be done is to extrapolate from current knowledge and trends in order to generate reasonable scenarios, while simultaneously remaining open to an array of other potential yet seemingly less likely futures. This balance of focus between the probable and possible is obviously a subjective practice.

Second, climate engineering – like many human endeavors – would pose both potential benefits and risks. It is essential to bear in mind that climate engineering techniques are not being considered and researched in isolation, but in response to anthropogenic climate change. That phenomenon is poised to cause enormous harms to humans, nonhuman species, and ecosystems, particularly those that are already highly vulnerable. When quantified in monetary terms, and discounted for the future, climate change is expected to do global harm in the order of tens of trillions of US dollars.[7] The irreversible loss of species and ecosystems, and the suffering of the poor, are difficult to capture in such terms. Because the leading conventional means to reduce these risks – adaptation and emissions abatement – are each expensive, then an economically optimal climate policy where these means are adopted until their marginal costs equal their marginal benefit would still result in significant harm from climate change. Moreover, optimal policies are unlikely to be implemented due to abatement's collective action problem: the need for large international transfers of financial resources to achieve genuinely effective adaptation, and the ability of economic interests that would be harmed by optimal policies to block action.

Some climate engineering proposals presently appear to be able to reduce climate risks, and in some ways that abatement and adaptation cannot. Yet climate engineering presents physical and social risks of its own. Most importantly for the purposes of this chapter, both climate engineering and the climate change (or global warming, or elevated atmospheric GHG concentrations) that it may reduce satisfy the definition of "pollution," "adverse effects," or "damage" that multilateral environmental agreements strive to minimize. This definitional coverage presents a challenging tension between climate change and climate engineering. Because of this, in some circumstances the net expected effects of climate change and climate engineering should collectively be taken into account. How this is done, which factors are considered, the relative weights given to these factors, and the thresholds of harm's magnitude and probability will depend on the particular legal environment as well as on the actual and anticipated impacts of the phenomena at hand. Given the high levels of uncertainty concerning climate engineering's effects as well as the limitations of indoor research and modeling, any such balancing would need to be a dynamic process that is capable of responding to new information.[8]

Finally, this chapter is an exercise in the interpretation of international law, relying on the text of explicit agreements, widely accepted formulations of custom and principles, and – where helpful – on the preparatory and other supporting documents that can provide insight into states' intention behind and interpretation of international law. In reality, international law is not so clear-cut, but instead operates in a political context of state and nonstate actors that have diverse interests and levels of power. These actors have therefore interpreted, applied, and enforced international law inconsistently, and will continue to so.

The next section briefly introduces international law. Readers who are already familiar with this topic may skip this discussion. From there, the chapter reviews existing international law that is applicable to climate engineering, organized in sections concerning multilateral environmental agreements, nonbinding environmental agreements, custom, principles of international environmental law, and related domains other than the environment. Emerging norms and codes of conduct, which may be highly relevant to climate change, are examined next. The chapter closes with a discussion of the scholarship of the international law of climate engineering, including a brief review of it, an attempt to identify the challenges that it poses for legal scholars, and possible future directions.

3.2. INTERNATIONAL LAW

Under international law, countries, often called states or nations, are sovereign. Here, words such as "state" are often anthropomorphized and meant to imply the governing institutions that exercise control over a territory and a group of people. Sovereignty means that countries are free to manage their internal affairs and to carry out relations with other sovereign countries as they see fit, free from unwanted external interference. Their territories are defined and mutually exclusive. All states are equally sovereign, at least legally. There are no institutions that can exert legally binding authority over sovereign states without the states' consent.[9] Of particular importance for this chapter is that sovereignty generally includes the right for countries to exploit their own natural resources as they deem appropriate, except when this has negative extraterritorial impacts.[10] Notably, some areas – such as the high seas, Antarctica, and outer space – are not within the jurisdiction of sovereign countries.[11]

Reality is not so distinct Nationals of one country travel to others. Some territories remain legally dependent on distant countries, sometimes in a quasi-colonial manner. Other areas are claimed by multiple states. A government might not be able to exert its authority effectively over certain regions or populations within its borders. Some regions might not have any formal or legal governing institutions. Activities in one state's territory or under its control sometimes harm the environment of another. State and nonstate actors regularly try to influence and interfere in the internal and external affairs of other states, not always peacefully. Some countries are more powerful than others and are consequently able to influence other states, calling into question the theoretical equality of legal sovereignty.

When states interact, they face problems that are analogous to those that individuals in society face. Both countries and individuals benefit by having expectations regarding others' behavior, and therefore make explicit and implicit mutual promises. In the international arena, these promises have coalesced into something that resembles law. In fact, through this, states collectively try to fill the same sorts of functions internationally as national law does domestically: prevent and resolve conflicts, provide beneficial public goods, reduce negative externalities, promote positive externalities, enforce contracts and property rights, and encourage behavior consistent with widely agreed-upon normative principles. The central and very important difference is that people and domestic institutions are subject to binding national law that is enforced through the threat of socially sanctioned appropriation of property, freedom, and sometimes even life, whereas sovereign countries are not. This lack of centralized or hierarchical enforcement gives international law certain distinct characteristics.

International law is typically described as coming from three sources. First, states can make explicit agreements with one another in which they promise to do, to not do, to try to do, or to try not to do specific activities. These treaties, agreements, or conventions resemble contracts, and are considered legally binding (inasmuch as international law is legally binding) on those countries – often called Parties – that ratify them. Prior to ratification, states might sign a treaty, signaling their intention to ratify. In the meantime, such non-Party signatories may not act contrary to the agreement's core objective, but are not legally bound by it.[12] The vast majority of treaties are between only two countries, but some count many more as Parties. Parties generally may withdraw from treaties.[13] Some treaties establish institutions, such as regular meetings of Parties, secretariats, and scientific advisory bodies, that perform various intergovernmental functions. Second, over the course of centuries of interaction, countries have developed customary behavior among them. Once such customary behavior is widely practiced and there is evidence that this conduct arose from a sense of legal obligation, it then is considered legally binding. Customary international law is not codified, although some authoritative institutions regularly suggest language. States may explicitly object, and be exempt from, particular tenets of customary international law. Third, general principles guide the interpretation and development of international law. These are not legally binding in themselves, but must be operationalized in a treaty or in custom. Beyond these three sources, others such as nonbinding agreements among countries, rulings of international tribunals and domestic courts, statements of intergovernmental organizations, and scholarly writing can influence international law.

Despite the absence of a centralized or hierarchical power, international law can be and is enforced.[14] First, states can sometimes enforce international law through reciprocation, in which violators are punished by others' equivalent violations, which are often legally sanctioned.[15] This tit-for-tat can be effective when the reciprocal violation is beneficial for the punisher, at least in the short term, such as in trade agreements. Second, states can punish violators through retaliation, either within the issue area at hand or, more often, in another one. However, retaliation, such as economic sanctions or military action, is costly to the punisher. Retaliation is thus itself often a public good and will consequently be undersupplied due to collective action problems. Third, violators will suffer reputational damage, and will find it subsequently more difficult to reap the benefits of international cooperation. Notably, states can experience reputational damage also for acting contrary to nonbinding international law, or even to unwritten expectations, blurring the definition of "binding." Finally, ex post renegotiation – more cynically called bribes – can be an additional mechanism to reduce noncompliance.[16] Here, other countries offer to pay the violator either directly or (more often) indirectly to cease the breach, sometimes under the guise of assistance with compliance. This exchange can amount to de facto renegotiation of the agreement, and presents the hazard of strategic future noncompliance to extract more payments.

Together, these four enforcement mechanisms are of limited effectiveness. For this reason, much of international law emphasizes preventing violations and conflicts in the first place. Furthermore, as noted in the introductory section, enforcement occurs in a political context of states with diverse interests and levels of power. Not surprisingly, actual enforcement depends on the identities of the violator and the victim.

Just as the reality of countries and their sovereignty is less clear than legal theory implies, so too are the external and internal boundaries of international law unclear. In recent decades there has been growing recognition of diverse arrangements by authoritative state, substate, intergovernmental, and nonstate institutions that seek to intentionally and explicitly influence various actors' actions. Such "transnational law" or "global governance" has advantages over more narrowly defined international law in some circumstances, such as when the regulated actors operate in transboundary manners, when the conditions are highly dynamic, and when political leaders have insufficient incentives or opportunities to adopt national or international policy within the issue area.

International law typically governs the actions of states, not of individual people, corporations, or other nonstate entities. Although there are a few exceptions, such as international criminal law, for the most part nonstate actors are governed indirectly. That is, countries might ratify a treaty in which they promise to require, prohibit, encourage, or regulate certain behaviors from nonstate actors that are within their jurisdiction or under their control. The state often implements international law through domestic law and administrative policy. If it does so in a manner consistent with the standards of international law, then it generally remains in compliance even when a private actor within its jurisdiction or under its control acts in a way contrary to international law.

Finally, most of climate engineering's legal issues relate to environmental concerns, such as avoiding climate change, reducing pollution, preventing transboundary harm, and conducting environmental impact assessments. International environmental law is thus central. Notably, international environmental law is generally not oriented toward protecting ecosystems and Earth systems for their own sake. Instead, "almost all justifications for international environmental protection are predominantly and in some sense anthropocentric."[17]

3.3. LEGALLY BINDING MULTILATERAL ENVIRONMENTAL AGREEMENTS

Multilateral environmental agreements contain explicit promises that states have made to each other regarding the environment. This section reviews legally binding ones, organized in subsections for those focused on the atmosphere, on oceans, on procedural duties, and on other topics. Section 3.5 discusses nonbinding environmental agreements.

Here, and in the other sections that review international law, when an agreement invokes a tenet of customary international law or a general principle of international environmental law, it will be briefly mentioned along with any relevant details, but discussed in more depth in the appropriate later section.

3.3.1. *Atmospheric Agreements*

Climate change is foremost an atmospheric phenomenon. Its cause is elevated atmospheric concentrations of GHGs, and the changes manifest primarily and most immediately in the atmosphere. Climate engineering would operate either directly in the atmosphere, by making it more reflective, by withdrawing carbon dioxide from it, or by allowing more heat to escape, or indirectly, by making the surface below it more reflective or by reducing the amount of solar radiation that it receives.

International law does not clearly define the atmosphere and responsibility for its quality. A state has sovereign rights over its airspace, which is the volume above its territory extending upwards to the undefined border with outer space. However, the air that constitutes the atmosphere moves and mixes, including across national boundaries and through airspaces. Pollutants consequently travel, and these movements are often addressed through various bilateral and regional agreements. Some atmospheric pollutants, notably ozone-depleting substances and GHGs, have global effects and therefore call for a global response. Proposals for a comprehensive "Law of the Atmosphere" have surfaced occasionally – especially when the scientific understanding concerning climate change risks first emerged – but these have not encountered warm receptions.[18] The International Law Commission (ILC) of the UN is presently developing draft guidelines, examined in more detail below, on the protection of the atmosphere that reflect current international environmental law, although the guidelines' scope excludes issues that existing international law already addresses.[19] The present draft of these guidelines defines the atmosphere as "the envelope of gases surrounding the Earth" and states that "the protection of the atmosphere from atmospheric pollution and atmospheric degradation is a pressing concern of the international community as a whole."[20]

3.3.1.1. UN Framework Convention on Climate Change

The UN Framework Convention on Climate Change (UNFCCC) is the central international legal instrument for multilaterally coordinated efforts to limit climate change and its impacts. As a framework, it is a general agreement with limited commitments, and sets out expectations for subsequent, more detailed protocols. All globally recognized countries, as the phrase is typically understood, are Parties to the UNFCCC.[21] Its objective is the

> stabilization of greenhouse gas concentrations in the atmosphere at a level that would prevent dangerous anthropogenic interference with the climate system. Such a level should be achieved within a time frame sufficient to allow ecosystems to adapt naturally to climate change, to ensure that food production is not threatened and to enable economic development to proceed in a sustainable manner.[22]

Because the excess quantity of the leading anthropogenic GHG – carbon dioxide – is naturally withdrawn from the atmosphere at very slow rates, it is essentially a cumulative pollutant. Genuine stabilization of its concentration implies something close to net zero emissions. This stabilized concentration should prevent "dangerous anthropogenic interference with the climate system," which is undefined in the UNFCCC itself. However, in a subsequent agreement under the UNFCCC – the Paris Agreement, discussed below – countries agreed that global average temperatures should stay well below 2°C above preindustrial levels, implying a stabilized carbon dioxide concentration of less than 460 parts per million.[23] The current emissions trajectory, while uncertain, implies that the atmospheric carbon dioxide concentration will cross this threshold around the year 2035 under "business-as-usual" assumptions."[24] The second half of the objective indicates that this stabilization must be done somewhat rapidly. This goal rules out a large overshoot of the carbon dioxide concentration limit followed by its drawdown through CDR. Together, a literal reading of the objective under present circumstances and forecast trajectories calls for the rapid development of and heavy reliance upon CDR.[25] Notably, of the four scenarios (called "Representative concentration pathways") currently used by the Intergovernmental Panel on Climate Change (IPCC), the most optimistic two, which are forecast to prevent 2°C warming and to delay it until the twenty-second century, respectively, each assume large quantities of CDR. In these projections, the CDR would be carried out through bioenergy with carbon capture and storage (BECCS). At the scales envisioned, this would constitute climate engineering. However, this approach would compete with agriculture for arable land, increasing food prices and running contrary to the UNFCCC's objective "that food production is not threatened." It would also degrade preserved and relatively natural land, reducing biodiversity.

The relationship between the UNFCCC's objective and SRM is unclear. These climate engineering techniques would not directly affect greenhouse gas concentrations, although they would have indirect effects. Moreover, SRM itself could be seen as "dangerous anthropogenic interference with the climate system," ex ante due to expected risks, or ex post due to actual negative impacts. On the other hand, certain SRM methods appear to be rapidly effective at reducing the most dangerous aspects of climate change. Moreover, compared to emissions abatement, SRM is more likely to be implemented at a level that is nearer to the optimum, due to their different problem structures.[26] Therefore, a hope is that SRM could be used to slow down or prevent most climate change impacts – including those on ecosystems, food production, and economic development, which are all emphasized in the objective – while society transitions to zero net GHG emissions. (SRM's large uncertainties and its inability to impede ocean acidification, are important dampers on this hope.) Notably, the UNFCCC does not prohibit or exclude any means to reduce climate risks.

The UNFCCC invokes several principles of international environmental law and tenets of customary international law. The former include the environment as a common concern of humankind, common but differentiated responsibilities, general equity, intergenerational equity, sustainable development, and precaution.[27] The latter tenets are the sovereign right to exploit domestic natural resources and the responsibility to prevent transboundary harm.[28] The Convention is also anthropocentrically and economically oriented. The anthropocentricism is apparent both in the objective, in which two of the three criteria for the speed of GHG stabilization are for the sake of the human race, and in its first principle: "The Parties should protect the climate system for the benefit of present and future generations of humankind."[29] That is, the UNFCCC's objective and commitments are to be pursued *not* with the goal of a planet or atmosphere that is less impacted by human activities, but instead with one that prioritizes humans' well-being primarily and ecosystems secondarily. The importance of economic activity, particularly economic growth in the developing countries, is clear throughout the UNFCCC. This is seen both in its firm commitments (e.g., "Parties … shall … employ appropriate methods … with a view to minimizing adverse effects on the economy, on public health and on the quality of the environment, of projects or measures undertaken by them to mitigate or adapt to climate change") and in its soft ones (e.g., "Parties should … tak[e] into account that policies and measures to deal with climate change should be cost-effective so as to ensure global benefits at the lowest possible cost").[30] Because some SRM methods presently appear to have very low financial costs of implementation, these passages could imply that these lower-cost means to reduce climate risks should be given some priority within a wider portfolio of policies. Again, such an approach should consider SRM's own risks to people and the environment.

Although the UNFCCC's commitments mainly focus on gathering and sharing information and on developing general national plans, four themes found therein have implications for climate engineering. First, and most specifically, Parties are committed to minimizing a range of adverse effects – on the economy, public health, and the environment – in the activities that they undertake to reduce climate change risks.[31] Therefore, they would need to practice due diligence, such as by carrying out prior impact assessments, in their climate engineering programs.[32]

Second, Parties are to achieve the Convention's objective of stabilizing GHG concentrations through both emissions abatement and the conservation and enhancement of sinks and reservoirs.[33] For example, "All Parties … shall … promote and cooperate in the conservation and enhancement, as appropriate, of sinks and reservoirs of all greenhouse gases not controlled by the Montreal Protocol, including biomass, forests and oceans as well as other terrestrial, coastal and marine ecosystems."[34] All CDR methods would utilize sinks ("any process, activity or mechanism which removes a greenhouse gas, an aerosol or a precursor of a greenhouse gas from the atmosphere") and some also rely on reservoirs ("a component or components of the climate system where a greenhouse gas or a precursor of a greenhouse gas is stored").[35] This indicates that, at least in theory, all UNFCCC Parties have implicitly agreed to promote and cooperate in CDR implementation, in ways appropriate and consistent with the other terms of the Convention.

The third relevant set of Parties' commitments concerns research and technology. Several commitments in the UNFCCC call for Parties to undertake research, to cooperate therein, and to share the results. For example,

> All Parties … shall … Promote and cooperate in scientific, technological, technical, socio-economic and other research, systematic observation and development of data archives related to the climate system and intended to further the understanding and to reduce or eliminate the remaining uncertainties regarding the causes, effects, magnitude and timing of climate change and the economic and social consequences of various response strategies.[36]

The term "various response strategies" is not defined but presumably could include climate engineering. Other commitments concern the development and diffusion of technologies, especially those that would "control, reduce or prevent anthropogenic emissions of greenhouse gases," which appears to include CDR techniques.[37] To this end, in 2010 the UNFCCC Conference of Parties (COP) created a Technology Mechanism, including a Technology Executive Committee, to implement the UNFCCC's commitments related to the development and transfer of technology.[38]

Adapting to a changing climate is the fourth and final set of UNFCCC commitments that might apply to climate engineering. Several commitments and one principle explicitly call for adaptation, for example through the development and implementation of national adaptation plans.[39] Others, such as those concerning "various response strategies" and technology transfer, are implicitly adaptive. Adaptation is undefined both in the UNFCCC itself as well as by the decisions of its COPs. The Convention's website and other documents often use a definition developed by the IPCC: "Adaptation refers to adjustments in ecological, social, or economic systems in response to actual or expected climatic stimuli and their effects or impacts. It refers to changes in processes, practices, and structures to moderate potential damages or to benefit from opportunities associated with climate change."[40] SRM climate engineering could fall within this unofficial definition's latter sentence, in that it would be a change in practices to moderate potential climate damages.

The Kyoto Protocol of 1997 is a protocol to the UNFCCC that is in force, having been ratified by almost all UNFCCC Parties. (The USA did not ratify, and Canada ratified and later withdrew.) It is dedicated to specific GHG emissions abatement targets, especially for industrialized countries. It thus has little implication for SRM, save for further commitments by industrialized country Parties to implement policies to inter alia research, promote, develop, and transfer "environmentally sound technologies," and to cooperate in research to reduce uncertainties concerning "various response strategies."[41] In contrast, CDR technologies are highly relevant. Explicitly, the industrialized country Parties' policies cited above are, among other things, to protect and enhance sinks and reservoirs and to research, promote, develop, and transfer "carbon dioxide sequestration technologies."[42] Furthermore, the industrialized countries are able to satisfy their GHG emissions targets by, among other methods, afforestation, which is sometimes considered a CDR method.[43] The Parties may choose to permit other activities involving agricultural soils, land use changes, and forestry to contribute to achieving industrialized Parties' targets.[44] This could include biochar, for example. However, the Kyoto Protocol's first round of targets ended in 2012; an amendment providing for a second round that would end in 2020 has been approved but is not yet in force.[45] CDR is present implicitly in other commitments toward abatement. For example, industrialized countries are to implement systems to estimate removal of all GHGs by sinks, including through the enhancement of sinks.[46] Furthermore, the Kyoto Protocol established several flexibility mechanisms in order for the industrialized country Parties to meet their emissions targets more efficiently. The Protocol leaves the criteria for these mechanisms somewhat general.[47] The scope for CDR within the COP's guidelines for flexibility mechanisms, such as joint implementation and the clean development mechanism, is presently limited to forestry activities such as afforestation, but could be broadened.

At the 2015 COP in Paris, international negotiators approved the second legally binding agreement under the UNFCCC, and it entered into force the following year. Compared with the Kyoto Protocol, the Paris Agreement offers a fundamentally different approach to emissions abatement. Here, Parties' first aim is to keep global warming "well below 2°C," and they are "to pursue efforts to limit the temperature increase to 1.5°C."[48] Limiting expected warming to 2°C will require net GHG emissions to peak almost immediately, to be zero by sometime soon after 2050, and to be negative thereafter.[49] Parties are to do this not only by abatement but also through "a balance between anthropogenic emissions by sources and removals by sinks of greenhouse gases in the second half of this century," indicating the growing importance of CDR.[50] All Parties are to submit their own nationally determined contributions toward the Agreement's abatement aim. Almost all Paris Agreement Parties have done so already, and none of these explicitly includes carbon dioxide removal.[51] The Paris Agreement replaces the Kyoto Protocol's various flexibility mechanisms with a new approach to voluntary cooperation in implementing Parties' contributions, including through "internationally transferred mitigation outcomes."[52] The UNFCCC institutions will need to further clarify which activities, including enhancements of sinks and reservoirs, qualify toward a Party's own contribution and those of others via internationally transferred mitigation outcomes. This process will presumably build upon their analogous work for the Kyoto Protocol, and could address a growing number of activities, including some that fall within common definitions of CDR. Such determinations will be simpler for the more contained CDR techniques, such as direct air capture and BECCS, where the quantity of removed GHG can be monitored, reported, and verified fairly well. In contrast, the amount of captured carbon dioxide through ocean fertilization would be very difficult to ascertain reliably.[53] In addition, the Paris Agreement places much more importance on adaptation than did the UNFCCC and the Kyoto Protocol. In the Agreement's second of two aims, adaptation is now an explicit goal on par with abatement.[54] It is to be pursued in "a manner that does not threaten food production" and "with a view to contributing to sustainable development."[55] If SRM were to be considered a component of adaptation, then Parties to the Agreement could include it as part of their obligatory adaptation plans and it would constitute part of the global stocktaking, which is to occur every five years.[56] Such an inclusion could be justified by the growing unlikelihood of keeping global warming to within 1.5°C or 2°C. Furthermore, Parties' adaptive SRM activities should satisfy a number of desiderata set out by the article: They should be participatory and transparent, take into consideration vulnerable groups, be based upon the best available science, and be integrated with socioeconomic and environmental policies and actions.[57]

Finally, somewhat independent of the precise wording of the objective, principles, and commitments of the UNFCCC and its protocols, its affiliated institutions are a logical home for some form of climate engineering assessment and governance, broadly defined. The UNFCCC is clearly the central international legal instrument to reduce climate change risks. It also has global participation and robust institutions, including annual COPs, a Secretariat, a Subsidiary Body for Scientific and Technological Advice, and an explicit – albeit weak – dispute settlement procedure.[58] This is not to assert that it should be the site of the international regulation of all climate engineering.[59] Instead, the UNFCCC institutions might eventually be able to offer a forum for international discussion of how various climate engineering methods could further the objectives of the UNFCCC and its related agreements. In addition, the elaboration of the rules and guidelines for the UNFCCC regime's reporting standards, the Kyoto Protocol's commitments and flexible mechanisms, and the Paris Agreement's voluntary cooperation could situate various CDR techniques within or outside the portfolio of internationally recognized emissions abatement methods.

3.3.1.2. Vienna Convention for the Protection of the Ozone Layer

Prior to the emergence of climate change as an international scientific, political, and legal issue, the destruction of stratospheric ozone by the emission of certain anthropogenic substances was the preeminent known environmental problem of the global atmosphere. This loss of stratospheric ozone increases the amount of ultraviolet radiation at the Earth's surface, especially at higher latitudes, posing risks to humans and the environment. Fortunately, the general alignment of state interests, the large net benefits of ending emissions of ozone-depleting substances, and the availability of substitute chemicals enabled the international community to respond rapidly and effectively. The Vienna Convention for the Protection of the Ozone Layer of 1985 is the central framework convention, has limited commitments, and enjoys universal participation. It is furthered by its Montreal Protocol of 1987, in which its 197 Parties commit to phase out specific ozone-depleting substances. The agreements are supported by a standing secretariat and other dedicated institutions at the UN Environment Programme, regular COPs, a robust compliance mechanism, three Assessment Panels that provide scientific and technical input, and procedures for amending the agreements and updating other policies.

Stratospheric ozone and climate engineering are potentially related via general and specific mechanisms. The general relationship is that elevated atmospheric GHG concentrations and climate change will affect stratospheric ozone concentrations as well as ground-level ultraviolet radiation in complex ways. However, the sign and magnitude of the net effect remain uncertain, and they will certainly vary by latitude.[60] Regardless, CDR climate engineering techniques, if effective, would lessen all of these effects, whereas SRM would do so for only some. One specific issue is that sulfate particles, the leading candidate for stratospheric aerosol injection SRM, might deplete stratospheric ozone. However, the sulfate aerosol's presence might partially or fully counteract this effect by blocking some incoming ultraviolet radiation.[61] Therefore, the net effect of sulfate stratospheric aerosol injection is also unclear. Notably, the potential impact of sulfate aerosols on stratospheric ozone is a leading area of SRM research, through both the exploration of alternative aerosol materials and a proposed field experiment.[62] A second specific issue is that cirrus clouds might naturally contribute to stratospheric ozone depletion yet might block the rise of water vapor, which in the stratosphere acts as both a GHG and an ozone destroyer.[63] Cirrus cloud thinning could thus affect stratospheric ozone, although again the net impact is uncertain.

The implications of the Vienna Convention and its Montreal Protocol for climate engineering implementation are therefore unclear and dependent upon the outcome of future research. If stratospheric aerosol injection (or some other form of climate engineering) were to cause or were likely to cause "adverse effects resulting from modification or likely modification of the ozone layer," then Parties would be committed to adopt policies to control, limit, reduce, or prevent these activities.[64] Note that "significant" implies a relatively moderate threshold for the magnitude of harm, greater than "detectable" but not necessarily "serious" or "substantial." The Parties to the Montreal Protocol could then choose to add the ozone-depleting substance to the list of controlled substances.[65] Because they are to "take appropriate measures … to protect human health and the environment against adverse effects resulting or likely to result from human activities which modify or are likely to modify the ozone layer," the Parties' decision should balance the protective and adverse effects that occur through impacts on the ozone layer.[66] However, given that sulfate aerosols are already produced, albeit unintentionally, through industrial processes (especially coal combustion) in the lower atmosphere at an annual rate roughly ten times that which would be needed to offset global warming, in this case the Parties would need to incorporate some qualifier regarding the location of emissions.[67]

In contrast to its approach to climate engineering implementation, the Vienna Convention is somewhat clearer regarding climate engineering research. Its Parties commit to

> "[c]o-operate by means of systematic observations, research and information exchange in order to better understand and assess the effects of human activities on the ozone layer" and "to initiate and co-operate in … research and scientific assessment on: The physical and chemical processes that may affect the ozone layer; … [c]limatic effects deriving from any modifications of the ozone layer; … [and] [s]ubstances, practices, processes and activities that may affect the ozone layer, and their cumulative effects."[68]

Climate engineering in general, as well as cirrus cloud thinning and stratospheric sulfate aerosols specifically, are activities and/or substances that may affect the ozone layer. Therefore, Parties implicitly committed themselves in this article to researching potential impacts on stratospheric ozone from climate engineering activities. The obligation to conduct research, which the Montreal Protocol reiterates and expands, also includes a corollary duty to transfer technology.[69]

3.3.1.3. Convention on Long-Range Transboundary Air Pollution

The third and final multilateral atmospheric agreement that is relevant to climate engineering is the Convention on Long-Range Transboundary Air Pollution (CLRTAP) of 1979. Like the UNFCCC and the Vienna Ozone Convention, it is a framework convention with operationalizing protocols. In contrast, though, CLRTAP is a regional agreement, developed under the UN Economic Commission for Europe (UNECE), addressing the problem of acid rain due to the transboundary movement of certain pollutants. Its 51 Parties include all fully industrialized countries and most emerging economies of North America, Europe, and central Asia. As in other framework conventions, Parties to CLRTAP itself have only general and often softly worded commitments to, for example, monitor, exchange information, consult with one another, and develop air quality management systems.[70] CLRTAP's principles use obligatory language, in which its Parties inter alia "shall endeavour to limit and, as far as possible, gradually reduce and prevent air pollution."[71] Here, the definition of air pollution – which is seen elsewhere in international environmental law – includes GHGs, perhaps global warming, and any harmful substances used for atmospheric climate engineering:

> "Air Pollution" means the introduction by man, directly or indirectly, of substances or energy into the air resulting in deleterious effects of such a nature as to endanger human health, harm living resources and ecosystems and material property and impair or interfere with amenities and other legitimate uses of the environment.[72]

Note that, here, deleterious effects must reach a certain threshold of magnitude, and must have actually already occurred, which is or will soon be the case with GHGs and global warming. This definition highlights the recurring tension in international environmental law between climate change and climate engineering. Clearly, CLRTAP and its current Protocols – like current legally binding international law – do not address climate engineering per se. However, given this definition of pollution, a new Protocol to govern climate engineering, especially those methods that would operate by introducing substances into the air, would be within the scope of CLRTAP. Finally, this agreement has institutional support from the UNECE, its own standing bodies, a noncompliance procedure, ongoing scientific support, and the capacity to be amended through protocols.

The legal implications of CLRTAP for the implementation of climate engineering, and especially of atmospheric SRM and cirrus cloud-seeding methods, are somewhat similar to those of the Vienna Ozone Convention. At the very least, source Parties must report their emissions of acid rain precursors – which include sulfates, presently the leading candidate substance for stratospheric aerosol injection SRM – and consult with other Parties "which are actually affected by or exposed to a significant risk of long-range transboundary air pollution."[73] CLRTAP also commits its Parties to the research and development of, among other things, technologies to reduce air pollution and "economic, social and environmental assessment[s] of alternative measures for attaining environmental objectives including the reduction of long-range transboundary air pollution."[74] Given that GHGs and perhaps also global warming are air pollution according to the CLRTAP definition, this provision implies a commitment to research, develop, and assess climate engineering methods that have a potential to reduce the deleterious effects of GHGs and/or global warming. If a climate engineering activity were known to actually cause deleterious effects, then CLRTAP Parties would be obligated to, among other things, endeavor to limit and to gradually reduce and prevent it, as far as possible. If these activities also reduced the deleterious effects of GHGs and/or global warming, then the Parties (likely operating through the Convention's Executive Body and Implementation Committee) should also to take this into account as appropriate. Furthermore, if particular climate engineering techniques were believed to offer "the best available technology which is economically feasible and low- and non-waste" to reduce harm from GHGs and/or global warming, then Parties arguably would be committed to include them as part of their air quality management systems that are designed to combat air pollution, though it is unclear whether the Parties would have to actually carry them out.[75]

The goals and commitments of CLRTAP are furthered by eight protocols, most of which establish emission limits for various long-range air pollutants. All protocols are in force, with differing constellations of participating states.[76] Three of these – the 1985 Helsinki Protocol, the 1994 Oslo Protocol, and the 1999 Gothenburg Protocol – include sulfate emissions. Research on stratospheric sulfate aerosol injection appears to be favored under the Oslo Protocol, wherein Parties commit to "encourage research, development, monitoring and cooperation related to … [t]he understanding of the wider effects of sulfur emissions on human health [and] the environment."[77] One reading of this language is that it is a commitment to improve the understanding of the potential health and environmental impacts of sulfate-based stratospheric aerosol injection. The sulfate emissions at the quantities necessary for climate response field tests or for full implementation of stratospheric aerosol injection, if conducted within the territory of a Party, would exceed the limits under these Protocols.

CLRTAP has a noncompliance procedure. Cases of possible noncompliance are first reviewed by the Implementation Committee, and decisions regarding how to respond are taken by the Executive Body. These bodies would be able to consider three mitigating factors when considering noncompliance with a CLRTAP Protocol due to stratospheric sulfate aerosol injection. First, these activities might reduce global warming and the harm therefrom. Second, the goal of CLRTAP and its Protocols is to reduce adverse effects on humans and ecosystems from air pollution, whose definition includes GHGs and perhaps global warming.[78] Although it would be in Parties' general interests to consider both the beneficial and adverse effects of atmospheric climate engineering activities, they are not obligated to consider the former under CLRTAP and its Protocols. Third, the sulfate particles injected into the stratosphere would be deposited across the globe, with a very small impact on the precipitation within Parties' territories. Therefore, although the sulfate emissions from SRM research or implementation might exceed a Party's limit, their effect on acid rain would be minimal.

3.3.2. *Marine Agreements*

Climate engineering and the world's oceans are deeply interrelated. Elevated atmospheric GHG concentrations will affect the oceans by warming and acidifying their waters and by raising the sea level. In turn, the oceans will moderate terrestrial climatic effects by serving as sinks for both GHGs and heat. Thus, regardless of their location, all climate engineering methods undertaken at sufficient scale would affect the oceans, at least indirectly, by reducing atmospheric carbon dioxide or incoming solar radiation and the resulting heat. Moreover, some climate engineering activities may actually take place in, on, or over the seas, either inherently – such as ocean upwelling or marine cloud brightening – or potentially – such as stratospheric aerosol injection performed from maritime vessels or platforms.

In some ways, the oceans resemble the atmosphere: physically in that they consist of a large body of a fluid that mixes through currents, and legally in that some portions of this body are demarcated as being within the sovereign territory of (some) states. The most important legal difference for present purposes is that there is a central comprehensive multilateral agreement that regulates most maritime activities, the UN Convention of the Law of the Sea. Most of this subsection consequently focuses on its lengthy text, but then looks at a pair of related agreements that govern dumping in the oceans. Notably, in the cases described here, either the multilateral agreement systems contain specific provisions for research, or their Parties have chosen to make such a distinction in their implementation. Consequently, they place climate engineering research and implementation in distinct legal lights.

3.3.2.1. United Nations Convention of the Law of the Sea

The world's oceans are historically the most important area beyond national jurisdiction, and have been the object of much international law.[79] The 1982 United Nations Convention on the Law of the Sea (UNCLOS) is a comprehensive multilateral agreement that established a legal and institutional setting for international cooperation to govern the activities of states and – indirectly – private actors in, on, and above the oceans. UNCLOS describes Parties' rights, duties, and other commitments in their maritime activities, including their obligations to protect the marine environment and their rights and duties in conducting marine scientific research. As such, it governs climate engineering activities that would take place in or be likely to impact the marine environment, including the atmosphere above the oceans. The agreement counts most countries – but not the USA – as Parties, and much of its content is considered to reflect customary international law, which applies to non-Party states as well.[80] UNCLOS is supported by regular meetings of its Parties; by dedicated bodies created by UNCLOS, including an International Tribunal of the Law of the Sea to resolve disputes; and by the International Maritime Organization (IMO), a specialized UN agency that predates UNCLOS.

Among its purposes, UNCLOS is an environmental agreement that provides that "States have the obligation to protect and preserve the marine environment" with neither qualification nor exception.[81] UNCLOS does not define the "marine environment," but the term is generally understood to include the entire space above (i.e., the atmosphere and the sea surface), within, and below (i.e., the seabed and subsoil) the oceans.[82] Parties' sovereign right to exploit their natural resources is explicitly subject to their duty to protect and preserve the marine environment.[83] In this context, climate engineering could be considered a means of exploiting the ocean's natural resources. For CDR, the natural resource would be the water's ability to absorb carbon. Applying this category to SRM is difficult but not impossible by, for example, considering the ocean's capacity to produce reflective clouds as a natural resource. Parties are also to cooperate in developing regulations for environmental protection.[84] Somewhat more specifically, the agreement addresses "pollution of the marine environment," defined in a manner very similar to that of CLRTAP as:

> the introduction by man, directly or indirectly, of substances or energy into the marine environment, including estuaries, which results or is likely to result in such deleterious effects as harm to living resources and marine life, hazards to human health, hindrance to marine activities, including fishing and other legitimate uses of the sea, impairment of quality for use of sea water and reduction of amenities.[85]

Pollution can come from elsewhere, such as the land or the terrestrial atmosphere, provided that it impacts the marine environment. Therefore, for example, terrestrial CDR methods such as enhanced weathering (in which the captured carbon dioxide and dissolved minerals enter surface waters and subsequently the oceans) or land-based stratospheric aerosol injection could cause pollution of the marine environment. Furthermore, GHGs and probably global warming qualify under UNCLOS as pollution of the marine environment.[86] Unlike CLRTAP, the UNCLOS definition includes substances or energy that are merely likely to cause deleterious effects, not only those that have already done so. Parties' commitments concerning pollution of the marine environment include the following:

- "to take … all measures consistent with this Convention that are necessary to prevent, reduce and control pollution of the marine environment from any source";

- to "ensure that activities under their jurisdiction or control are so conducted as not to cause damage by pollution to other States and their environment, and that pollution arising from incidents or activities under their jurisdiction or control does not spread beyond the areas where they exercise sovereign rights";
- to notify potentially affected Parties and competent international organizations when they become aware of actual or imminent pollution damage;
- to cooperate in eliminating the effects of pollution and in preventing or minimizing the resulting damage;
- to "take all measures necessary to prevent, reduce and control pollution of the marine environment resulting from the use of technologies under their jurisdiction or control";
- to monitor the risks or effects of pollution and to publish the results therefrom;
- to assess and to communicate the expected effects of potential "substantial pollution of or significant and harmful changes to the marine environment"; and
- to adopt and enforce laws and regulations to reduce pollution, including from their own vessels, those that enter their territorial waters or their quasi-territorial exclusive economic zones (see below), land-based sources, and the atmosphere.[87]

Additionally, Parties are obligated to undertake their own measures, and to cooperate with others, to conserve living resources and mammals and "to protect and preserve rare or fragile ecosystems as well as the habitat of depleted, threatened or endangered species and other forms of marine life."[88]

What these and other provisions mean with respect to climate engineering activities is uncertain. Their application would depend in part upon the extent to which the activity in question would reduce or would be likely to reduce the "pollution" of atmospheric GHG concentrations or global warming, as well as the extent to which the activity itself would result in or would be likely to result in deleterious effects. This dynamic is another example of the tension between climate change and climate engineering. At one extreme, if it were certain or likely that the activity would lessen the negative impacts from climate change on the marine environment and maritime activities, while posing little risk of its own, Parties with the capacity to do so might be obligated – at least theoretically – to undertake the climate engineering activity. The state would need to do so in a manner consistent with UNCLOS, including exercising due diligence by taking all measures necessary to minimize the deleterious effects of the climate engineering activity itself.[89]

At the other extreme, if the activity were unlikely to directly or indirectly reduce atmospheric GHG concentrations or global warming yet would cause or would be likely to cause large deleterious effects, then Parties would be committed to take all measures necessary to prevent, reduce, and control the climate engineering activity within their jurisdiction or under their control.[90] If substantial negative impacts were expected, the Party carrying out or overseeing the activity would need to assess and communicate these expected effects prior to undertaking it.[91] Because some of the greatest threats to the marine environment from elevated carbon dioxide concentrations come from the acidification of ocean waters, these considerations generally tilt in favor of CDR climate engineering methods, which could prevent, reduce, and/or control this effect, compared with SRM proposals, which would not directly reduce ocean acidification. Likewise, those climate engineering methods that would intervene more directly in the marine environment, such as ocean fertilization and microbubble injection, appear more likely to have deleterious effects on the marine environment than those that would do so less directly or only indirectly, such as stratospheric aerosol injection or terrestrial enhanced weathering. Here, UNCLOS would appear less favorable to the former (direct) than the latter (indirect) methods.

The tension and balance among the deleterious impacts of climate change, the potential for climate engineering to reduce these impacts, and climate engineering's own environmental risks are further complicated by a provision in UNCLOS: "In taking measures to prevent, reduce and control pollution of the marine environment, States shall act so as not to transfer, directly or indirectly, damage or hazards from one area to another or transform one type of pollution into another."[92] Climate engineering could transfer hazards from one area to another, such as from the atmosphere to the ocean, and might transform one type of pollution, such as global warming, into another, such as stratospheric sulfates that destroy ozone. (Notably, other responses to climate change, including emissions abatement and adaptation, as well as actions to reduce pollution in general done at sufficient scale, also sometimes transfer the location of damage or hazards and transform the type of pollution.) Notably, this article uses the strong verb "shall," and makes no explicit provision for the possibly lower relative magnitude and/or probability of the new damage, hazard, or pollution. Legal scholars have reached divergent conclusions regarding this article's implication for climate engineering. For example, James Edward Peterson concludes that "only if the process left harmful residues or byproducts would it appropriately be considered polluting."[93] Yet Philomene Verlaan calls it "a particularly difficult hurdle for geo-engineering projects," and asserts that "[p]roponents of geo-engineering projects must show, inter alia, why such projects do not violate Article 195."[94] However, note that the oceans presently absorb approximately one-third of anthropogenic carbon dioxide emissions.[95] Once emissions reduce and cease, this process will continue, eventually removing most of the remaining anthropogenic carbon dioxide until the atmosphere and ocean reach a new equilibrium. In this regard, some CDR methods would arguably not be a transfer and transformation, but instead merely the acceleration of a process that is already occurring in response to anthropogenic disturbance to the atmosphere.

Because UNCLOS contains numerous provisions regarding marine scientific research, climate engineering research would face different legal circumstances than would implementation. Although UNCLOS generally supports and encourages "marine scientific research," the term remains undefined. Most definitions considered during the drafting of UNCLOS and in legal scholarship emphasize the importance of the marine environment as the object of study, but diverge on whether the research must occur in the marine environment.[96] Climate engineering research is discussed in detail in Chapter 6 of this volume, but it is noted here that the situation is complicated by the fact that there is often no clear distinction between outdoor research and implementation for many climate engineering methods. Regardless, Parties and competent international organizations have a right to conduct marine scientific research, and are obliged, among other things, to

> "promote and facilitate the development and conduct of marine scientific research"; to "promote international cooperation in marine scientific research"; "to create favourable conditions for the conduct of marine scientific research in the marine environment"; and "to make available … information on proposed major programmes and their objectives as well as knowledge resulting from marine scientific research."[97]

In fact, Parties are committed "to observe, measure, evaluate and analyse, by recognized scientific methods, the risks or effects of pollution of the marine environment."[98] This obligatory category could include both climate change and climate engineering, as they each pose risks of pollution of the marine environment. However, Parties' right to conduct research is subject to some limitations, most generally "to the rights and duties of other States."[99] More specifically, marine scientific research shall

> "be conducted exclusively for peaceful purposes"; "be conducted with appropriate scientific methods and means"; "not unjustifiably interfere with other legitimate uses of the sea"; and "be conducted in compliance with all relevant regulations … including those for the protection and preservation of the marine environment."[100]

As discussed in more detail below, Parties and sponsoring international organizations are responsible for and might be held liable for damage resulting from scientific research undertaken.[101]

Parties' rights, obligations, and other commitments with respect to climate engineering will vary based upon the location of the activity. This factor is somewhat complicated in that the jurisdiction over ships and their activities is partially shared among states. Across the seas, ships must fly the flag of a particular state that has given them permission to do so and to which the ship has a genuine link.[102] The flag state is supposed to exercise its jurisdiction over its flagged ships, including by ensuring that its crew is familiar with regulations concerning marine pollution and by adopting laws and regulations to minimize pollution from its flagged ships.[103] Furthermore, UNCLOS divides the water, water surface, and atmosphere horizontally into three primary zones of jurisdiction of coastal states. The first 12 nautical miles from shore is the territorial sea of the coastal state, and is part of its sovereign territory.[104] Activities other than innocent passage are subject to the approval of the coastal state.[105] Climate engineering activities would not be innocent, at the very least because the foreign ship would be engaged in an "activity not having a direct bearing on passage."[106] Coastal states' jurisdiction in their territorial waters also includes "the exclusive right to regulate, authorize and conduct marine scientific research," and research there by foreign ships requires their express consent.[107] Coastal states have a right to enforce their laws and regulations in territorial waters, including through detention and physical inspection of ships and through punishment by expulsion from their territorial seas, monetary penalties, arrests (in some circumstances), and – in the case of willful and serious pollution – unstated nonmonetary penalties.[108] Coastal states arguably have an obligation to take such enforcement action.[109]

The second primary jurisdictional zone defined by UNCLOS is the coastal state's exclusive economic zone (EEZ), which is the first 200 nautical miles from the coast.[110] There, the coastal state has sovereign rights over "exploring and exploiting, conserving and managing the natural resources, whether living or non-living … and with regard to other activities for the economic exploitation and exploration of the zone," as well as jurisdiction over installations and structures, marine scientific research, and protection of the marine environment.[111] In exercising its limited jurisdiction in the EEZ, the coastal state and other states must have due regard for each other's rights and duties.[112] States and, consequently, their flagged ships in an EEZ must comply with the laws and regulations of the coastal state.[113] The enforcement rights of a coastal state and its jurisdiction over marine scientific research, such as climate engineering research, in its EEZ (which hereinafter refers to the EEZ beyond the coastal state's territorial waters) are somewhat similar to those in its territorial waters.[114] However, the coastal state should also grant its consent for marine scientific research "in normal circumstances."[115] In turn, the researching state must provide certain information to the coastal state, both before and during the research project. This obligation includes ensuring that the coastal state may participate in or be represented in the research project if it so wishes.[116] As noted above, climate engineering implementation might qualify as a means of exploiting or managing the ocean's nonliving natural resources. Under such an interpretation, coastal states would have sovereign rights over the climate engineering activities in question within their EEZs. Regardless, any dispute concerning activities in the EEZ between the flag state and the coastal state "should be resolved on the basis of equity and in the light of all the relevant circumstances, taking into account the respective importance of the interests involved to the parties as well as to the international community as a whole."[117] This proviso implies that the severity of climate risks that the states face and the potential for the climate engineering activity to reduce (or exacerbate) the risks are relevant factors in resolving the conflict.

The third and final marine zone is the high seas, which lie beyond the EEZs and are open to all states for peaceful purposes, including marine scientific research, provided that the states exercise their freedoms there "with due regard for the interests of other States."[118] Thus, climate engineering and other activities on the high seas should not inappropriately interfere with, inter alia, the navigation, overflight, fishing, and scientific research of other states.[119] In the event of an incident on the high seas that results in serious damage to the marine environment or other vessels, the flag state is to hold an inquiry.[120] As an aside, states' rights in the high seas and in the EEZs of other states include the right to overflight, and thus their various rights and duties in these zones described above would extend to atmospheric-based climate engineering activities undertaken there.[121]

One particular challenge on the high seas is the regulation of ships that bear the flag of a non-ratifying state or no flag at all. For example, a ship that was reportedly flying the flag of a Canadian indigenous people's village conducted rogue ocean fertilization in the high seas in 2012.[122] It is reasonable to conclude that Parties might be responsible for the activities of their nationals in such circumstances. However, UNCLOS is not entirely clear to what extent its provisions regarding the marine environment and marine scientific research extend to Parties' nationals. On one hand, several of the key articles regarding preventing, reducing, and controlling pollution commit Parties to take measures regarding "activities under their jurisdiction or control," whereas others refer to "their natural or juridical persons," suggesting that the former duties do not extend to the activities of the latter group.[123] On the other hand, Parties can exercise some control over their nationals, and their obligations to protect and preserve the marine environment in general and to take all measures to minimize pollution from any source are without any qualification regarding their jurisdiction or control.[124]

In four articles, UNCLOS establishes, or at least reaffirms, liability for damage from Parties' activities at sea, for the most part limited to damage from acts that are contrary to UNCLOS or to international law generally. Some of these provisions could apply to climate engineering activities that cause harm, depending on the circumstances at hand and on the interpretation of the articles. The first and most generally relevant article addresses liability for harm from pollution to the marine environment, which, as discussed above, could include climate engineering activities and/or the results of such activities if certain conditions were satisfied.[125] This article states that "concerning the protection and preservation of the marine environment [Parties] shall be liable in accordance with international law." This language implies the application of the existing customary international law of state responsibility, in which states are responsible – including but not limited to providing "reparation –" for harm to other states from their acts that are contrary to international law. More specifically, it only requires Parties to "ensure that recourse is available in accordance with their legal systems for prompt and adequate compensation or other relief in respect of damage caused by pollution of the marine environment by natural or juridical persons under their jurisdiction."[126] This provision establishes only a procedural standard for compensation – not a substantive one for state liability – for harm from pollution arising from acts that are not necessarily contrary to international law. Furthermore, under this article UNCLOS Parties are also to cooperate in implementing and further developing international law relating to liability for pollution damage. A second article in UNCLOS establishes similar standards for liability for pollution damage arising from marine scientific research.[127] Unlike the more general article regarding pollution described above, which merely refers to existing international law, this one explicitly states that Parties and international organizations will be liable for such damage due to research that they or their citizens undertake and that is somehow contrary to UNCLOS. In the third relevant UNCLOS article, Parties are liable for damages and loss from their enforcement of laws and regulations with respect to pollution of the marine environment when the enforcement measures are "unlawful or exceed those reasonably required in the light of available information."[128] This provision could apply, for example, either to Parties' efforts to prevent, reduce, or control the pollution arising from climate engineering, or more speculatively, to their own climate engineering activities if these were undertaken with the intent for the activities to function as an enforcement mechanism against the pollution by GHGs or global warming.[129] The fourth article establishes state liability for harm to the seabed, ocean floor, and subsoil thereunder, again only for harm arising from noncompliance with UNCLOS.[130]

A few additional notes regarding liability for marine activities are necessary. The four provisions in UNCLOS described above imply the existence of a harmed victim who can pursue compensation, yet marine life – which is among the possible victims in the definition of pollution of the marine environment – that is not a living resource and that is located in the high seas, cannot be represented by a state or private person in order to claim compensation.[131] Additionally, UNCLOS leaves the scope of "damage" and "loss" undefined. Finally, the International Convention on Liability and Compensation for Damage in Connection with the Carriage of Hazardous and Noxious Substances by Sea would create a two-tier civil liability system paid for through mandatory insurance and an industry-wide fund.[132] Shipowners would be held strictly liable for damage, including environmental damage, from specific substances that are shipped by sea. The most widely considered material for stratospheric aerosol injection – sulfur in various forms – is on the list of substances, and future climate engineering techniques could utilize other hazardous and noxious substances. However, this agreement is not in force due to the absence of ratifying states, and this appears unlikely to change soon.

In one part of UNCLOS, Parties make various commitments to develop and transfer marine technology, especially to developing states. This obligation could apply to climate engineering technologies, depending on how one interprets the undefined term "marine technology" and the particular climate engineering technology at hand. Most generally, Parties are "to promote actively the development and transfer of marine science and marine technology on fair and reasonable terms and conditions."[133] The commitment to the "development of the marine scientific and technological capacity of States which may need and request technical assistance in this field" is explicitly with regard to "the protection and preservation of the marine environment, marine scientific research and other activities in the marine environment," strengthening the case that some of these provisions could apply to climate engineering.[134] This is to be done with due regard for the rights and duties of the holders, suppliers, and recipients of marine technology.[135] Together, these provisions could reasonably be interpreted as committing industrialized UNCLOS Parties to the development of climate engineering methods and the transfer of these technologies to developing countries on terms that do not unduly harm patent holders and others with interests in the production of these technologies.

Four particular categories of climate engineering methods are examined more closely here due to their possible locations or means of operation. First, direct air capture and BECCS technologies, which would qualify as climate engineering depending upon their scale, could store the captured carbon dioxide on or under the seabed.[136] There are provisions in UNCLOS regarding the seabed and the subsoil thereunder.[137] These were crafted to govern the exploration and exploitation of natural resources – especially fossil fuels – there. However, some articles could apply to the seabed surface or subsurface storage of carbon dioxide. The zones, rights, and obligations at these deep locations differ somewhat from those of the superjacent water, water surface, and atmosphere described above, although most of the provisions in UNCLOS regarding the protection of the marine environment and the minimization of pollution apply to all locations. Coastal states have certain sovereign rights to the seabed and subsoil of their continental shelves, a zone that includes at least the seabed surface and subsurface under states' EEZs, but that can extend further into the area under the high seas depending on the topology of the seabed.[138] Among coastal states' rights there is the exclusive right "to authorize and regulate drilling on the continental shelf for all purposes," which would presumably include sub-seabed carbon dioxide storage.[139] Beyond the continental shelf lies "the Area," which is governed in a manner that is novel in international law.[140] It is designated in UNCLOS to be the common heritage of mankind, managed and regulated by the International Seabed Authority. Conduct there must be "in the interests of maintaining peace and security and promoting international cooperation and mutual understanding," and only for peaceful purposes.[141] The International Seabed Authority has the authority to develop regulations to protect the marine environment of the Area.[142] As stated above, Parties are explicitly liable for damage caused by their conduct in the Area that fails to comply with the UNCLOS.[143] Both Parties and the International Seabed Authority may carry out marine scientific research there, provided that the researching states promote cooperation.[144]

Numerous regional multilateral agreements supplement UNCLOS regarding maritime activities and protection of the marine environment. Although this chapter does not examine these in detail, it should be noted that the Convention for the Protection of the Marine Environment of the North-East Atlantic, which is in effect and has been ratified by 16 Parties, has been amended to permit and regulate the storage of captured carbon dioxide below the seabed of the northeastern Atlantic Ocean.[145]

The second category of climate engineering methods operate through the placement of diffuse substances into the marine environment. This group includes CDR proposals such as ocean fertilization, ocean alkalization, and direct air capture followed by injection of carbon dioxide into the water column. It could also include maritime SRM by injection of microbubbles, stratospheric aerosol injection, or marine cloud brightening as well as cirrus cloud thinning. Both sets of methods might (or might not) be considered dumping, which the UNCLOS defines as "any deliberate disposal of wastes or other matter from vessels, aircraft, platforms or other man-made structures at sea," but excludes "placement of matter for a purpose other than the mere disposal thereof, provided that such placement is not contrary to the aims of this Convention."[146] Based on an ordinary interpretation of these words, the purpose of placing fertilizing or alkalinizing matter in the ocean, or the injection of aerosols or fine ocean mist, would not be *mere* disposal of those substances, but instead would be to indirectly remove carbon dioxide from the atmosphere or to reduce climate change risks in some other manner. This logical exclusion from dumping would require that the practice not be contrary to UNCLOS's unstated aims. Based upon its Preamble and its numerous commitments, these implicitly include protection of the marine environment, raising the tension between and consequently the balance of climate change and climate engineering.[147] For example, if ocean fertilization were likely to have or actually had deleterious effects, it would qualify as pollution of the marine environment, especially if it sequestered little carbon dioxide relative to its own deleterious effects. If this impact were known or expected in advance, the activity appears to be contrary to the aims of UNCLOS, and would therefore be dumping.[148] There would be some question whether marine cloud brightening, microbubble injection, and ocean upwelling would also be considered dumping, because these practices would merely involve the movement of natural material from one location to another within the marine environment. In contrast, the injection of captured carbon dioxide into the water column *does* appear to be the mere disposal thereof. Note that the definition of dumping does not include matter such as carbon dioxide from land-based sources that is transported to the marine environment via pipelines or outfall structures; that would potentially be pollution from land-based sources.[149]

Regardless, Parties are to prevent, reduce, and control pollution of the sea by dumping.[150] This requires that the dumping be likely to cause deleterious effects, which may or may not be the case with these marine climate engineering methods, depending upon their modality, quantity or magnitude, and location. Coastal states have the right to regulate dumping in their EEZs, where the practice requires their prior consent.[151] The Parties to UNCLOS are to establish global and regional regulation regarding dumping in order to reduce the pollution that arises from it.[152] Moreover, Parties' national laws "shall be no less effective … than the global rules and standards."[153] International regulation has been developed and implemented largely through the London Convention and London Protocol, considered below, although participation in these agreements is not as broad as that of UNCLOS. A possible interpretation, albeit an unsettled and controversial one, is that UNCLOS Parties that are not parties to the London Convention and/or the London Protocol are nevertheless committed to some or all aspects of those dumping-specific agreements.[154]

With regard to the third category of climate engineering methods, some researchers believe that marine cloud brightening SRM could be carried out by numerous unmanned boats. Although the interchangeable words "ship" and "vessel" remain undefined in UNCLOS, they are widely understood to include unmanned ships. They would therefore be required to bear the flag of a state to which they have a genuine link. That flag state would need to issue laws and regulations to prevent, reduce, and control the ships' potential pollution of the marine environment, and to conduct an inquiry in the event that the ship causes serious damage to the marine environment.[155] Nevertheless, some legal uncertainties arise with the deployment of unmanned ships, whether for climate engineering or for other purposes.[156] Specifically, UNCLOS gives certain responsibilities to the ship's master, officers, and crew, none of whom would be present on unmanned ships. For example, the flag state of a ship must take measures

> "to ensure … that each ship is in the charge of a master and officers who … are fully conversant with and required to observe the applicable international regulations concerning the safety of life at sea, the prevention of collisions, the prevention, reduction and control of marine pollution, and the maintenance of communications by radio."[157]

To some extent, the shore-based vessel operator(s) could be considered as fulfilling these roles. Other questions surround who would be liable for accidents or violations of international or coastal state laws and regulations. Another set of concerns regards how to ensure that unmanned ships would navigate the seas in such a manner that they would have due regard for the rights and interests of other states. This includes not hindering other ships' navigation and not entering other states' territorial water or – in some circumstances such as unauthorized marine scientific research – their EEZs.

In the fourth category, some marine climate engineering methods, particularly ocean upwelling, could function through objects that are placed and left in the ocean.[158] Parties' rights and obligations under UNCLOS regarding this are somewhat confusing in that the agreement usually – but not always – uses the undefined terms "artificial islands, installations and structures" in the general context and "installations and equipment" in the context of marine scientific research, yet in some passages omits one or either of these terms.[159] Regardless, throughout the ocean, Parties' obligation to take measures to minimize pollution of the marine environment explicitly extends to "pollution from other installations and devices operating in the marine environment."[160] Installations or equipment for research must bear the identifying markings of the state of registry or the international organization to which they belong.[161] In the EEZs or on the continental shelves, coastal states have jurisdiction over these placed objects, including the exclusive right to construct, authorize, and regulate artificial islands, installations, and structures.[162] Further, due notice must be given of their construction.[163] Although coastal states are to grant permission under normal circumstances for other states to conduct marine scientific research in their EEZs and on their continental shelves outside their territorial waters, "the construction, operation or use of artificial islands, installations and structures" is one of four justifications for them to deny such permission.[164] For marine scientific research in all locations, unused installations, structures, and equipment must be removed; artificial islands, installations, and structures may have a designated zone of safety; and all categories of placed objects may not interfere with existing sea lanes.[165] On the high seas, all Parties have the "freedom to construct artificial islands and other installations permitted under international law," provided that they exercise due regard for the exercise of high seas freedoms by other states.[166] Moreover, because objects placed in the water for climate engineering might be (and are presently imagined mostly to be) free floating, the Party that oversees their placement would need to ensure that they do not migrate into the EEZ of another state.

Although the possible application of UNCLOS to particular scenarios of climate engineering remains uncertain in many ways, the agreement might be able to provide a legal and institutional home for future governance of a wide array of climate engineering activities due to a handful of remarkable characteristics. UNCLOS contains numerous commitments that use legally binding language for the prevention, reduction, and control of pollution. This pollution can include GHGs, global warming, and potentially harmful climate engineering activities if their actual or expected negative effects occur in the marine environment. This encompasses the ocean's water and atmosphere, and cases in which the pollution originates from the atmosphere or land. The threshold of "likely to result in such deleterious effects" extends potential governance of climate engineering to an anticipatory stage. Furthermore, most countries are Parties to UNCLOS, and the major one that is not – the USA – explicitly considers most of it to reflect customary international law, including the parts regarding protection of the marine environment and marine scientific research.[167] UNCLOS Parties generally meet annually and can adopt amendments to the Convention. There is a standing International Tribunal for the Law of the Sea to resolve disputes, and the IMO offers an institutional home.[168] Of course, UNCLOS is not perfect for regulating climate engineering: Many of its commitments are vague, and its institutions' exercise of authority over climate engineering activities could lead to conflicts with the provisions of the UNFCCC.

3.3.2.2. London Convention and London Protocol

The London Convention on the Prevention of Marine Pollution by Dumping of Wastes and Other Matter and the London Protocol thereto are a pair of multilateral treaties that focus upon reducing marine pollution from dumping. They are both in effect and institutionally supported by the IMO. As noted above, they might represent the global rules and standards to which UNCLOS refers, and could therefore be legally binding to some extent on those Parties to UNCLOS that are not Parties to the London Convention and London Protocol.[169] The 1972 London Convention has 87 Parties, including all major industrialized maritime countries, as well as a handful of countries – including the USA – that are not Parties to UNCLOS. In contrast, the 1996 London Protocol – which is intended to replace the London Convention and supersedes the conventions for those states that are Parties to both – presently has 45 Parties, a cohort that lacks major states such as the USA and Russia.[170] Their specific objectives, stated in the obligatory language of "shall," refer to the control of *all* sources of pollution of the marine environment, especially that from dumping.[171] The two agreements apply to all maritime waters, including the high seas, EEZs, and non-inland territorial waters, as well as to the Parties' flagged ships and loading that occurs in their ports. The Protocol uses somewhat stronger language than the Convention and, unlike the Convention, actually defines pollution.[172] Although that definition is mostly similar to that of UNCLOS, it is limited to matter – not including energy – and thus could encompass elevated atmospheric and dissolved greenhouse gases and possibly some climate engineering methods, but not global warming. Among other things, Parties to the London Protocol are committed to promote scientific research on pollution from dumping and from "other sources of marine pollution relevant to this Protocol." The latter presumably includes research into marine CDR, especially those forms that may constitute dumping.[173] Furthermore, Parties may not "transfer, directly or indirectly, damage or likelihood of damage from one part of the environment to another or transform one type of pollution into another."[174] Although this clause resembles that in UNCLOS, its application is limited to Parties' actions in implementing the Protocol.

Most commitments in these two agreements are specific to pollution from dumping, despite their broader objectives. The definitions of dumping in the agreements are very similar to those in UNCLOS, in particular retaining the exception for purposes "other than the mere disposal thereof, provided that such placement is not contrary to the aims of" the agreement.[175] The older Convention uses a "black list" of prohibited substances that may not be dumped (with the dumping of non-prohibited substances subject to impact assessment) and a "gray list" of substances that require a special permit, while all other substances to be dumped require a general permit.[176] Most substances that are presently considered for placement into the water for climate engineering purposes, such as iron and carbon dioxide, are found on neither the "black list" nor the "gray list." However, the latter does categorically include those substances "which, though of a non-toxic nature, may become harmful due to the quantities in which they are dumped, or which are liable to seriously reduce amenities."[177] Consequently, climate engineering activities that were considered to be dumping and were of a sufficient scale would require a special permit under the terms of the London Convention. In contrast, the newer London Protocol generally prohibits dumping except for a "gray list" of substances that require a permit, while "being mindful of the Objectives and General Obligations of this Protocol."[178] That "gray list" includes "inert, inorganic, geological material" and "organic material of natural origin," terms that are not defined and whose applicability to substances used in climate engineering could be clarified in a permitting process. Furthermore, Parties to the London Protocol commit to applying "a precautionary approach to environmental protection from dumping of wastes or other matter whereby appropriate preventative measures" are taken, even in the absence of "conclusive evidence to prove a causal relation between inputs and their effects."[179]

The Parties to the two agreements, which meet jointly, have taken several actions regarding ocean fertilization.[180] In response to the stated intentions of for-profit enterprises to fertilize the oceans and subsequently market carbon credits, the Parties approved a resolution in 2008 stating that ocean fertilization ("any activity ... with the principle [*sic*] intention of stimulating primary productivity in the oceans [excluding] conventional aquaculture, or mariculture, or the creation of artificial reefs") falls within the scope of the two agreements.[181] The nonbinding resolution also concluded that ocean fertilization should generally not be allowed with the exception of legitimate scientific research, which should not be considered to be dumping and should be assessed on a case-by-case basis.[182] The Parties to the two agreements later developed a framework – also nonbinding – for such assessment for scientific research involving ocean fertilization.[183] Under this, Parties are to first determine whether the planned activity would indeed be ocean fertilization and have the attributes of legitimate scientific research: adding to scientific knowledge, not having direct financial gain and the influence therefrom, being subject to peer review, and committing to publish in a peer-reviewed outlet and to make data publicly available.[184] If this is the case, the agency would subsequently conduct an environmental impact assessment that should include, inter alia, site selection and description, exposure assessment, risk characterization, and risk management.[185] The latter process, as well as the ultimate approval decision, should follow a "precautionary approach," which is not elaborated but can be interpreted as being consistent with that of the London Protocol.[186] Prior to any approval, the researching Party should consult with all stakeholders and notify potentially affected countries. Notably, the assessment framework is adaptive, in that the approval can be modified or revoked based upon the content of the required reporting.

In 2013 the Parties to the London Protocol approved a broader amendment to that agreement, which would regulate "marine geoengineering" in general. The amendment – not yet in force – defines this as:

> a deliberate intervention in the marine environment to manipulate natural processes, including to counteract anthropogenic climate change and/or its impacts, and that has the potential to result in deleterious effects, especially where those effects may be widespread, long lasting or severe.[187]

When and if this amendment comes into effect, the Parties would collectively maintain a new Annex 4 to the Protocol, listing specific marine geoengineering activities.[188] Notably, a two-thirds majority of its Parties that are present and voting can modify the Protocol's Annexes.[189] The Parties would then be committed to not allowing the placement of matter into the sea for these activities, unless the Annex's particular listing allows for case-by-case authorization by the Party.[190] Placement of matter for marine geoengineering activities that are not listed in the Annex is implicitly permitted, provided that this placement is neither dumping nor contrary to the Protocol's aims. Presently, the list in the Annex includes only ocean fertilization, which Parties should permit only if the activity is legitimate scientific research.[191] Parties would need to adopt administrative or legislative measures to ensure that pollution to the marine environment from these listed activities is, as far as practicable, prevented or minimized and that the activities are not contrary to the Protocol's aims.[192]

The case-by-case authorizations should follow both a general assessment framework as well as any specific assessment framework for that particular activity. The former is intended to be legally binding upon the Parties that ratify the amendment and is detailed in another new Annex, whereas the latter will be nonbinding and approved by the meetings of Parties.[193] Under the general, legally binding assessment framework, the Party in whose jurisdiction or under whose control the proposed activity would occur is to require a detailed description of the activity, to notify potentially affected countries, and to develop a consultation plan.[194] The activity's proponents must demonstrate that the activity is not mere disposal, that it would fulfill its purpose, that its "rationale, goals, methods, scale, timings and locations as well as predicted benefits and risks" are justified, and that they have the financial resources to carry out the proposed activity adequately.[195] The responsible Party is also to encourage consultation with all stakeholders, and consent – while not required – "should be sought from all countries with jurisdiction or interests in the region of potential impact."[196] Furthermore, the Party and any potentially affected countries should seek expert advice, including peer review of proposals.[197] Ultimately, a permit requires that the proposal has satisfactorily completed all assessments, impact evaluations, and consultation requirements; that it would fulfill its purpose; that the risk management and monitoring requirements have been determined; that the proposal's environmental harm would be minimized while benefits would be maximized; and that pollution would be minimized as far as practicable.[198]

The approved Annex also offers a non-exhaustive list of reasons that some potential marine geoengineering techniques may require specific marine scientific research, and describes the characteristics that constitute such research as well as the conditions that should be imposed upon it.[199] These required conditions are similar to those found in the nonbinding assessment framework for ocean fertilization – i.e., adding to scientific knowledge, having no direct financial gain and the influence therefrom, being subject to peer review, and committing to publish in a peer-reviewed outlet and to make data publicly available. The Annex adds to these the requirement of an appropriate research methodology and sufficient financial resources to carry out the proposed research activity. The two paragraphs concerning marine scientific research are not related explicitly to any particular obligation, but instead appear to be intended to guide the assessment of listed marine geoengineering activities that are limited to or might have alternative assessment criteria for legitimate scientific research.

This amendment to the London Protocol, although not yet in force, is notable in several regards. First, it is the only instrument of international law that is intended to be legally binding upon states and is specifically concerned with climate engineering, although the definition of marine geoengineering is not limited to climatic purposes. In fact, this definition – likewise the first of its type in a binding legal instrument – includes SRM and cirrus cloud-thinning activities, at least those that take place at sea.[200] Second, the amendment relies upon an expansive interpretation of the Protocol's scope, which is not limited to dumping. Instead, as noted, it also includes a commitment to "protect and preserve the marine environment from all sources of pollution," which is recalled in the recitals found in the resolution's Preamble.[201] Furthermore, the amendment's regulatory scope concerns all placement of matter for marine geoengineering, not only dumping. Indeed, the listed marine geoengineering activities will be exempt from the Protocol's central article that prohibits dumping.[202] Third, as with marine geoengineering, criteria for marine scientific research are laid out in a legally binding international legal instrument for the first time.[203] These characteristics of and conditions for marine scientific research, which are also found in similar form in the earlier nonbinding assessment framework, could help clarify that concept in the context of UNCLOS, where the phrase is oft-repeated – including in a freedom to conduct it on the high seas – yet is undefined. Although the Parties to the London Protocol lack the authority to modify UNCLOS, these characteristics and conditions are consistent with UNCLOS's rights and obligations regarding marine scientific and could provide rough contours as to how some members of the international community delineate marine scientific research. Finally, the amendment explicitly adopts a balancing approach to the tension between climate change and climate engineering, calling for "conditions [to be] in place to ensure that, as far as practicable, environmental disturbance and detriment would be minimized and the benefits maximized."[204]

As a final additional note, direct sequestration in the water column, or on or under the seabed, of carbon dioxide that has been captured at sea would qualify as dumping under the London Convention and the London Protocol. In 1999 the Parties to the former received a report from its Scientific Group, which concluded that "fossil fuel derived CO_2 was considered an industrial waste" – a prohibited category. However, they did not reach consensus on this issue, and the legal status of the placement of captured carbon dioxide into the water column consequently remains unclear under the London Convention.[205] The Parties to that agreement did, however, later adopt and subsequently revise guidelines on the storage of carbon dioxide in sub-seabed geological formations.[206] Parties amended the newer London Protocol – whose definition of dumping explicitly includes matter placed in the seabed and the subsoil thereof – almost immediately after it entered into force, in order to allow marine sequestration of carbon dioxide streams only below the surface of the seabed, prohibiting them in the water column or on the seabed surface.[207] A further amendment to the Protocol was approved in 2009, but is not yet in force, that would allow for the international export of captured carbon dioxide for sub-seabed storage.[208]

3.3.3. *Procedural Agreements*

States' obligations under international environmental law are often procedural in nature. Indeed, most multilateral agreements reviewed in this section rely heavily on procedural duties, as does the customary international law discussed below.[209] Two agreements developed under the UNECE solely rely upon procedural duties, and could apply to climate engineering activities.

3.3.3.1. Convention on Environmental Impact Assessment in a Transboundary Context

The Convention on Environmental Impact Assessment in a Transboundary Context (the Espoo Convention) is the first of two procedure-based multilateral environmental agreements developed through UNECE. Finalized in 1991, its current 45 Parties include the European Union, several former Soviet countries, and Canada. If a proposal, such as one for climate engineering activity, is subject to a decision of a national competent authority, then it is a "proposed activity" per the Convention.[210] Although the Espoo Convention's definition of "impact" has a very low threshold ("any effect caused by a proposed activity on the environment … includ[ing] effects on cultural heritage or socio-economic conditions resulting from alterations to" the environment), the obligations in the Convention are limited to cases of actual or likely significant adverse transboundary impacts.[211] The qualifier "transboundary" refers only to those impacts that are experienced in the jurisdiction of one Party and originate from that of another. Most outdoor climate engineering activities will have impacts; many will be subject to regulatory decisions by national authorities; and some of those activities of sufficient scale may cause significant adverse transboundary impacts on other Parties.

The only commitments that apply to all significant adverse transboundary impacts require Parties to "take all appropriate and effective measures to prevent, reduce and control significant adverse transboundary environmental impact from proposed activities."[212] The Parties of the Espoo Convention are also to "give special consideration to" establishing and expanding research programs for better understanding environmental impacts, a soft obligation that could encourage climate engineering research programs.[213] The other commitments mostly apply to only the specific proposed activities listed in an Appendix, none of which could be reasonably construed to include any of the climate engineering techniques that are currently suggested.

However, a proposed activity that is not listed but is deemed likely to have significant adverse transboundary impacts can still be subject to the obligations of the Espoo Convention. For this, a concerned Party (presumably the state of origin of the unlisted proposed activity) is obligated to enter into discussions with another Party (presumably a potentially affected state), and if they both so agree, they are to treat the proposed activity as if it were listed.[214] The Convention recommends some criteria of size, location, and effects that the Parties can use in this consideration.[215] Large-scale climate engineering research and certainly its deployment could satisfy these. When and if Parties agree to this, then the Party of origin is subject to a number of procedural obligations. These include, prior to a decision to authorize, conducting an environmental impact assessment that provides particular information, such as a description of the proposed activity, details of its likely impacts, possible steps to mitigate adverse impacts, potential alternative activities, and uncertainties.[216] The Party of origin is obligated to notify and consult with other affected Parties.[217] In particular, the publics in the likely affected areas are to have the opportunity to participate.[218] The Party of origin should take the outcome of the assessment, the public comments, and the international consultations into account in making the final decision regarding whether and how to proceed with the proposed activity.[219] The Espoo Convention also calls on its Parties, "to the extent possible," to conduct environmental assessments on their relevant policies, plans, and programs, implicitly only those that concern activities covered by the Convention.[220]

The Espoo Convention is supplemented by its Protocol on Strategic Environmental Assessment, which entered into force in 2010, and presently has 26 European Parties, including the European Union. They commit to undertake strategic environmental assessments of certain listed categories of official draft plans and programs that are likely to have significant environmental or health effects.[221] This obligation might extend to plans and programs for climate engineering, particularly for some CDR techniques.[222] Notably, the definitions of likely effects to be considered include, among other things, those on climate.[223] Parties to this Protocol also commit to ensuring the public availability of relevant information, and to providing opportunities for public participation and consultation.[224] If a plan or program would likely have significant transboundary environmental and/or health effects, the Party of origin is to notify and consult with the potentially affected Party, including providing an opportunity for public comment in the affected state.[225]

3.3.3.2. Convention on Access to Information, Public Participation in Decision-Making and Access to Justice in Environmental Matters

The 1998 Convention on Access to Information, Public Participation in Decision-making and Access to Justice in Environmental Matters (the Aarhus Convention) is another UNECE multilateral environmental agreement that relies upon procedural obligations. In this case, its 47 Parties pursue the three objectives given in the Convention's title, which they generally guarantee as rights.[226] In this sense, the Aarhus Convention resembles a human rights agreement more than a traditional environmental one. In contrast to most multilateral environmental agreements, the effects of concern need not be transboundary. Thus, Parties are to offer these rights to the public of the state of origin of the environmental information or proposed activity, as well as to the publics of other Parties, in a nondiscriminatory manner.[227]

Concerning the first objective, "environmental information" is information that concerns the state of the environment and its elements as well as factors – including public policies, plans, and programs, and the analyses upon which they are based – that affect or are likely to affect the environment.[228] Climate engineering activities such as outdoor tests and implementation clearly fall within this definition. Parties' public authorities must make such relevant environmental information available to the public in accordance with requirements, such as timeliness, and with limitations, such as preventing adverse effects on intellectual property rights.[229] Requests for information do not need to demonstrate a particular interest, such as actual or potential harm. The public authorities are also to proactively establish and maintain systems for the collection and dissemination of environmental information.[230]

Regarding the second objective, the public that might be affected, as well as environmental nongovernmental organizations, have the right to participate in decision-making regarding whether to permit proposed activities that could affect the environment, which would include many climate engineering activities. This right could include any activities specifically listed in the Convention's Annex, any activity for which domestic impact assessment legislation provides public participation, or unlisted activities that might have a significant effect on the environment.[231] Once this obligation is triggered, Parties must ensure that the affected public and environmental groups are informed and given access to relevant information and the opportunity to participate, including by providing comments. This participation and comments are to be taken into account by public authorities in their decision-making. Some of these provisions extend, with qualifications, to plans, programs, and policies.[232]

The third objective of the Aarhus Convention, access to justice, is intended to address deficiencies in how Parties have carried out the first two objectives. It is achieved by setting minimum standards for redress for members of the public who have been denied environmental information or who wish to challenge a prior decision concerning environmental matters in which they have a sufficient interest.[233] In this, the Parties are to ensure that courts or other independent tribunals can enforce the rights granted in the Aarhus Convention. The Convention also has relatively strong, novel noncompliance procedures.[234]

The Aarhus Convention is furthered by its Kiev Protocol on Pollutant Release and Transfer Registers, which is in effect through the participation of 32 states as well as the European Union. Its Parties are to ensure the public availability of information regarding transfer and release of pollutants.[235] In this context, a pollutant is any substance that may be harmful to human health or the environment due to its introduction into the environment, and includes both accidental and deliberate release.[236] Thus, substances intended for use in climate engineering might qualify as pollutants. The Protocol's Parties are to establish and maintain a publicly accessible register of such information, which is to include information from private actors acquired via mandated reporting.[237] The Protocol's Parties are to be guided by the precautionary approach in their implementation of the Protocol.[238]

3.3.4. *Other Multilateral Environmental Agreements*

3.3.4.1. Convention on Biological Diversity

The Convention on Biological Diversity (CBD) is one of the most important and far-reaching multilateral environmental agreements. Agreed at the 1992 UN Conference on Environment and Development in Rio de Janeiro, it now includes almost all of the world's countries – except the USA – among its Parties. The CBD can be described as a framework treaty whose commitments are broad, general, and often weakened through qualifying language. Its objective extends beyond the conservation of biological diversity to encompass the sustainable uses of biological resources and the equitable sharing of benefits from genetic resources.[239] With respect to its conservation objective, climate engineering as a response to the risks of climate change could have positive, negative, or mixed impacts on biological diversity.[240] Furthermore, because many human activities, such as climate engineering activities, undertaken at sufficient scale will affect biodiversity, the CBD's broad scope and robust institutional support have led it to function as a vehicle for environmental protection in general. Indeed, its implementation has touched upon issues as diverse as economic development, trade, agriculture, tourism, and climate change.

Two principles and two commitments of the CBD are, in general, relevant for climate engineering activities that might affect biodiversity. The Convention's singular explicit guiding principle is a restatement of states' sovereign right to exploit their own natural resources and their concomitant responsibility to prevent transboundary harm.[241] The CBD also invokes precaution, but only by "noting" it in the document's Preamble. In terms of commitments, the CBD Parties are, "as far as possible and as appropriate," to identify activities that have or are likely to have significant adverse impacts on biodiversity, and to monitor the effects thereof.[242] Article 14 of the CBD, the second relevant commitment, has provisions for three contexts.[243] In terms of domestic effects, Parties are, once again "as far as possible and as appropriate," to require environmental impact assessments for proposed activities that are likely to have significant adverse effects on biological diversity, and to ensure that the impacts are "duly taken into account." In the case of likely significant transboundary impacts, the Parties to the CBD are to promote notification, exchange of information, and consultation, by encouraging multilateral arrangements as appropriate. Finally, if there is grave and imminent danger to biological diversity, Parties should have arrangements for emergency responses and encourage international cooperation. If these dangers are transboundary, then immediate notification and action are required.

The CBD COPs have issued four decisions concerning climate engineering. These are nonbinding on Parties but have been influential.[244] The first, in 2008, was limited to ocean fertilization, under which the COP requested

> Parties and urges other Governments, in accordance with the precautionary approach, to ensure that ocean fertilization activities do not take place until there is an adequate scientific basis on which to justify such activities, including assessing associated risks, and a global, transparent and effective control and regulatory mechanism is in place for these activities; with the exception of small scale scientific research studies within coastal waters. Such studies should only be authorized if justified by the need to gather specific scientific data, and should also be subject to a thorough prior assessment of the potential impacts of the research studies on the marine environment, and be strictly controlled, and not be used for generating and selling carbon offsets or any other commercial purposes;[245]

Parties expanded this statement two years later to include all climate engineering, albeit in softer language:

> The Conference of the Parties … Invites Parties and other Governments … to consider the guidance below …
>
> Ensure … in the absence of science based, global, transparent and effective control and regulatory mechanisms for geo-engineering, and in accordance with the precautionary approach and Article 14 of the Convention, that no climate-related geo-engineering activities that may affect biodiversity take place, until there is an adequate scientific basis on which to justify such activities and appropriate consideration of the associated risks for the environment and biodiversity and associated social, economic and cultural impacts, with the exception of small scale scientific research studies that would be conducted in a controlled setting in accordance with Article 3 of the Convention, and only if they are justified by the need to gather specific scientific data and are subject to a thorough prior assessment of the potential impacts on the environment.[246]

At that time, the COP adopted a tentative definition for geoengineering:

> Any technologies that deliberately reduce solar insolation or increase carbon sequestration from the atmosphere on a large scale and that may affect biodiversity (excluding carbon capture and storage from fossil fuels when it captures carbon dioxide before it is released into the atmosphere).[247]

A later report refined the definition to "a deliberate intervention in the planetary environment of a nature and scale intended to counteract anthropogenic climate change and its impacts."[248] Two later decisions by the CBD COP reaffirmed the previous decisions, made general observations, and requested Parties to address gaps in the understanding of the impacts of climate engineering on biodiversity.[249] The most recent one, from 2016, notably called for more research "in order to better understand the impacts of climate-related geoengineering on biodiversity and ecosystem functions and services, socio-economic, cultural and ethical issues and regulatory options."[250]

These decisions by the CBD COP are important, as they represent the only negotiated consensus concerning climate engineering in general (and ocean fertilization specifically) from representatives of most of the world's states. The two substantive statements are ones of concern, calling upon all states to ensure that climate engineering activities of a certain type or scale do not take place until explicit criteria are met. In the case of ocean fertilization, small-scale scientific research studies within coastal waters – a location that ironically would not be scientifically useful – are the only exception. In order for climate engineering in general to be covered by these decisions, the activity must be at a large enough scale that it would affect biodiversity, and an exception is made for "small scale scientific research studies that would be conducted in a controlled setting." Because that final term has not been defined, it remains unclear whether a "controlled setting" is limited to indoor activities or could include low-risk or fairly well-contained outdoor ones. Each decision describes the conditions under which the request for preventing activities would be lifted, and these are highly congruous: There must be an adequate scientific basis, an assessment or consideration of risks, and (science-based) global, transparent, and effective control and regulatory mechanisms. Appropriate assessment procedures prior to a climate engineering activity could satisfy the first two conditions. In contrast, the need for global regulation is a challenging criterion to satisfy. In its 2012 general decision on climate engineering, the CBD COP noted the continued lack of adequate regulation, while reports issued by the CBD Secretariat have acknowledged that ocean fertilization experiments and sub-seabed carbon dioxide storage are "possible exceptions" to this absence and that "[m]any ocean-based potential geoengineering approaches are already covered under the" London Convention and London Protocol."[251]

At the same time, the CBD COP decisions are limited in their effects, legal and otherwise. Most importantly, as COP decisions they are nonbinding. Indeed, reports issued by the CBD Secretariat call the 2010 decision a "nonbinding normative framework."[252] Furthermore, the CBD itself and the COP decisions consistently use soft and highly qualified language, and that of the decisions is vague. For example, the 2010 decision concerning "climate-related geo-engineering" in general merely "invites" states to "consider the guidance." That decision's requirement that it be interpreted in accordance with the CBD's Article 14 indicates that the Parties intended the decision to be further limited to activities that are likely to have *significant adverse* effects on biological diversity.[253] Although they are sometimes referred to as moratoria, that is an inaccurate characterization of both their substance and intention. Indeed, the COP explicitly rejected the word "moratorium" in the case of the ocean fertilization decision.[254]

Furthermore, the decisions have also been controversial. A first-hand report of the negotiations toward the 2010 COP decision concluded that "the delegates were not well informed about geoengineering, and negotiations were conducted in haste without proper scientific consideration."[255] Moreover, the ad hoc Consultative Group on Ocean Fertilization of the Intergovernmental Oceanographic Commission of the UN Educational, Scientific and Cultural Organization issued a statement in response to the 2008 CBD COP decision on ocean fertilization. In this, the Group stated that it was "concerned that [the decision] … places unnecessary and undue restriction on legitimate scientific activities," that the limitation of scientific research to coastal waters was "new, arbitrary, and counterproductive … [with] no scientific basis," and that scientific information regarding preservation of marine diversity could be obtained through further research.[256]

As a final note, the CBD Secretariat and Subsidiary Body on Scientific, Technical and Technological Advice have issued a number of reports in order to inform its COPs. The most thorough one was published in 2012, and updated in 2016.[257]

3.3.4.2. Antarctica Treaty System

The continent of Antarctica is the second of three primary areas that are typically considered beyond national jurisdiction, and the area south of 60° latitude is governed by a number of interrelated treaties.[258] Some climate engineering activities could occur there, in either the terrestrial or the marine environments. In particular, the Southern Ocean is a leading location for potential ocean iron fertilization. Furthermore, although SRM would generally be less effective there due to the angle of incoming sunlight and the already high albedo, some researchers are interested in regional SRM as a means to preserve ice sheets.[259] Generally speaking, the treaties concerning Antarctica call upon their Parties both to protect the environment and to conduct scientific research. Unlike UNCLOS, these agreements prioritize neither of these goals over the other, emphasizing a potential tension between them.

The central Antarctic Treaty counts 53 Parties, including the major economic powers, and has been in effect since 1961. It establishes a freedom of scientific investigation – presumably including that of climate engineering research – and encourages international cooperation in this area.[260] At their meetings, Parties are to discuss, inter alia, the facilitation of scientific research and the "preservation and conservation of living resources."[261] All activities in Antarctica must be for peaceful purposes.[262]

The 1991 Protocol on Environmental Protection of the Antarctic Treaty (the Madrid Protocol) further explicates states' commitments, obligations, and rights concerning scientific research and environmental protection in Antarctica. It is in effect, with 33 participating states. All of these goals remain high priorities in the Protocol. For example, in its objective, Parties commit to "the comprehensive protection of the Antarctic" and "designate Antarctica as a natural reserve, devoted to peace and science."[263] Its first principle emphasizes this, providing that both environmental protection and "scientific research, in particular research essential to understanding the global environment" are to be "fundamental considerations in the planning and conduct of all activities."[264] Another principle is that "[a]ctivities shall be planned and conducted so as to accord priority to scientific research … including research essential to understanding the global environment."[265] The upshot of this is that climate engineering activities that would help protect the Antarctic environment and climate engineering research that would be free of adverse environmental impacts are each to be prioritized, but it is unclear how Parties should respond to proposals for climate engineering activities – especially research – that might have adverse impacts.

Regarding their substantive commitments, Parties to the Madrid Protocol are to plan their activities in the Antarctic to limit adverse environmental impacts.[266] This limitation includes avoiding, among other things, "adverse effects on climate or weather patterns; significant adverse effects on air or water quality; significant changes in the atmospheric, terrestrial (including aquatic), glacial or marine environments"; detrimental changes to the populations of plants and animals; putting endangered or threatened species at further risk; and degrading or putting at risk "areas of biological, scientific, historic, aesthetic or wilderness significance."[267] These obligations all emphasize the tension between climate change and climate engineering, which respectively will and might have adverse effects on climate, the environment, species, and significant areas. Furthermore, Parties are to plan their activities in Antarctica based on sufficient information to allow a prior assessment of and informed judgments about the potential impacts on the continent's environment and its value as a site of scientific research.[268] In line with this, Parties must cooperate as well as monitor, assess, and report the environmental impacts of their activities.[269] Scientific research programs are explicitly subject to prior environmental impact assessment. If an activity such as climate engineering "results in the significant adverse modification of habitat," or if a research activity were planned to take place in a "Specially Protected or Managed Area," then a permit from the Party's appropriate regulatory authority would be required.[270] Finally, if ocean fertilization or other marine geoengineering were to introduce substances into the sea "in quantities or concentrations that are harmful to the marine environment," then it would be prohibited, although noncommercial government ships are exempt from this prohibition.[271]

3.3.4.3. Convention on the Prohibition of Military or Any Other Hostile Use of Environmental Modification Techniques

The Convention on the Prohibition of Military or Any Other Hostile Use of Environmental Modification Techniques (ENMOD) was completed in 1976 in order to end the use of weather modification techniques in warfare and other hostile situations. It is now in effect through the participation of its 77 Parties, which include almost all major industrialized states.[272] It is highly relevant to climate engineering because its definition of "environmental modification" clearly encompasses almost all climate engineering proposals.[273]

Centrally, the ENMOD Parties agree "not to engage in military or any other hostile use of environmental modification techniques having widespread, long-lasting or severe effects as the means of destruction, damage or injury to any other State Party."[274] That is, prohibition requires meeting three criteria: to be military or hostile, to satisfy one of three criteria of scale, and to directly harm another ENMOD Party. The trio "widespread, long lasting or severe," which was later used in the amendment to the London Protocol regarding marine geoengineering, is not defined in ENMOD itself but was in a nonbinding Understanding thereto. This Understanding describes them as:

> "widespread": encompassing an area on the scale of several hundred square kilometres; "long lasting": lasting for a period of months, or approximately a season; "severe": involving serious or significant disruption or harm to human life, natural and economic resources or other assets.[275]

The first two criteria of scale can be determined somewhat objectively, and climate engineering large-scale field research projects or implementation would most likely satisfy them. The criterion of severity, which notably is limited to life, resources, and assets, is less certain.

Simultaneously, ENMOD recognizes and is rhetorically supportive of peaceful environmental modification. The agreement explicitly "shall not hinder the use of environmental modification techniques for peaceful purposes."[276] Furthermore, in its Preamble, Parties recognize "that the use of environmental modification techniques for peaceful purposes could improve the interrelationship of man and nature and contribute to the preservation and improvement of the environment for the benefit of present and future generations."[277] To that end, Parties are to facilitate the exchange of information regarding such peaceful uses, and those Parties "in a position to do so shall contribute … to international economic and scientific co-operation in the preservation, improvement and peaceful utilization of the environment."[278] If climate engineering is able to counter climate change risks, then it would be such "preservation, improvement and peaceful utilization," and Parties with the capacity to do so are obligated to contribute to cooperation.

A challenge to the implementation and enforcement of ENMOD is its weak institutional support. It has neither a standing secretariat nor regular meetings of its Parties. Such meetings are infrequently proposed. The ENMOD Parties have held two meetings, but declined to do so again in 2013.[279] If the Parties were to choose to do so, a meeting could provide a potential institutional vehicle to clarify the relationship among climate engineering, international law, and the preferences of much of the international community.

3.4. OTHER DOMAINS OF INTERNATIONAL LAW

3.4.1. *Civil Aviation*

Some climate engineering techniques, such as stratospheric aerosol injection and cirrus cloud thinning, could be researched or implemented using aircraft, which would emit particular substances. States would need to comply with international aviation law, which governs the international aspects of such vehicles. The central multilateral agreement, the Convention on International Civil Aviation (the Chicago Convention), grants sovereignty to Parties over their airspace and enjoys global participation.[280] This airspace extends upward from their territory – from both land and territorial waters – to the undefined upper border with outer space.[281] Parties are to ensure that aircraft bear a nationality mark and comply with other standards, and under a related treaty, the state of registration can exercise jurisdiction over aircraft.[282] States are to permit unscheduled flights through their airspace provided that the foreign aircraft are not military, customs, or police aircraft, and that the aircraft operate with a purpose consistent with the aims of the Chicago Convention.[283] Scheduled flights as well as pilotless, military, customs, and police aircraft require prior authorization.[284] Flights over the high seas must also follow the Convention's rules.[285]

Regarding aircraft emissions, which would presumably include substances intentionally injected into the atmosphere for climate engineering, Parties may establish their own regulations for all aircraft operating in their territory, as long as these rules are enforced without distinction to the aircraft's country of registration.[286] However, Parties are expected the keep the rules relatively uniform and consistent with guidelines established by the International Civil Aviation Organization (ICAO), a UN body established by the Chicago Convention.[287] Deviations from the ICAO guidelines are to be reported by Parties.[288] The ICAO has guidelines regarding specific pollutants from aircraft emissions, but these do not address materials such as sulfates that are presently considered for climate engineering purposes.[289] Therefore, atmospheric climate engineering – even in the airspace of countries other than that of the aircraft's nation of registry – appears at first glance to be compliant with international aviation law. However, the state in whose airspace the climate engineering activity occurred could assert that such flights are inconsistent with the aims of the Chicago Convention or that they are contrary to other international law.[290]

If a Party has reasonable grounds to believe that an aircraft in its territory is being used for purposes inconsistent with Convention's aims, then it may require that the aircraft land and take any appropriate means that are consistent with international law, as well as giving instructions that any such violations must cease.[291] Under certain circumstances – including to ensure compliance with any other treaty – a state may "interfere with" an aircraft in the case of a criminal offense.[292] Parties are to take steps to ensure that aircraft in their territory comply with these requests, and that they do not use their own aircraft for purposes inconsistent with the aims of the Chicago Convention.[293] Disputes between Parties that negotiation cannot resolve may be submitted to the ICAO Council.[294]

3.4.2. *Space Law*

SRM climate engineering could, at least in principle, be done in outer space, by placing objects in orbit or at the L1 Lagrangian point between the Earth and the Sun.[295] Although these proposals presently appear to be prohibitively expensive, this might not always be the case. States' activities in outer space, an area that remains undefined, are governed by a set of multilateral agreements. The 1967 Treaty on Principles Governing the Activities of States in the Exploration and Use of Outer Space, including the Moon and Other Celestial Bodies (Outer Space Treaty) is foundational, and counts as Parties all states with space programs. In it, Parties are to conduct space activities "for the benefit and in the interests of all countries … in accordance with international law … in the interest of maintaining international peace and security and promoting international cooperation and understanding" and "with due regard to the corresponding interests of all other States Parties."[296] Parties are to inform the UN, the international scientific community, and the public as to their space activities and the results thereof.[297] All space activities must be conducted or authorized by the Party's government, and the Party is responsible for those activities.[298] The Treaty established a "freedom of scientific investigation in outer space," and Parties should facilitate and encourage cooperation in carrying out research.[299] These commitments would apply to climate engineering activities in space.

The most interesting – and perhaps most relevant – provisions under space law are those regarding liability for harm from space activities. The Outer

Space Treaty establishes, and the Convention on International Liability for Damage Caused by Space Objects details the liability provisions. Under these, the Party (or Parties) that launches, procures the launching, or provides the launching site has absolute liability (i.e., with no need to demonstrate fault) for damage from their space objects to other Parties in outer space, in the atmosphere, and on the Earth.[300] Multiple liable Parties are jointly and severally liable.[301] This liability implicitly includes harm from accidents as well as from expected operations, and from direct contact as well as from remote effects. Therefore, space-based SRM is the only suggested climate engineering technique under which the state that undertakes or authorizes it would clearly be liable for transboundary harm through a treaty's provisions.

3.4.3. *Human Rights*

Human rights law might be a somewhat ambiguous domain of international law with respect to climate engineering, but it can provide guidance as to how states and even private actors should proceed with climate engineering research, development, and possible implementation. Human rights may shape large-scale outdoor climate engineering activities, scientists' conduct of research, people's enjoyment of the benefits of scientific research, and decision-making procedures.

Before going forward, some clarification may be helpful. Human rights are a subcategory of rights – that is, claims of entitlement by someone to something from someone else – that all people have by the mere fact of being human. The duty bearers of human rights are usually, but not always, states, which have committed themselves to ensuring human rights through a series of binding and nonbinding international legal instruments. So-called first-generation or negative human rights are political and civil rights such as those to life (i.e., to not be arbitrarily deprived of one's life), to equality before the law, and to freedom of thought, conscience, and religion. States generally have duties to ensure that people within their territory and subject to their jurisdiction can enjoy these rights. Second-generation or positive human rights are economic, social, and cultural rights such as those to food, housing, the highest attainable standard of health, education, and participation in cultural life. States have progressive duties first to respect the right by not directly interfering with its enjoyment, second to protect it by preventing third parties from interfering with its enjoyment, and third to fulfill it by taking action toward its full realization.[302]

In the case of large-scale outdoor climate engineering activities, interpretation of human rights could point in contradictory directions depending upon the circumstances. If the field research or implementation of particular climate engineering proposals were expected to have environmental impacts severe enough to threaten inter alia the availability of food, housing, and standards of health for a state's population, then carrying out these activities might undermine human rights. For example, BECCS is a CDR technique that the IPCC's more optimistic scenarios assume will be implemented at very large scales, requiring very large areas of arable land. The resulting competition for land could cause food prices to increase, which could put at risk the right to food.[303] Whether this would actually be the case would depend on factors including the impact on food prices, the magnitude of this price impact on people's ability to access food, the strength of the climate engineering activity's causal effect on food prices, and the climate change risks – particularly to food production – that the CDR activities prevented or reduced. Under circumstances in which the threat to a second-generation human right – such as that to food – is clear, states would be obligated to respect the right by not undertaking the climate engineering activity themselves and to protect the right by trying to prevent others from doing so. On the other hand, if the existing evidence was that certain forms of climate engineering research or implementation could greatly reduce grave climate risks, then the state might arguably have a duty to authorize, encourage, or undertake these activities in order to fulfill rights such as those to food, housing, and the highest attainable standard of health.[304] Given the complexities of climate change and of climate engineering, as well as the diversity of the conditions and interests of the human race, climate engineering is likely to have the capacity to undermine the human rights of some people, while fulfilling those of others. The uncertainty of actual outcomes from climate engineering further exacerbates this challenge.

Scientific research is the object of human rights as well. Some writers assert that academic freedom is a human right based upon the rights to freedom of opinion and expression, to an education, and to freedom of scientific research and creative activity.[305] Any such academic or scientific freedom would be subject to balancing with and constraints from other fundamental rights, funding, safety, research subjects' rights, and environmental harm. An additional human right of scientists is the protection of their "moral and material interests" from their research activities.[306] This indicates a duty of states to progressively realize inventors' rights to benefit from their work, for which patents are the leading vehicle. Thus, states are to respect this right by refraining from infringing upon patents themselves, to protect it by preventing infringement by third parties, and to fulfill it by implementing laws and administrative regulations that award and defend patents. Again, this should be done with due consideration for potentially conflicting legitimate interests such as other human rights.

International law also provides for a human right to enjoy the benefits of scientific research. In fact, enjoying these benefits may be a precursor to the fulfillment of other human rights such as those to life, food, and health. A Special Rapporteur to the UN Human Rights Council describes the right to enjoy the benefits of scientific progress as having four normative tenets.[307] First, everyone should be able to access science's benefits, without discrimination. In the case of climate engineering, this indicates that governments should ensure that data, results, publications, and materials are freely available, such as through data repositories and open access publication, in a nondiscriminatory manner. Second, everyone should have the opportunity to contribute to scientific research. This tenet is roughly congruous with freedom of scientific research, described above. Third, people have a right to participate in science-related decision-making, particularly regarding setting major research priorities and policies. Here, states should provide their residents and citizens with such opportunities to participate in relevant decision-making concerning climate engineering research. Fourth, states should ensure the conservation, development and diffusion of science and technology. In particular, industrialized countries should internationally collaborate in scientific research, and should prioritize technologies that would be of the greatest benefit to poor and otherwise marginalized people. This implies that they should support research intro climate engineering methods that could reduce climate risks to vulnerable populations, and that developing countries should take steps toward the appropriate development, importation and dissemination of those climate engineering technologies that may assist their own residents and citizens.

Procedural rights are also sometimes considered among human rights. For example, populations that are likely to be affected by climate engineering might have the rights to be notified, to access relevant information, to participate in decision-making processes, and to have access to the appropriate administrative and judicial processes, especially in seeking remedies for harm.[308] Both the Universal Declaration of Human Rights and the International Covenant on Civil and Political Rights might have established a right to access information.[309] However, participation in decision-making in environmental matters is not consistently recognized as a human right per se. Instead, particular legal instruments such as the Aarhus Convention and the customary international law of transboundary harm advance the participatory right of potentially affected publics.[310] Regardless, the large, if not global scale of climate engineering's effects complicates procedural rights. This would particularly be the case if the "affected public" were to include those whom climate engineering would affect in nonphysical ways.

Finally, because law that is specific to climate engineering is largely absent, private actors could play important roles in ensuring that research proceeds in a manner that is consistent with human rights. For example, professional societies, funders, publishers, and scholars could develop and promulgate codes of conduct that are informed by international human rights law.[311]

3.4.4. *Intellectual Property*

In general, countries promote innovation and new technologies through laws that grant inventors temporary exclusive rights to use their inventions.[312] This intellectual property law is primarily a national matter, but is harmonized through a number of multilateral agreements and international institutions, especially the Agreement on Trade-Related Aspects of Intellectual Property Rights (TRIPS) of the World Trade Organization. At least three particular issues arise for possible climate engineering technologies and inventions.[313] First, given the stakes of climate risks, a patent for a climate engineering invention might be essential for the protection of humans and the environment, yet the patent holder might fail to utilize or license the patent, or might demand excessive licensing terms. If so, states may intervene and, for example, compel the patent holder to license the invention on reasonable terms. TRIPS permits such compulsory licensing pursuant to procedural limitations.[314] Second, some people consider some climate engineering methods to be objectionable or unacceptably risky. Under TRIPS, if a particular climate engineering technique were shown to put at risk "*ordre public* or morality, including to protect human, animal or plant life or health or to avoid serious prejudice to the environment," then a state may exclude it from patentability.[315] Third, holders of actual or potentially valuable patents might attempt to influence public bodies' decision-making regarding climate engineering. This concern might warrant special measures.

Regardless, international policy will remain the purview of domestic law and possibly innovative arrangements by private actors.

3.4.5. *International Disputes*

States sometimes disagree, and might do so over climate engineering. Most often, they resolve their disputes through nonlegal and political means. Dispute resolution occasionally moves to more legalized approaches, such as negotiation, mediation, and arbitration. Some multilateral agreements have specialized mechanisms, such as compulsory arbitration or dedicated special tribunals, to try to resolve disputes within their scope. Four international legal fora have much broader ranges of issues that they can address. First, the UN General Assembly has universal participation and can consider almost any matter, but its resolutions are nonbinding.[316] Second, the UN Security Council is limited to the "maintenance of international peace and security," which could include climate engineering activities if they were to greatly raise international tensions. The Security Council can issue legally binding, majoritarian resolutions, although five of the world's most powerful countries have veto power.[317] Even if it can agree on a binding resolution, any enforcement such as sanctions or military action must be performed by one or more willing states. Third, the International Court of Justice can resolve international disputes. However, in order for its rulings in contentious issues to be legally binding, each state must consent to the jurisdiction of the court prior to the trial.[318] By majority vote of the UN General Assembly, a legal question can be referred to the International Court of Justice (ICJ) for an advisory opinion; such an opinion could speak to the obligations of states that have not consented to the ICJ's jurisdiction for such matters, but it would not be binding on them. Finally, states may submit a dispute to the Permanent Court of Arbitration (PCA). However, it lacks standing judges and is thus better considered as an institution for facilitating arbitration. After years of little to no activity, states have been turning to the PCA, perhaps due to its confidentiality, lack of precedent-setting, and generally faster resolution.

3.5. NONBINDING MULTILATERAL ENVIRONMENTAL AGREEMENTS AND INTERGOVERNMENTAL INSTITUTIONS

Although the phrase "international law" generally refers to those instruments, either explicit agreements or custom, that are intended to be legally binding, nonbinding multilateral agreements and the role played by intergovernmental institutions are also important. These agreements can influence the behavior of states that wish to avoid any potential reputational costs that might result from acting contrary to a nonbinding agreement or the advice of an authoritative intergovernmental body. Furthermore, nonbinding agreements and the decisions of intergovernmental institutions often provide a sense of the preferences, norms, aspirations, and interests of various states and of the international community as a whole. They can also establish initial terms for future negotiations toward binding agreements. Some of them can trigger certain obligations under domestic laws, such as reporting requirements. International institutions can play essential coordinating roles. Their authority often derives both from their intergovernmental character and from the expertise – often scientific or technical – of their decision-makers.

A handful of nonbinding multilateral legal instruments and intergovernmental institutions' decisions and reports might have a bearing on the future governance of climate engineering. Particular attention is given here to those nonbinding agreements that have been globally endorsed, or nearly so. Given that these agreements, by their nature, use vague and hortatory language, inferring what they might mean for climate engineering is a subjective practice.

3.5.1. *UN Environmental Summits*

A series of UN-organized global summits on the environment and development have shaped the principles of international environmental. The first two – held in Stockholm in 1972 and Rio de Janeiro in 1992 – are the most significant.

3.5.1.1. Declaration of the UN Conference on the Human Environment

Contemporary international environmental law arguably began with the 1972 UN Conference on the Human Environment, which 113 states attended in Stockholm. The resulting Declaration contains proclamations and principles that sound decidedly anthropocentric to twenty-first-century ears, and that regularly refer to the benefits of improving nature. Therefore, to the extent that the Stockholm Declaration still reflects states' preferences regarding international environmental law, it generally supports interventions in the natural world to improve humanity's well-being.

For example, its purpose is to

> "inspire and guide the peoples of the world in the preservation and enhancement of the human environment," and it proclaims that "[i]n our time, man's capability to transform his surroundings, if used wisely, can bring to all peoples the benefits of development and the opportunity to enhance the quality of life … For the purpose of attaining freedom in the world of nature, man must use knowledge to build, in collaboration with nature, a better environment."[319]

The anthropocentric tone continues in the principles, the first of which includes the assertion that humanity "bears a solemn responsibility to protect and improve the environment for present and future generations."[320] Consistent with this, science and technology are to be "applied to the identification, avoidance and control of environmental risks and the solution of environmental problems and for the common good of mankind."[321] The Declaration also calls for the minimization of transboundary harm and for international cooperation in protecting and improving the environment.[322]

3.5.1.2. Rio Declaration on Environment and Development

Twenty years after the Stockholm Conference and Declaration, the UN hosted the Conference on Environment and Development in Rio de Janeiro. The resulting Declaration, which the UN General Assembly later endorsed, reflects how the priorities of the international community had changed since Stockholm.[323] It gives greater, albeit qualified, emphasis to the protection of natural world itself and it elevates the importance of economic development of poorer countries. These environmental and economic goals are often folded together under the rubric of sustainable development. The implications of the Rio Declaration for climate engineering are less certain compared with those of the Stockholm Declaration. Climate engineering could help protect the natural world from climate change and enable greater economic development, and it could cause significant harm to the environment and to vulnerable populations.

Some anthropocentrism remains: The first sentence of the first principle is "Human beings are at the centre of concerns for sustainable development."[324] Furthermore, the Rio Declaration refers to only two rights: that of states' "sovereign right to exploit their own resources pursuant to their own environmental and developmental policies" (coupled with their responsibility to prevent transboundary harm), and a right to development, which remains undefined.[325] To the extent that the latter right indeed exists in international law, then climate engineering might help enable it, given that climate change risks will be disproportionately borne by poor states and that GHG emissions and economic development remain coupled.

More specifically, the Rio Declaration calls for "improving scientific understanding" and for developing "new and innovative technologies."[326] In one of its principles, states are to avoid transferring sources of environmental harm to other states.[327] If climate engineering merely moved the location of climate risks, it would violate this principle. Finally, the Rio Declaration reiterates a number of procedural duties, such as states' obligations to conduct environmental impact assessments, notify potentially affected states, and provide public access to information, as well as several principles of international environmental law, including common but differentiated responsibilities, precaution, polluter pays, intergenerational equity, cooperation, and sustainable development.[328]

3.5.1.3. From the Johannesburg Declaration to Sustainable Development Goals

The UN organized further summits on sustainable development in Johannesburg in 2002 and again in Rio de Janeiro in 2012. Instead of a couple dozen new, clarified, or expanded principles, their reports offer long lists of recognitions, aspirations, and goals. These serve mostly to reaffirm and, by their nature, add little to international environmental law.[329] The report from the 2012 summit, *The Future We Want*, reads like a "laundry list" of nearly 300 such goals. Notably, one of these expresses concern about the potential negative environmental impacts of ocean fertilization.[330]

The UN Member States agreed at the 2012 summit to establish a working group to develop a new set of overarching goals to replace the UN Millennium Development Goals, which had established targets for 2015. The UN General Assembly approved the resulting 17 Sustainable Development Goals in September 2015.[331] In turn, these goals consist of 169 targets to be reached by 2030. Combating climate change not only has its own goal, but is also integrated into several others and is emphasized in the Preamble. Three months later, the Paris Agreement incorporated much of the Sustainable Development Goals' climate-specific language.

Few implications specific to climate engineering can be gleaned from the Sustainable Development Goals. Their Preamble does "note with grave concern the significant gap between … Parties' mitigation pledges … and … having a likely chance of holding the increase in global average temperature below 2°C or 1.5°C," implying a potential important role for CDR.[332] Further, human and institutional capacity is to be improved on, among other things, reducing the impacts of climate change.[333] One could argue that SRM is a means to reduce climate impacts. Finally, the combination of the climate goal with others, such as ending poverty, achieving food security, and sustainable development more generally, imply that climate change risks must be reduced rapidly and at low cost.

3.5.2. *UN Environment Programme*

Since its launch soon after the 1972 Stockholm Conference, the UN Environment Programme (UNEP) has played an important role in coordinating other international bodies and national governments, and in laying the foundation for the development of international environmental law. Examples are given below.

3.5.2.1. UNEP Provisions for Co-operation between States in Weather Modification

In 1980 UNEP approved Provisions for Co-operation between States in Weather Modification, in cooperation with the World Meteorological Organization

(WMO). As in ENMOD, the term "weather modification" is defined in a manner that would include SRM climate engineering, and perhaps CDR methods as well.[334] Also like ENMOD, it is generally supportive of states' use of weather modification in appropriate circumstances, in this case for "the benefit of mankind and the environment."[335] To this end, "States should encourage and facilitate international co-operation in weather modification activities, including research."[336] As its name implies, the document calls on states to cooperate through, for example, information exchange, notification, and consultation.[337] States are to collect relevant information regarding their weather modification activities and to share it with the WMO.[338] Finally, they are further encouraged in the document to undertake prior environmental assessment of their weather modification activities that might have transboundary impacts, and to conduct those activities in a manner that prevents environmental damage in other countries or in areas beyond national jurisdiction.[339] Note that only the UNEP Governing Council, which has 58 states as members, approved the Provisions.

3.5.2.2. Recent Activity

UNEP has demonstrated some institutional interest in climate engineering in recent years. It supported a workshop led by the UN Education, Social and Cultural Organization (UNESCO), described below. Along with the Intergovernmental Oceanographic Commission (IOC) of UNESCO and the WMO, UNEP cosponsored a 2011 International Scientific Conference on Problems of Adaptation to Climate Change, hosted by the Russian Academy of Sciences. There, climate engineering was among the topics discussed.[340]

3.5.3. *UN Education, Social and Cultural Organization*

UNESCO is a UN specialized agency with the mandate to promote international collaboration in the domains of education, science, and culture. It hosted an expert meeting on climate engineering in 2010.[341] The 20 participants recommended an international research program modeled on the World Climate Research Programme.[342] This would be sponsored by UNESCO, the IOC, the International Council for Science, and the WMO, and "could address the technological and scientific challenges of geoengineering and ensure that legitimate scientific research into this controversial issue may proceed."[343] A policy brief was later published, which bore the endorsements of UNEP and the Scientific Committee on Problems of the Environment.[344]

The IOC of UNESCO commissioned a report on ocean fertilization.[345] The Surface Ocean Lower Atmosphere Study helped prepare it, and the International Geosphere-Biosphere Programme, the Scientific Committee on Oceanic Research, the World Climate Research Programme, and the International Commission on Atmospheric Chemistry and Global Pollution sponsored it. The IOC issued the report in 2010.

3.5.4. *World Meteorological Organization*

The WMO is a UN agency that fosters international coordination and cooperation in diverse issues of weather and climate, including the atmosphere, oceans, and water resources. The WMO was referenced above in the contexts of the 1980 UNEP Provisions for Co-operation between States in Weather Modification and the 2010 UNESCO expert meeting. Under the guidance of its Expert Team on Weather Modification, the WMO also appears to be in the process of developing a statement on climate engineering. The draft statement "closely follows" the previous such statement from the American Meteorological Society.[346] The most recent plan within the WMO is for its Commission for Atmospheric Sciences to work toward an assessment of climate engineering, in cooperation with the World Climate Research Programme, the IOC of UNESCO, the IMO, and others.[347]

3.5.5. *The Intergovernmental Panel on Climate Change*

UNEP and the WMO established the IPCC in 1988 in order to assess the state of climate change knowledge, the impacts of climate change, and possible response strategies. Its primary output has been the occasional publication of comprehensive assessment reports, which collect and summarize the most recent scientific information. Beginning with the Third Assessment Report, published in 2001, these have devoted limited – but increasing – attention to climate engineering. The most recent Fifth Assessment Report considers climate engineering methods in a number of contexts. It concluded, for example, that:

> Several CDR techniques could potentially reduce atmospheric greenhouse gas (GHG) levels. However, there are biogeochemical, technical and societal limitations that, to varying degrees, make it difficult to provide quantitative estimates of the potential for CDR …
>
> SRM is untested, and is not included in any of the mitigation scenarios, but, if realisable, could to some degree offset global temperature rise and some of its effects. It could possibly provide rapid cooling in comparison to CO_2 mitigation …

> SRM technologies raise questions about costs, risks, governance and ethical implications of development and deployment. There are special challenges emerging for international institutions and mechanisms that could coordinate research and possibly restrain testing and deployment.[348]

In 2011 the IPCC convened an expert meeting on climate engineering, and later published a meeting report.[349]

3.5.6. *UN General Assembly Resolution on Oceans and the Law of the Sea*

In 2007 the UN General Assembly approved a Resolution containing a passage relating to climate engineering. The Resolution on Oceans and the Law of the Sea urged states to exercise caution when considering proposals for ocean fertilization.[350] It further stated that, while large-scale ocean fertilization activities were not justified at that time, states should encourage research in that area.[351]

3.6. CUSTOMARY INTERNATIONAL LAW

Customary international law is a set of legally binding rules derived from states' repeated behavior and evidence that they do so out of a sense of legal requirement. Because it is not centrally transcribed, the precise obligations that customary international law imposes upon states are often somewhat unclear. Nevertheless, it provides an essential gap-filling function, preventing, resolving, and governing international disputes over environmental rights and duties in the absence of a specific multilateral agreement. Three domains of customary international law will be relevant for climate engineering: the right to exploit natural resources, the duty to prevent transboundary harm, and state responsibility for harm. This section also discusses a new document that attempts to gather together existing custom regarding the atmosphere.

3.6.1. *Sovereign Right to Exploit Natural Resources*

If sovereignty is the foundation of international relations and international law, then states' sovereign right to exploit their own natural resources is one of the two cornerstones of international environmental law. In theory, this right could be considered an implicit default condition within the system of sovereign states, at least prior to the rise of environmental law in the twentieth century.[352] Developing countries that were asserting their independence in the wake of decolonization made the right for a state to exploit its natural resources explicit. This right was initially recognized globally, beginning with the 1952 UN General Assembly Resolution on the Right to Exploit Freely Natural Wealth and Resources and the 1962 Declaration on Permanent Sovereignty over Natural Resources.[353] Within international environmental law, many (if not most) multilateral agreements, beginning with the 1972 Stockholm Declaration, provide for or restate it.[354] The implication of this tenet of customary international law is that states have a presumptive right to conduct climate engineering activities within their own territory, provided that they do so in a manner consistent with their other rights and obligations.

3.6.2. *Prevention of Transboundary Environmental Harm*

The second of the two cornerstones of international environmental law is that states have a responsibility to prevent transboundary environmental harm – including that to areas beyond the limits of national jurisdiction – arising from activities within their jurisdiction or under their control. Such transboundary harm may occur due to outdoor climate engineering activities of sufficient scales. Although this responsibility arose in the early twentieth century distinct from and in parallel with the sovereign right to exploit natural resources, the two fused in the 1972 Stockholm Declaration and since then have generally been presented as two sides of a single coin.[355] Many multilateral environmental agreements reiterate this responsibility.[356] In 1996 the International Court of Justice stated that the responsibility is part of customary international law,[357] citing a principle of the Rio Declaration, in which

> States have, in accordance with the Charter of the United Nations and the principles of international law, the sovereign right to exploit their own resources pursuant to their own environmental and developmental policies, and the responsibility to ensure that activities within their jurisdiction or control do not cause damage to the environment of other States or of areas beyond the limits of national jurisdiction.[358]

This responsibility is not an inviolable obligation to ensure that absolutely no transboundary environmental harm occurs. Instead, the objective is the prevention of transboundary harm, while the obligation of states is one of practicing due diligence regarding activities, such as outdoor climate engineering activities, that occur within their jurisdiction or control and that pose a risk of significant transboundary harm.[359] The due diligence standard is roughly proportional to the probability and the magnitude of the risk. In its Draft Articles on the Prevention of Transboundary Harm from Hazardous Activities, the ILC of the UN attempted to clarify the responsibilities based upon state practice and the rulings of international tribunals. It concluded that the responsibility to prevent transboundary harm arises when an activity could pose a "risk of causing significant transboundary harm through their physical consequences … [including] risks taking the form of a high probability of causing significant transboundary harm and a low probability of causing disastrous transboundary harm."[360] Each of these two formulations of risky activities could apply to various suggested climate engineering activities. Once that threshold has been met, states must undertake "all appropriate measures" to prevent the potential harm and to reduce its risk. These measures include requiring authorization for the activity, performing an environmental impact assessment, notifying and cooperating in good faith with potentially affected states, informing the public, and developing contingency plans for an emergency.[361] The precise steps are subject to consultations between the countries, and are to be "based on an equitable balance of interests."[362] Factors to consider in this balancing of interests include:

> the importance of the activity, taking into account its overall advantages of a social, economic and technical character for the State of origin in relation to the potential harm for the State likely to be affected; … [and] the economic viability of the activity in relation to the costs of prevention and to the possibility of carrying out the activity elsewhere or by other means or replacing it with an alternative activity.[363]

A balancing of interests should thus take into account both the benefits and risks of the proposed climate engineering activity for all concerned states.[364] An activity that posed great risk to another state with little likely benefit for the state of origin would be interpreted quite differently than a modest, low-risk one undertaken by a state whose great risks from climate change could subsequently be lowered through the proposed climate engineering activity. If the consultations fail to produce a consensus among the states, then the state of origin is to take into account the interests of potentially affected states in its decisions, such as whether to authorize the activity.[365]

Because large-scale outdoor climate engineering activities could pose risks of transboundary impacts, the state of jurisdiction or control is obligated to undertake these procedural duties. Beyond that, scholars have divergent opinions as to the potential of these customary responsibilities to regulate the risks of climate engineering effectively. Reichwein and coauthors argue that the risk of abrupt termination of SRM is great enough that due diligence may call for SRM not to be implemented in the first place.[366] However, this conclusion requires that sudden termination has a high risk probability, which is not self-evident. Furthermore, the responsibility to prevent transboundary harm requires states to comply with certain procedural duties and, in the absence of consensus, to take other states' interests into account. Customary international law does not prohibit activities that are known to cause transboundary harm. In fact, a passage was removed from a previous version of the ILC's articles that said that the state of origin shall refuse authorization of an activity that would cause unavoidable transboundary harm.[367] Reichwein and coauthors, as well as Bodansky, cite the difficulty in demonstrating specific risks with sufficient confidence as a barrier to the regulatory effectiveness of the customary international law regarding transboundary harm.[368] Elsewhere, Bodansky highlights the challenges and likely controversy in interpreting this responsibility in the case of climate engineering, given that climate engineering itself is intended to prevent transboundary harm.[369]

A final, unexplored question is whether the responsibility to prevent transboundary harm could be interpreted as obligating states to research, or even implement climate engineering, if it is found to be effective and without undue adverse impacts. Because all states emit GHGs, this customary responsibility obligates them to adopt, implement, and enforce policies that aim to abate their GHG emissions.[370] Countries, especially those with greater emissions, might bear an analogous responsibility to enact policies to consider reducing their transboundary harm of climate change through the research – and perhaps implementation – of climate engineering.

3.6.3. *State Responsibility and Liability*

Customary international law also has provisions regarding how states should respond to acts that have caused transboundary harm, including environmental damage that could result from climate engineering activities. Although these provisions are less well developed than those concerning the prevention of transboundary harm, the ILC has tried to capture them in two documents. The division between responsibility for acts that are contrary to international law and compensation for transboundary harm from hazardous activities that are consistent with international law has been useful, in large part because the legal status of the latter group – which arguably does not reflect customary international law – is contested.

As described in the ILC Draft Articles on Responsibility of States for Internationally Wrongful Acts, a state's action or omission is wrongful if it breaches international law and is attributable to the state.[371] Regarding the latter criterion, the Draft Articles focus on whether the actor who actually undertook or authorized the action is part of "the state" or not. In a somewhat different vein, in the context of climate engineering, some writers have expressed concern that the doctrine of state responsibility would require the attribution of specific environmental harms to particular climate engineering activities, which would be a challenging endeavor.[372] However, many (if not most) of the obligations in international environmental law are ex ante procedural duties that countries should or must carry out prior to a risky activity. A state that failed to comply with these obligations is still responsible, independent of any ex post manifestation of harm. However, demands for reparation by an injured state, discussed immediately below, do require injuries that can be attributed to the state's wrongful acts.

If a state has committed a wrongful act, it should cease the activity, assure that the act will not recur, and make full reparations for any injuries.[373] Reparations can take the form of restitution (reestablishing the situation that existed before the wrongful act), compensation (providing something of value, usually money, to compensate for the harm), and satisfaction ("an acknowledgement of the breach, an expression of regret, a formal apology, or another appropriate modality"), in that order of priority.[374] Restitution is often difficult for environmental damage. Thus, states are generally liable for compensation from transboundary harm that results from their legally wrongful acts, including failing to follow due diligence in preventing transboundary harm.[375] Although under the ILC Draft Articles reparations extend to injury, including material and moral, compensation is limited to "financially assessable damage."[376]

States face certain additional consequences if they engage in a "serious breach" of a peremptory norm of international law.[377] A serious breach is one that "involves a gross or systematic failure" of obligations.[378] A peremptory norm (or *jus cogens*) is a fundamental, core value of the international order, one from which no violation is permitted. Unlike other tenets of customary international law, countries are automatically bound by peremptory norms and therefore may not ratify treaties that are contrary to them. There is no definitive list of peremptory norms, but they typically include only the most egregious activities such as genocide, slavery, and crimes against humanity. The ILC declined to list them in the final version of its Draft Articles. A draft version 25 years earlier described an analogous category, that of "international crime," which included "a serious breach of an international obligation of essential importance for the safeguarding and preservation of the human environment, such as those prohibiting massive pollution of the atmosphere or of the seas."[379] The fact that the ILC removed this provision and the absence of the relevant state practice imply that "massive pollution of the atmosphere" would likely not violate peremptory norms of international law. At the same time, the possibility that maintaining the planet's climate might be regarded as a peremptory norm, and that a state's failure to prevent, reduce, and control GHG emissions and/or harmful climate engineering should not be completely ruled out.[380]

The ILC Draft Articles also describe under what circumstances acts contrary to international law should not be considered wrongful. Among these is necessity, under which a state's action is not wrongful if it is the only means for the state "to safeguard an essential interest against a grave and imminent peril," and if it does not put the essential interests of other states at risk.[381] The ICJ has ruled that grave and imminent environmental risks can constitute a state of necessity.[382] For countries such as small island states that could face existential risks from climate change, necessity might operate as a legal preclusion from wrongfulness for climate engineering activities that would otherwise be contrary to international law. However, a state may not invoke necessity if it has contributed to the situation that gave rise to the necessity.[383] All countries have emitted GHGs, which are the cause of climate change and its risks, and have thus all contributed to the state of necessity. At the same time, historical contributions to anthropogenic GHG emissions and policies to control them vary dramatically among countries, and those countries with very low historical emissions or aggressive emissions abatement policies might be able to successfully invoke the necessity defense for climate engineering activities that would otherwise breach international law.

As with other international environmental legal instruments, climate engineering must be considered in the context of the likely environmental harm from climate change. States might have committed wrongful acts, such as failing to adopt effective GHG emissions abatement policies in contravention of the UNFCCC, its Kyoto Protocol, its Paris Agreement, or the customary responsibility to prevent transboundary harm. In this context where GHG abatement failure is the wrongful act, climate engineering could have two roles pursuant to the customary law of responsibility, one for the responsible state and one for the injured state. Regarding the former, as described above, states that are responsible for a wrongful act must, among other things, cease the activity and make reparations to the injured states. Climate engineering could contribute to fulfilling these obligations. If CDR were to reduce a responsible state's annual net emissions to zero, then it would arguably have stopped its wrongful activity. Furthermore, although restitution of diffuse environmental damage, including that from climate change, is typically difficult, climate engineering might be able to contribute to this. For example, large-scale CDR might restore the atmosphere to its former condition by counteracting a country's historical GHG emissions. Furthermore, if SRM climate engineering could be implemented safely and with climatic impacts consistent with current models, then it could help restore the climate, including offsetting effects that might injure affected states.

Injured states, by contrast, have a right to undertake countermeasures. These are acts that are otherwise contrary to international law that are directed at the responsible state in order to induce it to comply.[384] Countermeasures can be taken by an injured state in response to a typical breach, or by any state in response to a breach of an obligation that is owed to the international community as a whole.[385] Considering that the UNFCCC recognizes the prevention of climate change as a common concern of humanity, the latter may indeed be the case.[386] Therefore, in this scenario a state might be able to carry out or authorize climate engineering activities that would otherwise be contrary to international law as a countermeasure. In particular, SRM could be effective in this context due to its high leverage, speed of action, and the reversibility of its direct climatic effects. Countermeasures are subject to several somewhat strict limitations. Before taking them, the injured state must provide the responsible state with the opportunity to fulfill its obligations. If the latter then fails to do so, the injured state must notify the responsible state of any intentions to perform countermeasures.[387] Countermeasures must be proportionate to the injury suffered, and may not be among certain prohibited acts, such as the use of force or violations of fundamental human rights.[388] The countermeasure must end as soon as the responsible state has ceased its wrongful activity, made sufficient assurances that the activity will not recur, and provided full reparations to the injured states.[389] One significant legal barrier to the use of climate engineering as a countermeasure in response to states' failure to abate GHG emissions is that "[a]n injured State may only take countermeasures against a State which is responsible."[390] Citing a ruling of the ICJ, the accompanying commentary of the ILC Draft Articles clarifies that the injured state may act in a manner that is otherwise contrary to international law with respect to *only* the responsible state; other states' rights under international law may not be violated.[391] On the one hand, SRM as it is presently understood, could be global or perhaps regional in its effect, and would consequently serve poorly as a targeted countermeasure. On the other hand, a desperate state – perhaps one that is also claiming a necessity defense – might argue that all countries have failed to abate their GHG emissions sufficiently and are therefore subject to countermeasures.

Compensation for harm from hazardous activities that were *not* contrary to international law is the subject of the ILC's second document regarding ex post obligations of states. Notably, given the lack of consensus on this matter, these provisions were released as Draft Principles instead of Draft Articles, and do not necessarily reflect customary international law. Nevertheless, the document can help to guide leading scholars' thinking and customary law's possible future direction. The Draft Principles address only transboundary damage, including "impairment of the environment" and any "reasonable response measures," that resulted from hazardous activities, defined as those "which [involve] a risk of causing significant harm."[392] The country in whose jurisdiction or under whose control the hazardous activity was carried out is considered the state of origin.

In the event of an activity involving a hazardous activity that is likely to cause transboundary damage, the state of origin must notify, consult with, and seek the cooperation of potentially affected states, as well as take appropriate response measures.[393] At the same time, the potentially affected state must "take all feasible measures" to minimize the damage. If a hazardous activity did cause transboundary damage, then the state of origin "should take all necessary measures to ensure that prompt and adequate compensation is available for victims."[394] Furthermore, it should impose strict liability on the operator who carries out the hazardous activity, and must ensure that its domestic courts and other public institutions can provide "prompt, adequate and effective remedies" to the victims of transboundary harm, without discrimination regarding victims' country of residence.[395] In other words, the Draft Principles do not claim that the state of origin is liable for damages, only that it should take steps to ensure that victims of transboundary harm have access to compensation and other remedies. This arrangement could occur through a variety of mechanisms including explicit and preferably strict liability for harm on the part of the operator, nondiscriminatory access to courts and other legal avenues of redress, mandatory insurance, industry-wide and international compensation funds, as well as possibly vicarious state liability. Finally, states should make all efforts to establish compensation regimes for particular categories of hazardous activities.[396] These regimes should rely, as appropriate, on industry-wide and/or state funds to supplement the resources and insurance cover of the operator.

Some scholars have explored whether a category of ultrahazardous or abnormally dangerous activities exists that would render states or operators strictly liable under the customary international law for transboundary harm arising from them.[397] These terms are used only in the ILC's commentary to the Draft Principles (which do not necessarily reflect customary international law in any case). There, the ILC notes that strict liability for ultrahazardous activities is the "most proper technique."[398] Indeed, that is the tradition within many domestic jurisdictions, and an economic analysis generally supports it. However, within international law, strict liability for seemingly ultrahazardous activities has so far been limited to those three classes that are the subjects of dedicated multilateral agreements or regimes: nuclear energy, activities in space, and the maritime transport of oil.

3.6.4. *UN International Law Commission Draft Articles on the Protection of the Atmosphere*

In 2011 the ILC began work toward a set of guidelines that would offer an international legal framework for the protection of the atmosphere, and in 2016 it provisionally approved such a draft set. One of these guidelines states, "Activities aimed at intentional large-scale modification of the atmosphere should be conducted with prudence and caution, subject to any applicable rules of international law."[399] The accompanying commentary notes that this includes but is not limited to "what is commonly understood as 'geo-engineering,'" and that it aims neither to authorize nor prohibit such activities.[400] The phrase "activities aimed at intentional large-scale modification of the atmosphere" is based on ENMOD, and "prudence and caution" on rulings of the International Tribunal for the Law of the Sea. The commentary further says that "[t]he draft guideline is cast in hortatory language, aimed at encouraging the development of rules to govern such activities, within the regimes competent in the various fields relevant to atmospheric pollution and atmospheric degradation."[401] Beyond this, the draft guidelines largely consist of restatements of existing principles and customary international law, including the obligations to exercise due diligence, to conduct environmental impact assessments, and to utilize the atmosphere equitably. However, the draft guidelines apply these to the atmosphere, with no explicit reference to transboundary risks in the guidelines themselves.

The drafting process is not complete, and further changes are likely.

3.7. PRINCIPLES OF INTERNATIONAL ENVIRONMENTAL LAW

General principles are the third main source of international law. They are not themselves legally binding, but instead must be operationalized in custom or a treaty. Their primary purpose is to provide guidance for the interpretation and further development of international law. That is, they can be thought of as the spirit, as opposed to the letter, of the law. Even more so than customary international law, the identity and substance of principles is not fully agreed-upon. This section introduces some of the leading principles of international environmental law, and what they could mean for the international law of climate engineering.

3.7.1. *The Environment as a Common Concern of Humankind*

Some multilateral agreements explicitly designate certain aspects of the environment – including the conservation of biological diversity in the

CBD and climate change in the UNFCCC – as the common concerns of humankind.[402] Furthermore, international environmental law implicitly treats other components such as stratospheric ozone, Antarctica, and the marine environment, as well as the global environment in general, as common concerns. As a principle, this designation does not have clear legal consequences, but for the most part, it has resulted in all states having individual and collective legitimate interests in and responsibilities for the maintenance of the common concern, independent of any direct harm to other Parties. Legally speaking, these responsibilities are akin to *erga omnes* obligations, which countries owe to the international community as a whole.

Climate engineering, especially SRM, can be reasonably inferred to affect a common concern of humankind, given "that change in the Earth's climate and its adverse effects are a common concern of humankind" per the universally ratified UNFCCC. Thus, unilateral or minilateral implementation of global climate engineering in the absence of notification and consultation with other states, whether or not they would be likely to experience deleterious effects, as well as other customary obligations would be contrary to this principle of international environmental law. Furthermore, its probable impacts on humanity's common concern should guide the development of any international law specific to climate engineering.

3.7.2. *Common but Differentiated Responsibilities*

In contrast to most international legal obligations, states are not to share some environmental responsibilities equally. Instead, restoring stratospheric ozone, preventing climate change, the conservation of biological diversity, the preservation of the marine environment, and environmental protection in general are explicitly or implicitly treated in international environmental law as common but differentiated responsibilities.[403] This principle recognizes that, despite the equality of sovereign states under international law, their environmental responsibilities need not be uniform. All Parties have some responsibilities toward the goals and in the specific commitments at hand, but those countries with the greater capacity – i.e., the wealthier ones – must bear disproportionate obligations to satisfy them. For example, the principle of the UNFCCC regarding common but differentiated responsibilities concludes: "Accordingly, the developed country Parties should take the lead in combating climate change and the adverse effects thereof."[404] Both the Kyoto Protocol and the Paris Agreement continue and affirm this approach.

The principle of common but differentiated responsibilities implies that wealthier countries should carry most of the burdens of researching, developing, and implementing (if appropriate) climate engineering. In the specific case of the UNFCCC and its related agreements, this implication extends most clearly to CDR, given that their substantive content largely works toward the stabilization of GHG concentrations. However, one of the Paris Agreement's goals is the limitation of planetary warming, which states might be able to achieve – at least in part – through SRM.[405] In addition, the UNFCCC also calls on Parties to cooperate in research concerning undefined "various response strategies," and the Parties note, in the Paris Agreement, "the importance of technology for the implementation of mitigation and adaptation actions" and commit to "cooperative action on technology development and transfer."[406] Both of these terms – various response strategies and technology development, including for adaptation actions – might include both CDR and SRM climate engineering.

3.7.3. *Precaution*

Precaution, expressed as a principle or an approach, has been an increasingly common feature of international environmental law in recent decades. It is a legal tool to manage risk and uncertainty that is frequently cited when confronting issues of emerging technologies. Its particular formulations in the Rio Declaration, the CBD, the UNFCCC, CLRTAP's Oslo Protocol, the London Protocol, and the Kiev Protocol to the Aarhus Convention vary slightly.[407] The formulation in the UNFCCC is both typical and most relevant for climate engineering:

> The Parties should take precautionary measures to anticipate, prevent or minimize the causes of climate change and mitigate its adverse effects. Where there are threats of serious or irreversible damage, lack of full scientific certainty should not be used as a reason for postponing such measures, taking into account that policies and measures to deal with climate change should be cost-effective so as to ensure global benefits at the lowest possible cost. To achieve this, such policies and measures should take into account different socio-economic contexts, be comprehensive, cover all relevant sources, sinks and reservoirs of greenhouse gases and adaptation, and comprise all economic sectors.[408]

Scholars have debated precaution's meaning for climate engineering, given that climate change does and climate engineering might pose "threats of serious or irreversible damage" while demonstrating some "lack of full scientific certainty." As with the other instances where the risks and benefits of these two climatic phenomena seem to be in tension, any interpretation of this principle will depend on the specific technique, circumstances, and evidence at hand. This author has elsewhere asserted that precaution, as embodied in the UNFCCC, calls at least for the consideration, such as through research, of all means to reduce climate change risks, including that of climate engineering.[409] This argument relies upon the UNFCCC's call for precautionary measures to mitigate climate change's adverse effects, for measures and policies to be cost-effective, and for comprehensive consideration in developing response measures. Furthermore, "precautionary measures" is neither defined nor restricted, and the term could include not only CDR, which is clearly within the UNFCCC's scope and commitments, but also SRM. This position appears to be consistent with the decisions of the Parties to the London Protocol regarding marine geoengineering, which also requires a precautionary approach. In that case, both the nonbinding assessment framework for ocean fertilization and the approved amendment to regulate all marine geoengineering permit legitimate scientific research.[410]

3.7.4. *Polluter Pays*

A fourth principle of international environmental law is that the polluter, not the victim, should pay for environmental damage. However, despite the normative appeal of requiring the party who caused the harm to bear the burden of reparations, its implementation in international environmental law has been inconsistent, has been mainly in regional treaties, and often has utilized highly qualified language.[411] The principle was written into both the Rio Declaration and the London Protocol.[412] It is noticeably absent in the UNFCCC, despite this convention being drafted at the same time as the Rio Declaration.

To the extent that a principle of such limited acceptance were to apply to climate engineering, it could have two modalities of application. The primary cause of climate change is GHGs – a type of pollution – and climate engineering may offer a means to reduce their occurrence and/or effect. If so, then those states that have contributed most to the elevated atmospheric GHG concentrations should bear the costs of the means to reduce the risks from climate change, including any climate engineering research and implementation. The secondary source of "pollution" could be climate engineering itself, which might cause deleterious effects. It is less clear who should pay for any resulting damage, such as through environmental restoration and compensation for victims of harm, under the polluter pays principle. The states or other actors that implemented or undertook large-scale research or implementation of climate engineering are the most proximal source, yet the GHG-emitting states are the ultimate source of most climate change. This matter is clearly related to that of liability for harm from climate engineering.[413]

3.7.5. *Cooperation*

The obligation for states to cooperate in good faith is one of the cornerstones of international law. It was embodied in the Charter of the UN, and has since been explicitly stated or relied upon in most multilateral environmental agreements and in customary international law.[414] Note that cooperation does not demand agreement among states, only that they act in good faith and with due diligence regarding actions that might affect other countries. Because of its widespread application, the principle of cooperation is already legally binding in most climate engineering contexts through the UNFCCC or the customary law of preventing transboundary harm. Any future new legal instruments that would be specific to climate engineering should, at the very least, call upon states to cooperate in research and to share the results thereof. Moreover, they should require states, especially those engaged in climate engineering that would pose risks to other states, to notify, to share information, and to consult with one another.

3.7.6. *Equity*

A final principle of international environmental law is that of equity: acting with due regard for others' interests. Equity can include less direct effects than those captured by other legal concepts such as transboundary harm, and is particularly relevant when the potentially affected party is somehow disenfranchised or unable to effectively consult and negotiate with the decision-maker. Equity can be divided into that within and that between generations.

In the context of international environmental law, intragenerational equity can be thought of as a general normative framework that can guide the division of rights and obligations among states, underlying the related principles of cooperation and common but differentiated responsibilities. It considers the substantially uneven distribution of the impacts of states' actions. For example, ocean fertilization CDR could enable the acting state to meet its commitments to abate its net GHG emissions, yet might reduce marine resources on which other states depend for food and income. Likewise, the implementation of stratospheric aerosol injection SRM could reduce climate risks for many states, yet might change precipitation in some of them, putting their agricultural systems at risk. In this way, intragenerational equity resembles the prevention of transboundary harm, although the former can take a global perspective while the latter focuses primarily on bilateral relations. Equity is especially important in the management of shared resources as seen, for example, in the first principle of the UNFCCC:

> The Parties should protect the climate system for the benefit of present and future generations of humankind, on the basis of equity and in accordance with their common but differentiated responsibilities and respective capabilities.[415]

The second form of equity – intergenerational equity – concerns future populations who inherently lack self-representation in present discussions. Given the long length of time required before elevated atmospheric carbon dioxide concentrations will naturally lower, as well as the potentially high opportunity costs of foregone investments due to emissions abatement and adaptation, intergenerational equity is critical in the formulation of climate policy, yet it remains ambiguous and elusive.

For climate engineering, effects of both climate change and climate engineering on populations who lack an effective voice in governance will require particular attention. The principle of intergenerational equity implies that implementation of any climate engineering with global effects, such as SRM, should be subject to careful consideration and a high degree of scrutiny, especially in the absence of widespread consensus in the international community. The implications of intergenerational equity are complicated by the fact that, under conditions of elevated atmospheric GHG concentrations, SRM would need to be maintained for durations that would cross generations. On the one hand, transferring a burden to maintain such systems may be contrary to the principle of intergenerational equity.[416] At the same time, all other things being equal and assuming that climate engineering methods would function as planned, it appears contrary to intergenerational equity to fail to research, develop, and potentially implement a potential additional means to reduce climate risks – which will be borne almost entirely by future generations. For now, the cautious exploration of various climate engineering options through research appears to be consistent with the principle of intergenerational equity.

3.8. THE SCHOLARSHIP OF INTERNATIONAL ENVIRONMENTAL LAW AND CLIMATE ENGINEERING: THE PAST, CHALLENGES, AND THE FUTURE

The legal scholarship concerning climate engineering, particularly its international aspects, is now large and diverse enough that it can genuinely be called a body of scholarship. Although the first article to be published on this subject is more than 20 years old, the majority of these publications, which number well over 100, have appeared since 2013.[417] Some of the early essays offered introductory overviews to climate engineering proposals and the legal challenges that they posed.[418] Many of the writings – like this chapter – explored how existing international law could apply to climate engineering's various proposals.[419] Others offered suggestions for how climate engineering should be internationally governed.[420] These groups are not mutually exclusive.

More recent research has become more focused, and in some ways, more sophisticated. Often, it disaggregates the legal issues based upon the various climate engineering methods,[421] by where they would take place,[422] or by distinguishing research from implementation.[423] Ocean fertilization was a notable forerunner to these trends due to the controversial proposals in 2007 and the responses by the institutions of the CBD and of the London Convention and London Protocol.[424] Some scholars have looked to existing analogous technologies from which they could draw lessons,[425] including exploring to what extent the emerging international regulation of ocean fertilization could serve as a model for climate engineering more generally.[426] Instead of specific or detailed proposals for regulation, some articles offer modest suggestions for governance, such as principles or key issues, which may be more useful at this early stage.[427]

The most fruitful outputs from the last few years have been those that delve more deeply into particular legal aspects of and challenges posed by climate engineering. These include liability for harm,[428] intellectual property,[429] dispute settlement,[430] problematic uni- or minilateral implementation,[431] the concern of lessened GHG emissions abatement,[432] and its relationship with particular tenets of customary international law and general principles of international environmental law.[433] Finally, some recent scholarship uses climate engineering as a lens through which the authors can examine legal phenomena.[434]

Nevertheless, climate engineering continues to present challenges for the scholarship of international law, and especially that of environmental law. The first source of this difficulty is that climate engineering, due to its plural potential modalities, locations, and effects, invokes numerous international legal instruments. These were crafted at various times, by various parties, for various reasons, and are consequently often in tension with one another. Legal scholars must confront these tensions.

A second source of difficulty is the many ways in which climate engineering could both decrease and increase environmental risks – a phenomenon that this chapter has emphasized. Yet environmental law is generally oriented toward decreasing risks, and is poorly equipped to balance difficult and uncertain risk–risk tradeoffs.[435]

Third, SRM climate engineering presents a distinct problem structure, in that it would be a single best effort global public good.[436] The present evidence is that, in a world of inadequate efforts to reduce GHG emissions, SRM could be a global public good with large net benefits, although some groups and regions would lose in relative terms, and possibly in absolute terms.[437] These single best effort global public goods call for coordination, mutual restraint, and prevention of misuse. In contrast, environmental law generally strives to prevent and reduce negative externalities, such as pollution, and to manage common pool resources, such as the oceans and the atmosphere. These problems – which SRM also presents – are best controlled through mechanisms such as rules, legally binding commitments, liability, Pigouvian taxes, information disclosure, mandatory consultation and negotiation, and impact assessment.

Fourth, modern environmental law arose out of the justified recognition that humans were not properly accounting for all of the effects of our actions on the environment and on humans with the environment as the medium of harm. This lack of full accounting of impacts seemed most evident in cases of large-scale technological endeavors. Most responses to this, both within and beyond legal scholarship, have focused on reducing interventions in the environment, and on humility in our collective endeavors. Climate engineering runs counter to this cultural milieu. Indeed, the proposals are more congruent with the increasing recognition that humanity now influences the natural world to such a degree that a new geologic era – the Anthropocene – may be warranted.[438] The result of the latter two challenges to environmental legal scholarship listed here – those of problem structure and historical culture – is that many writers approach climate engineering proposals solely as risky, if not hubristic, interventions in the natural world that will have negative impacts on species, ecosystems, and humans, and which will damage our shared resources. This approach offers a picture that is far from complete.

Looking forward, scholars of international law should reflect upon these challenges. Those who work in the environmental vein should strive to understand and integrate diverse problem structures into their analyses, and use climate engineering as an opportunity to examine and make explicit their objectives, which are often left unstated. Furthermore, the examination of the international law of climate engineering has, understandably, been dominated by experts of environmental law. Yet a broader range of specialties is now needed. Chief among these are those of intellectual property, given the importance of patents as a means to regulate technologies, and of human rights, considering the potential yet unclear capacity for climate engineering to impact their provision. Within international environmental law, more attention is warranted on how its principles could and should guide the interpretation of existing law and the development of any future law of climate engineering. Besides precaution, other general international legal principles remain largely unexplored. Finally, research on the law of climate engineering could benefit through integration with the growing legal scholarship of the Anthropocene.

NOTES

1 *See* Durwood Zaelke and James Cameron, *Global Warming and Climate Change: An Overview of the International Legal Process*, 5 AMERICAN UNIVERSITY JOURNAL OF INTERNATIONAL LAW AND POLICY 249 (1989).

2 Cirrus cloud thinning, in which more heat is allowed to escape the Earth, seems to offer a third category of climate engineering. Because it resembles SRM in terms of its problem structure and atmospheric means of operation, the term "SRM" in this chapter generally includes cirrus cloud thinning. At the same time, cirrus cloud thinning is like CDR in that it influences the Earth's balance of long-wave radiation.

3 Daniel Bodansky, *May We Engineer the Climate?*, 33 CLIMATIC CHANGE 309 (1996). For a review and some history, particularly of the relevant international relations aspects, see Joshua B. Horton and Jesse L. Reynolds, *The International Politics of Climate Engineering: A Review and Prospectus for International Relations*, 18 INTERNATIONAL STUDIES REVIEW 438 (2016).

4 "Nonexcludable" means that the producer cannot effectively exclude others from enjoying the benefits of the good. "Nonrivalrous" means that enjoyment of the good by one party does not diminish the ability of others to enjoy it. Note that a "good" is meant here in the descriptive sense of a product or service, not in any normative sense. A public good is often desired by some and not by others, and is thus normatively neither good nor bad. Also note that a public good is an ideal type. In reality, goods have characteristics that place them along a spectrum. *See* Scott Barrett, *Why Cooperate? The Incentive to Supply Global Public Goods* (Oxford: Oxford University Press, 2007); Daniel Bodansky, *What's in a Concept? Global Public Goods, International Law, and Legitimacy*, 23 EUROPEAN JOURNAL OF INTERNATIONAL LAW 651 (2012). Adapting society to a changed climate is not a global public good.

5 Note that CDR (as well as many forms of abatement) would also have negative externalities, in some cases possibly negating at least some of its global benefits. Hypothetical ultra-cheap CDR could resemble SRM, allowing single actors at the country-level scale to experience net direct benefits, even in the absence of global cooperation. This presently appears to be an unlikely scenario. Many forms of GHG abatement also have positive externalities that provide direct and immediate local benefits, in particular a reduction in levels of conventional air pollutants.

6 Likewise, the negative impacts of SRM might be great enough to outweigh its benefits.
7 William Nordhaus, *A Question of Balance: Weighing the Options on Global Warming Policies* (New Haven: Yale University Press, 2008), 196.
8 Some legal scholars reject trying to balance the expected effects of climate engineering and climate change. *See, e.g.*, Ralph Bodle et al., *Options and Proposals for the International Governance of Geoengineering*, 14/2014 CLIMATE CHANGE, UMWELTBUNDESAMT/FEDERAL ENVIRONMENT AGENCY (GERMANY) 98, 136 (2014); Wilfried Rickels et al., *Large-Scale Intentional Interventions into the Climate System? Assessing the Climate Engineering Debate* (Kiel: Kiel Earth Institute, 2011).
9 States sometimes consent to external authorities, such as the European Union and the UN Security Council, having such authority. States may withdraw such consent, albeit with consequences that might be negative.
10 This right is coupled with a state's responsibility to ensure that its activities do not cause harm to another state or to areas beyond national jurisdiction. *See* The Rio Declaration on Environment and Development (1992), Principle 2. *See* section 3.6.2.
11 There are jurisdictional claims in Antarctica, but these are not widely recognized.
12 Vienna Convention on the Law of Treaties (1969) [hereinafter Vienna Treaties Convention], Art. 1.
13 *Id.*, Art. 56.
14 Andrew T. Guzman, *How International Law Works: A Rational Choice Theory* (Oxford: Oxford University Press, 2008).
15 Vienna Treaties Convention, Art. 60.
16 Eric A. Posner and Alan O. Sykes, *Economic Foundations of International Law* (Cambridge, MA: Harvard University Press, 2013), 32.
17 Patricia W. Birnie, Alan E. Boyle, and Catherine Redgwell, *International Law and the Environment* (Oxford: Oxford University Press, 2009), 7.
18 *The Changing Atmosphere: Implications for Global Security*, statement of the World Conference on the Changing Atmosphere, Toronto, Canada, June 27–30, 1988, para. 30; *Protection of the Atmosphere: International Meeting of Legal and Policy Experts: Statement of the Meeting of Legal Policy Experts*, Ottawa, Canada, February 20–22, 1989. Both are reprinted in Center for International and Environmental Law, *Selected International Legal Materials on Global Warming and Climate Change*, 5 AMERICAN UNIVERSITY JOURNAL OF INTERNATIONAL LAW AND POLICY 513 (1990). *See also* James K. Sebenius, *Designing Negotiations Toward a New Regime: The Case of Global Warming*, 15 INTERNATIONAL SECURITY 110, 115–16 (1991).
19 *See* section 3.4.4.
20 Report of the International Law Commission, Sixty-eighth Session (A/71/10) (2016), Ch. VIII.
21 The Holy See and Palestine are not Parties but only observer states.
22 United Nations Framework Convention on Climate Change (1992) [hereinafter UNFCCC], Art. 2.
23 Paris Agreement (2015), Art. 2.1(a). The carbon dioxide limit assumes a climate sensitivity of 3°C.

24 "Business as usual" here means the RCP8.5 scenario. Detlef P. van Vuuren et al., *The Representative Concentration Pathways: An Overview*, 109 CLIMATIC CHANGE 5 (2011).

25 *See* Sabine Fuss et al., *Betting on Negative Emissions*, 4 NATURE CLIMATE CHANGE 850 (2014); T. Gasser et al., *Negative Emissions Physically Needed to Keep Global Warming Below 2°C*, 6 NATURE COMMUNICATIONS 7958 (2015).

26 As described above, both emissions abatement and SRM are global public goods. However, the benefits of the former to a single country generally do not outweigh the costs, whereas they might in the case of the latter. *See* Jesse Reynolds, *A Critical Examination of the Climate Engineering Moral Hazard and Risk Compensation Concern*, 2 THE ANTHROPOCENE REVIEW 174, 175–83 (2015).

27 UNFCCC, Preamble para. 1, Arts. 3.1, 3.3, 4.

28 *Id.*, Preamble para. 8.

29 *Id.*, Art. 3.1.

30 *Id.*, Arts. 4.1(f), 3.3. *See also* Preamble paras. 10, 16, 21, 22, Arts. 3.4, 3.5, 4.1(g), 4.1(h), 4.2(a), 4.7, 4.10, 7.2(a).

31 *Id.*, Art. 4.1(f).

32 *Id.*

33 *Id.*, Arts. 4.1(b), 4.1(d), 4.2(a).

34 *Id.*, Art. 4.1(d).

35 *Id.*, Arts. 1.7, 1.8. Indeed, any line that tries to separate CDR methods from traditional abatement activities would be unclear.

36 *Id.*, Art. 4.1(g). *See also* Arts. 4.1(h), 5.

37 *Id.*, Art. 4.1(c). *See also* Arts. 4.3, 4.5, 4.7, 4.8, 4.9.

38 Report of the Conference of the Parties on its Sixteenth Session, Held in Cancún from 29 November to 10 December 2010, Addendum Part Two: Action Taken by the Conference of the Parties at its Sixteenth Session (2011), paras. 113–127.

39 UNFCCC, Arts. 3.3, 4.1(b), 4.1(e), 4.1(f), 4.4.

40 *See* UNFCCC, *Focus: Adaptation*, http://unfccc.int/focus/adaptation/items/6999.php.

41 Kyoto Protocol to the United Nations Framework Convention on Climate Change (1997) [hereinafter Kyoto Protocol], Arts. 2.1(a)(iv), 10(c), 10(d). *See also* Arts. 10(b)(i), 11.2(b).

42 *Id.*, Arts. 2.1(a)(ii), 2.1(a)(iv).

43 *Id.*, Art. 3.3. Developing means to account for the removal of GHGs through land use change and forestry activities has been a long, contentious, and still unresolved process.

44 *Id.*, Art. 3.4.

45 Report of the Conference of the Parties Serving as the Meeting of the Parties to the Kyoto Protocol on Its Eighth Session, Addendum Part Two (2012), Decision 1/CMP.8.

46 *Id.*, Arts. 5, 7.1, 10, especially 10(b)(ii).

47 *Id.*, Arts. 6, 12.

48 Paris Agreement (2015), Art. 1.2(a).

49 United Nations Environment Programme, *The Emissions Gap Report 2016* (Nairobi: United Nations Environment Programme, 2016).
50 Paris Agreement, Art. 4.1.
51 *Id.*, Art. 4.2.
52 *Id.*, Art. 6.
53 Phillip Williamson et al., *Ocean Fertilization for Geoengineering: A Review of Effectiveness, Environmental Impacts and Emerging Governance*, 90 PROCESS SAFETY AND ENVIRONMENTAL PROTECTION 475 (2012).
54 *Id.*, Arts. 1.2(b), 7.
55 *Id.*
56 *Id.*, Arts. 7.9, 7.11, 7.14.
57 *Id.*, Art. 7.5.
58 UNFCCC, Arts. 7, 8, 9, 14. *See* Meinhard Doelle, *Geo-engineering and Dispute Settlement under UNCLOS and the UNFCCC: Stormy Seas Ahead?*, in CLIMATE CHANGE IMPACTS ON OCEAN AND COASTAL LAW: U.S. AND INTERNATIONAL PERSPECTIVES (Randall S. Abate, ed., Oxford: Oxford University Press, 2014).
59 Compare Matthias Honegger, Kushini Sugathapala, and Axel Michaelowa, *Tackling Climate Change: Where Can the Generic Framework Be Located?*, 7 CARBON AND CLIMATE LAW REVIEW 125 (2013); Jesse Reynolds, *Why the UNFCCC and CBD Should Refrain from Regulating Solar Climate Engineering*, in GEOENGINEERING OUR CLIMATE? ETHICS, POLITICS AND GOVERNANCE (Jason Blackstock and Sean Lo, eds., Abingdon: Routledge, 2018).
60 *See* A. F. Bais et al., *Ozone Depletion and Climate Change: Impacts on UV Radiation*, 14 PHOTOCHEMICAL & PHOTOBIOLOGICAL SCIENCES 19 (2015).
61 Giovanni Pitari et al., *Stratospheric Ozone Response to Sulfate Geoengineering: Results from the Geoengineering Model Intercomparison Project (GeoMIP)*, 119 JOURNAL OF GEOPHYSICAL RESEARCH: ATMOSPHERES 2629 (2014).
62 John A. Dykema et al., *Stratospheric Controlled Perturbation Experiment: A Small-scale Experiment to Improve Understanding of the Risks of Solar Geoengineering*, 372 PHILOSOPHICAL TRANSACTIONS OF THE ROYAL SOCIETY A: MATHEMATICAL, PHYSICAL AND ENGINEERING SCIENCES 0059 (2014); David W. Keith et al., *Stratospheric Solar Geoengineering without Ozone Loss*, 113 PROCEEDINGS OF THE NATIONAL ACADEMY OF SCIENCES 14910–14 (2016).
63 S. Roumeau et al., *Tropical Cirrus Clouds: A Possible Sink for Ozone*, 27 GEOPHYSICAL RESEARCH LETTERS 2233 (2000); Eric J. Jensen et al., *Ice Nucleation and Dehydration in the Tropical Tropopause Layer*, 110 PROCEEDINGS OF THE NATIONAL ACADEMY OF SCIENCES 2041 (2013).
64 Vienna Convention for the Protection of the Ozone Layer (1985) [hereinafer Vienna Ozone Convention], Art. 2.2(b).
65 Montreal Protocol on Substances That Deplete the Ozone Layer (1987) [hereinafter Montreal Protocol], Art. 2.10.
66 Vienna Ozone Convention, Art. 2.1.
67 Annual global lower atmospheric sulfur pollution is roughly 100 megatons of sulfur dioxide (50 megatons of sulfur) and declining. Offsetting the warming effect of the doubling of the atmospheric carbon dioxide concentration would require roughly 3 to 10 megatons of sulfur dioxide (1.5 to 5 megatons of sulfur). *See* Z. Klimont et al., *The Last Decade of Global Anthropogenic Sulfur Dioxide: 2000–2011 Emissions*,

8 ENVIRONMENTAL RESEARCH LETTERS 014003 (2013); Naomi E. Vaughan and Timothy M. Lenton, *A Review of Climate Geoengineering Proposals*, 109 CLIMATIC CHANGE 791, 810 (2011). Sulfate aerosols in the lower atmosphere presently offset approximately a quarter of climate change but are much less effective due to their shorter residence time and suboptimal particle size. *See* Olivier Boucher et al., *Clouds and Aerosols*, in *CLIMATE CHANGE 2013: THE PHYSICAL SCIENCE BASIS* (Thomas F. Stocker et al., eds., Cambridge: Cambridge University Press, 2013); Ben Kravitz, *Stratospheric Aerosols for Solar Radiation Management*, in *GEOENGINEERING RESPONSES TO CLIMATE CHANGE: SELECTED ENTRIES FROM THE ENCYCLOPEDIA OF SUSTAINABILITY SCIENCE AND TECHNOLOGY* (Tim Lenton and Naomi Vaughan, eds., New York: Springer, 2013), 24.

68 Vienna Ozone Convention, Arts. 2.2(a), 3.1.

69 Montreal Protocol, Arts. 9, 10A.

70 Convention on Long-Range Transboundary Air Pollution (1979) [hereinafter CLRTAP], Arts. 3–6, 8.

71 *Id.*, Art. 2.

72 *Id.*, Art. 1(a). *See* Philippe Sands and Jacqueline Peel, *Principles of International Environmental Law* (Cambridge: Cambridge University Press, 2012), 247.

73 CLRTAP, Arts. 5, 8. Note the lower threshold for consultation, in which source states must consult, upon request by states that are merely at risk or affected by long-range transboundary air pollution. *See also* Protocol to the 1979 Convention on Long-Range Transboundary Air Pollution on the Reduction of Sulphur Emissions or their Transboundary Fluxes by at Least 30 per Cent (1985), Art. 4; Protocol to the 1979 Convention on Long-Range Transboundary Air Pollution on Further Reduction of Sulphur Emissions (1994) [hereinafter Oslo Protocol], Art. 5; Protocol to the 1979 Convention on Long-Range Transboundary Air Pollution to Abate Acidification, Eutrophication and Ground-level Ozone (1999) [hereinafter Gothenburg Protocol], Art. 7.

74 CLRTAP, Art. 7.

75 *Id.*, Art. 6.

76 Note that several larger states, including Canada, Russia, the United Kingdom, and the USA, that have ratified CLRTAP are not Parties to one or more of these three Protocols.

77 Oslo Protocol, Art. 6.

78 CLRTAP, Arts. 2–3; Oslo Protocol, Art. 2.1; Gothenburg Protocol, Art. 2.1.

79 The other areas beyond national jurisdiction are outer space and (arguably) Antarctica.

80 The primary objections of the USA were to the agreement's provisions regarding the sharing of the resources of the seabed. A minority in the Senate presently blocks its ratification.

81 United Nations Convention on the Law of the Sea (1982) [hereinafter UNCLOS], Art. 192.

82 Mark J. Valencia and Kazumine Akimoto, *Guidelines for Navigation and Overflight in the Exclusive Economic Zone*, 30 MARINE POLICY 704, 708 (2006); Veronica Frank, *The European Community and Marine Environmental Protection in the International Law of the Sea: Implementing Global Obligations at the Regional Level* (Leiden: Martinus Nijhoff, 2007), 12.

83 UNCLOS, Art. 193.
84 *Id.*, Art. 197.
85 *Id.*, Art. 1.1(4).
86 *See* Alan Boyle, *Law of the Sea Perspectives on Climate Change*, 27 THE INTERNATIONAL JOURNAL OF MARINE AND COASTAL LAW 831, 832–33 (2012).
87 UNCLOS, Arts. 194, 196, 198–199, 204–22. Note that a coastal state may regulate foreign vessels in its exclusive economic zone (EEZ) outside of its territorial waters in order to minimize pollution only to operationalize generally accepted international standards, whereas inside its territorial waters it may do so to higher standards, provided that this does not interfere with innocent passage. *See id.*, Art. 211.4–5.
88 *Id.*, Arts. 61, 117–120, 194.5.
89 *Id.*, Arts. 194, 212.2.
90 *Id.*, Arts. 194, 196.
91 *Id.*, Arts. 204–206.
92 *Id.*, Art. 195.
93 James Edward Peterson, *Can Algae Save Civilization?: A Look at Technology, Law, and Policy Regarding Iron Fertilization of the Ocean to Counteract the Greenhouse Effect*, 6 COLORADO JOURNAL OF INTERNATIONAL ENVIRONMENTAL LAW AND POLICY 61, 92 (1995).
94 Philomene Verlaan, *Geo-engineering, the Law of the Sea, and Climate Change*, 3 CARBON AND CLIMATE LAW REVIEW 446, 458 (2009).
95 Monika Rhein et al., *Observations: Ocean*, in CLIMATE CHANGE 2013: THE PHYSICAL SCIENCE BASIS (Thomas F. Stocker et al., eds., Cambridge: Cambridge University Press, 2013), 292.
96 For example, proposals discussed during negotiations included "any study or related experimental work designed to increase man's knowledge of the marine environment." Informal Single Negotiating Text, Part III, Third United Nations Conference on the Law of the Sea, Volume IV, Summary Records, Plenary, General Committee, First, Second and Third Committees, as well as Documents of the Conference, Third Session (1975), Art. 1. *See* UN Division for Ocean Affairs and the Law of the Sea, Office of Legal Affairs, *The Law of the Sea: Marine Scientific Research: A Revised Guide to the Implementation of the Relevant Provisions of the United Nations Convention on the Law of the Sea* (New York: United Nations, 2010); Philomene Verlaan, *Marine Scientific Research: Its Potential Contribution to Achieving Responsible High Seas Governance*, 27 THE INTERNATIONAL JOURNAL OF MARINE AND COASTAL LAW 805 (2012). In contrast, the International Law Association (American Branch) Law of the Sea Committee requires that it be "undertaken in ocean space." George K. Walker, ed., *Definitions for the Law of the Sea: Terms Not Defined by the 1982 Convention* (Leiden: Martinus Nijhoff, 2011), 241.
97 UNCLOS, Arts. 238–239, 242–244.
98 *Id.*, Art.204.1.
99 *Id.*, Art. 238.
100 *Id.*, Art. 240. *See* Philomene A. Verlaan, *Experimental Activities that Intentionally Perturb the Marine Environment: Implications for the Marine Environmental Protection and Marine Scientific Research Provisions of the 1982 United Nations Convention on the Law of the Sea*, 31 MARINE POLICY 210 (2007); Anna-Maria

Hubert, *The New Paradox in Marine Scientific Research: Regulating the Potential Environmental Impacts of Conducting Ocean Science*, 42 OCEAN DEVELOPMENT AND INTERNATIONAL LAW 329 (2011).

101 UNCLOS, Art. 263.

102 *Id.*, Arts. 91–92. Flag states do not consistently comply with the requirements regarding the oversight of ships that bear their flag, including the one that there be a genuine link. *See* Alan Khee-Jin Tan, *Vessel-Source Marine Pollution: The Law and Politics of International Regulation* (Cambridge: Cambridge University Press, 2005), 47–62.

103 *Id.*, Arts. 94, 211.2.

104 *Id.*, Arts. 2–3.

105 *Id.*, Art. 17.

106 *Id.*, Arts. 18–19. Other causes for a ship engaged in climate engineering to not be in innocent passage include that the passage itself might not be "continuous and expeditious," the activity could be an "act of willful and serious pollution contrary to" UNCLOS, or the ship could be "carrying out ... research or survey activities."

107 *Id.*, Art. 245.

108 *Id.*, Arts. 27, 220, 230.2. *See* International Law Association Committee on Coastal State Jurisdiction Relating to Marine Pollution, *Final Report: London Conference (2000)*, *in* VESSEL-SOURCE POLLUTION AND COASTAL STATE JURISDICTION: THE WORK OF THE ILA COMMITTEE ON COASTAL STATE JURISDICTION RELATING TO MARINE POLLUTION *(1991–2000)*, (Erik Franckx, ed., The Hague: Kluwer, 2001), Conclusions 7, 9.

109 UNCLOS, Arts. 117, 194, 196.

110 *Id.*, Arts. 55, 57. The 24 nautical miles from land is the contiguous zone of the coastal state, which is not relevant for our purposes here. *See id.*, Art. 33.

111 *Id.*, Art. 56.1.

112 *Id.*, Arts. 56.2, 58.3.

113 *Id.*, Art. 58.3.

114 Its right of the physical inspection and potential detention of the foreign ship are dependent upon the severity of the suspected pollution, and penalties are limited to monetary ones. *See id.*, Arts. 220, 230.

115 *Id.*, Art. 246. *See also* Arts. 252–253.

116 *Id.*, Arts. 248–249.

117 *Id.*, Art. 59.

118 *Id.*, Arts. 86–88, 257. Archipelagic Parties have sovereignty over archipelagic waters, which are excluded from the high seas. *See id.*, Arts. 46–49.

119 *Id.*, Art. 87.1.

120 *Id.*, Art. 94.7

121 *Id.*, Arts. 58.1, 87.1(b). *See* section 3.4.1 below. This provision, coupled with Art. 56.1, seems to grant coastal states jurisdiction over atmospheric SRM in their EEZs. For example, a group of "senior officials and analysts" developed a set of "Guidelines for Navigation and Overflight in the Exclusive Economic Zone." They concluded that "States exercising the freedoms of navigation and overflight in a coastal State's EEZ should not interfere with or endanger the rights of the coastal State to protect and manage its own resources and their environment." Valencia and Akimoto, *supra* note 82, at 709.

122 *See* Neil Craik, Jason J. Blackstock, and Anna-Maria Hubert, *Regulating Geoengineering Research through Domestic Environmental Protection Frameworks: Reflections on the Recent Canadian Ocean Fertilization Case*, 7 CARBON AND CLIMATE LAW REVIEW 117 (2013).

123 *Compare* UNCLOS, Arts. 194.2, 196, 206 *with* Arts. 139, 153, 235, 263.

124 *Id.*, Arts. 192, 194.1.

125 *Id.*, Art. 235.

126 *Id.*

127 Id., Art. 263.

128 *Id.*, Art. 232.

129 For the latter, *see id.*, Arts. 213, 222, in which Parties are to take "measures necessary to implement applicable international rules and standards established through competent international organizations or diplomatic conference to prevent, reduce and control pollution of the marine environment from land-based sources [and] … from or through the atmosphere."

130 *Id.*, Art. 139.

131 UNCLOS uses "living resources" and "marine life" distinctly, although both are undefined. Provisions regarding the former consistently refer to conservation, management, utilization, exploitation, and allowable catch, implying that they are organisms that are extracted for economic gain. The sole article that invokes only the latter is "to protect and preserve rare or fragile ecosystems as well as the habitat of depleted, threatened or endangered species and other forms of marine life," implying a broader category. *Id.*, Arts. 61–68, 116–120, 194.5.

132 International Convention on Liability and Compensation for Damage in Connection with the Carriage of Hazardous and Noxious Substances by Sea (1996).

133 UNCLOS, Art. 266.1.

134 *Id.*, Art. 266.2.

135 *Id.*, Art. 267.

136 It may be possible to store captured carbon dioxide on the seabed, because at the low temperature and high pressure there, carbon dioxide is denser than water and could consequently remain as lake-like pools of liquid and/or solid hydrates in the valleys and crevices of the seabed.

137 UNCLOS, Arts. 76–85, 133–191.

138 *Id.*, Arts. 76–81.

139 *Id.*, Art. 82. For a legal analysis of sub-seabed storage of carbon dioxide off the US Atlantic coast, *see* Romany M. Webb & Michael B. Gerrard, *Policy Readiness for Offshore Carbon Dioxide Storage in the Northeast* (Sabin Center for Climate Change Law, June 2017), available at http://columbiaclimatelaw.com/files/2017/06/Webb-and-Gerrard-2017-06-Offshore-Carbon-Storage.pdf.

140 UNCLOS, Art. 1.1(1).

141 *Id.*, Arts. 136, 138, 141, 156–185.

142 *Id.*, Art. 145. It has done so for prospecting and exploration for certain materials, including polymetallic nodules, polymetallic sulfides, and cobalt-rich ferromanganese crusts.

143 *Id.*, Art. 139.1.

144 *Id.*, Art. 143.

145 Convention for the Protection of the Marine Environment of the North-East Atlantic (1992); Amendments to Annex II and Annex III to the OSPAR Convention for the Protection of the Marine Environment in the North-East Atlantic in Relation to the Storage of Carbon Dioxide Streams in Geological Formations (2007).

146 Note that disposal need not be into the sea itself. *See* UNCLOS, Art. 1.1(5).

147 *Id.*, Preamble para. 4.

148 Scholarly opinion diverges on this point regarding ocean fertilization. Proelss and Güssow roughly agree (as set forth in Rickels et al.). Freestone and Rayfuse argue that ocean fertilization is dumping, because the carbon dioxide that is indirectly put into the water column is merely being disposed of. Scott asserts that it is dumping because it is very likely to cause pollution and is thus contrary to the aims of UNCLOS. *See* David Freestone and Rosemary Rayfuse, *Ocean Iron Fertilization and International Law*, 364 MARINE ECOLOGY PROGRESS SERIES 227, 229 (2008); Rickels et al., 94; Karen N. Scott, *Regulating Ocean Fertilization under International Law: The Risks*, 7 CARBON AND CLIMATE LAW REVIEW 108, 112 (2013).

149 UNCLOS, Art. 207.1.

150 *Id.*, Art. 194.3(a).

151 *Id.*, Art. 210.

152 *Id.*, Art. 210.4.

153 *Id.*, Art. 210.6.

154 Verlaan, INTERNATIONAL JOURNAL OF MARINE AND COASTAL LAW 810–11.

155 UNCLOS, Arts. 94.7, 211.2.

156 *See* Eric Van Hooydonk, *The Law of Unmanned Merchant Shipping: An Exploration*, 20 THE JOURNAL OF INTERNATIONAL MARITIME LAW 403 (2014).

157 UNCLOS, Art. 94.4.

158 *See* Chapter 2. *See also* Alexander Proelss and Chang Hong, *Ocean Upwelling and International Law*, 43 OCEAN DEVELOPMENT AND INTERNATIONAL LAW 371 (2012).

159 An ordinary reading of these terms implies that, within each of the contexts, they are listed in order of decreasing size.

160 UNCLOS, Art. 194.3(d). *See also* Art. 208.1 for an extension of this duty to artificial islands, installations, and structures that operate on the seabed.

161 *Id.*, Art. 262.

162 *Id.*, Arts. 56.1(b), 60, 80.

163 *Id.*, Art. 60.3. Scholars have concluded that due notice is "through appropriate authorities within a reasonable amount of time in a suitable manner … through diplomatic or other designated channels … [and perhaps] national hydrographic offices." Walker, *supra* note 96, at 176.

164 *Supra* note 8, at Art. 246.5(c).

165 *Id.*, Arts. 60.3–60.7, 249.1(g), 260–261.

166 *Id.*, Art. 87. UNCLOS has no provision requiring that non-research artificial islands, installations, and structures bear the identifying markings of the state of registry or the international organization to which they belongs. This implies that Parties may build or place them on the high seas without such markings, after which time they could persist without an identifying state. However, this could

interfere with other states exercising their own high seas freedoms. Furthermore, allowing such placement of unmarked objects would, if they created pollution of the marine environment, cause Parties to be noncompliant with their commitment to take all measures to ensure that "pollution from other installations and devices operating in the marine environment" is minimized. *See id.*, Art. 194.3(d).

167 R. R. Churchill and A. V. Lowe, *The Law of the Sea*, Third edn (Manchester: Manchester University Press, 1999), 24; Yoshifumi Tanaka, *The International Law of the Sea* (Cambridge: Cambridge University Press, 2012), 264.

168 UNCLOS, Annex VI. *See* Doelle, *supra* note 58.

169 UNCLOS, Art. 210.6. See Verlaan, supra note 96.

170 Protocol to the Convention on the Prevention of Marine Pollution by Dumping of Wastes and Other Matter, 1972 [hereinafter London Protocol] (1996), Art. 23. The USA is making "progress towards joining." Report of the Thirty-sixth Consultative Meeting of the London Convention and the Ninth Meeting of Contracting Parties of the London Protocol (2014), para. 2.7.

171 Convention on the Prevention of Marine Pollution by Dumping of Wastes and Other Matter [hereinafter London Convention] (1972), Art. I; London Protocol, Art. 2.

172 London Protocol, Art. 1.10.

173 *Id.*, Art. 14.1.

174 *Id.*, Art. 3.3.

175 London Convention, Art. III.1; London Protocol, Art. 1.4.

176 London Convention, Art. IV.

177 *Id.*, Annex II, para. D.

178 London Protocol, Annex 1.

179 *Id.*, Art. 3.1.

180 *See* Bettina Boschen, *The Regulation of Ocean Fertilization and Marine Geoengineering under the London Protocol*, in CLIMATE CHANGE IMPACTS ON OCEAN AND COASTAL LAW: U.S. AND INTERNATIONAL PERSPECTIVES (Randall Abate, ed., Oxford: Oxford University Press, 2014); Harald Ginzky and Robyn Frost, *Marine Geo-engineering: Legally Binding Regulation under the London Protocol*, 8 CARBON AND CLIMATE LAW REVIEW 82 (2014).

181 Resolution LC-LP.1 on the Regulation of Ocean Fertilization (2008). This definition would also include fertilization via upwelling pipes. The resolution did not explicitly conclude that ocean fertilization was or was not dumping, only that ocean fertilization activities other than legitimate scientific research are contrary to the aims of the agreements and did not qualify for any exemption from the definition of dumping.

182 *Id.*, paras. 3–8.

183 Resolution LC-LP.2 on the Assessment Framework for Scientific Research Involving Ocean Fertilization (2010).

184 *Id.*, Sec. 2.

185 *Id.*, Sec. 3.

186 *Id.*, paras. 3.6.1, 4.3.

187 Resolution LP.4(8) on the Amendment to the London Protocol to Regulate the Placement of Matter for Ocean Fertilization and Other Marine Geoengineering Activities (2013), Annex, Art. 1, adding new para. 5*bis*.

188 *Id.*, Annex, Art. 1, adding new Annex 4. The Parties are developing a guidance document for considering the inclusion of new activities in this Annex. *See* Report of the Thirty-sixth Consultative Meeting, *supra* note 170. In November 2015, a new working group of the Joint Group of Experts on the Scientific Aspects of Marine Environmental Protection, which advises the UN and its related bodies on the scientific aspects of pollution of the marine environment, was formed. It will explore the potential environmental impacts of marine geoengineering, in part to explore potential future forms of marine geoengineering to be listed in the amendment's annex.
189 London Protocol, Art. 22. The modifications then apply to each Party that does not declare "that they are not able to accept the amendment at that time." *Id.*
190 Res. LP.4(8), Annex, Art. 1, adding new Art. 6*bis*.
191 *Id.*, Annex, Art. 1, adding new Annex 4, para. 1.
192 *Id.*, Annex, Art. 1, adding new Art. 6*bis*.2.
193 *Id.*, Annex, Art. 1, adding new Annex 5. The existing ocean fertilization assessment framework is to serve as the specific assessment framework for that activity.
194 *Id.*, Annex, Art. 1, adding new Annex 5, para. 10.
195 *Id.*, Annex, Art. 1, adding new Annex 5, paras. 5, 8.
196 *Id.*, Annex, Art. 1, adding new Annex 5, para. 11.
197 *Id.*, Annex, Art. 1, adding new Annex 5, para. 12.
198 *Id.*, Annex, Art. 1, adding new Annex 5, para. 26.
199 *Id.*, Annex, Art. 1, adding new Annex 5, paras. 7–8.
200 With some creative interpretation, the definition could include even those activities that occur on or over land but impact in the marine environment. However, the regulatory scope of the amendment is limited to placement of matter into the sea, thus excluding atmospheric activities.
201 Res. LP.4(8), Preamble para. 1.
202 *Id.*, Annex, Art. 1, adding new Art. 6*bis*.3.
203 Verlaan, *supra* note 96; Ginzky and Frost, *supra* note 180, at 90.
204 Res. LP.4(8) Annex, Art. 1, adding new Annex 5, para. 26.5.
205 Report of the Twenty-first Consultative Meeting of Contracting Parties to the Convention on the Prevention of Marine Pollution by Dumping of Wastes and Other Matter (2007), paras. 5.18–5.27. *See* Interpretation of the London Convention 1972: Mitigating the Environmental Impacts on the Oceans of Climate Change: Carbon Capture and Sequestration in the Marine Environment (submitted by the United Kingdom 2004).
206 2012 Specific Guidelines for the Assessment of Carbon Dioxide for Disposal into Sub-seabed Geological Formations (2012).
207 This amendment entered into force on February 10, 2007. The carbon dioxide stream must also be overwhelmingly pure. *See* London Protocol Annex 1, Art. 4.3.
208 Resolution LP.3(4) on the Amendment to Art. 6 of the London Protocol (Adopted on 30 October 2009), the Fourth Meeting of Contracting Parties to the 1996 Protocol to the Convention on the Prevention of Marine Pollution by Dumping of Wastes and Other Matter 1972. Such export would otherwise be prohibited by Article 6 of the Protocol.
209 *See* section 3.6.

210 Convention on Environmental Impact Assessment in a Transboundary Context (1991), Art. 1(v).
211 *Id.*, Art. 1(vii).
212 *Id.*, Art. 2.1.
213 *Id.*, Art. 9.
214 *Id.*, Art. 2.5.
215 *Id.*, App. II.
216 *Id.*, Arts. 2.2–2.3, 4, App. II.
217 *Id.*, Arts. 2.4, 3, 5.
218 *Id.*, Arts. 2.6, 3.8.
219 *Id.*, Art. 6.
220 *Id.*, Art. 2.7.
221 Protocol on Strategic Environmental Assessment to the Convention on Environmental Impact Assessment in a Transboundary Context (2003), Art. 4.1.
222 General categories include forestry, industry, waste management, and land use. Specific projects include large-diameter gas pipelines (presumably encompassing carbon dioxide, as elsewhere "natural gas" is stated), "[p]rojects for the use of uncultivated land or seminatural areas for intensive agricultural purposes," "[i]nitial afforestation and deforestation for the purposes of conversion to another type of land use," "[i]ndustrial installations for carrying gas," and deep drillings. *Id.*, Art. 4.2, Annexes I–II.
223 *Id.*, Art. 2.7.
224 *Id.*, Arts. 1(d), 2.6, 3.2, 5.3, 5.4, 6.3, 8, 11.2, 12.2.
225 *Id.*, Art. 9.
226 Convention on Access to Information, Public Participation in Decision-making and Access to Justice in Environmental Matters [hereinafter Aarhus Convention] (1998), Art. 1.
227 *Id.*, Art. 3.9.
228 *Id.*, Art. 2.3.
229 *Id.*, Art. 4.
230 *Id.*, Art. 5.
231 *Id.*, Art. 6, Annex I.
232 *Id.*, Arts. 7–8. *See* Svitlana Kravchenko, *The Aarhus Convention and Innovations in Compliance with Multilateral Environmental Agreements*, 18 COLORADO JOURNAL OF INTERNATIONAL ENVIRONMENTAL LAW AND POLICY 1 (2007).
233 Aarhus Convention, Art. 9.
234 *Id.*, Art. 15. Members of the public and environmental organizations may bring complaints before a noncompliance committee, whose members are nominated by, among others, environmental organizations. The "non-confrontational, non-judicial and consultative" procedure results in recommendations for the noncompliant Party. The Meeting of Parties may suspend the treaty rights of a noncompliant Party.
235 Protocol on Pollutant Release and Transfer Registers to the Convention on Access to Information, Public Participation in Decision-Making and Access to Justice in Environmental Matters [hereinafter Kiev Protocol] (2003), Art. 1.
236 *Id.*, Art. 2.
237 *Id.*, Arts. 3–11.
238 *Id.*, Art. 3.4.

239 Convention on Biological Diversity [hereinafter CBD] (1992), Art. 1.

240 Reports issued by the CBD Secretariat have said as much, highlighting the tension between climate change and climate engineering: "If effective, geoengineering would reduce the impacts of climate change on biodiversity at the global level." Phillip Williamson and Ralph Bodle, *Update on Climate Geoengineering in Relation to the Convention on Biological Diversity: Potential Impacts and Regulatory Framework* (Montreal: Secretariat of the CBD, 2016), 13.

241 CBD, Art. 3.

242 *Id.*, Art. 7.

243 *Id.*, Art. 14.

244 See the description of the LOHAFEX debate within Germany in Chapter 6. For more on nonbinding multilateral environmental agreements, see section 3.5 below.

245 Report of the Conference of the Parties to the Convention on Biological Diversity on the Work of its Ninth Meeting (2008), Decision IX/16 C.4.

246 Report of the Tenth Meeting of the Conference of Parties to the Convention on Biological Diversity (2010), Decision X/33.8(w). Art. 3 is the restatement of a state's right to exploit its natural resources and its responsibility to prevent transboundary harm.

247 *Id.*, fn. 76.

248 Secretariat of the CBD, *Geoengineering in Relation to the Convention on Biological Diversity: Technical and Regulatory Matters*, (Montreal: Secretariat of the CBD, 2012), 8.

249 Report of the Eleventh Meeting of the Conference of Parties to the Convention on Biological Diversity (2012), Decision XI/20; Conference of the Parties to the Convention on Biological Diversity (2016), Climate-related geoengineering, Decision XIII/14.

250 Conference of the Parties to the Convention on Biological Diversity (2016), *supra* note 249, para. 5.

251 *Id.*, Decision XI/20, para. 8; Regulatory Framework for Climate-related Geoengineering Relevant to the Convention on Biological Diversity (2012), 6; Williamson and Bodle, *supra* note 240, at 12. Deference to the London Convention and London Protocol is reinforced by CBD, Art. 22, in which Parties agree to implement the CBD "with respect to the marine environment consistently with the rights and obligations of States under the law of the sea."

252 Williamson and Bodle, *supra* note 240, at 144.

253 CBD, Art. 14.1.

254 However, the COP report noted "that the call for a moratorium on ocean fertilization would be implicit." Report of the COP to the CBD at its 9th Meeting, *supra* note 170, para. 250. *See* Jesse L. Reynolds, Andy Parker, and Peter Irvine, *Five Solar Geoengineering Tropes that Have Outstayed Their Welcome*, 4 EARTH'S FUTURE 562 (2016).

255 Masahiro Sugiyama and Taishi Sugiyama, *Interpretation of CBD COP10 Decision on Geoengineering* (Socio-economic Research Center, Central Research Institute of Electric Power Industry, 2010), 1.

256 Statement by the IOC ad hoc Consultative Group on Ocean Fertilization, Submitted by the Intergovernmental Oceanographic Commission (IOC) (2008), Addendum (June 14, 2008).

257 Secretariat of the CBD, *supra* note 248; Williamson and Bodle, *supra* note 240.
258 More accurately, the land area is subject to claims by seven states, but these are not fully recognized and further claims are prohibited by the Antarctic Treaty.
259 Michael C MacCracken, *On the Possible Use of Geoengineering to Moderate Specific Climate Change Impacts*, 4 ENVIRONMENTAL RESEARCH LETTERS 045107 (2009); John Latham et al., *Marine Cloud Brightening: Regional Applications*, 372 PHILOSOPHICAL TRANSACTIONS OF THE ROYAL SOCIETY A: MATHEMATICAL, PHYSICAL AND ENGINEERING SICENCES 20140053 (2014).
260 The Antarctic Treaty (1959), Arts. II–III.
261 *Id.*, Art. IX.
262 *Id.*, Art. I.
263 Protocol on Environmental Protection to the Antarctic Treaty (1991), Art. 2.
264 *Id.*, Art. 3.1.
265 *Id.*, Art. 3.3.
266 *Id.*, Art. 3.2(a).
267 *Id.*, Art. 3.2(b).
268 *Id.*, Art. 3.2(c).
269 *Id.*, Arts. 3.2(d)–(e), 6, 8, 17, Annex I.
270 *Id.*, Annex II, Arts. 1, 3; Annex V, Art. 4.
271 *Id.*, Annex IV, Arts. 4, 11.
272 For example, France is the nonparty state with the largest economy.
273 Convention on the Prohibition of Military or Any Other Hostile Use of Environmental Modification Techniques [hereinafter ENMOD] (1977), Art. II.
274 *Id.*, Art. I.1.
275 Understandings Regarding the Convention, Report of Conference of the Committee on Disarmament, General Assembly Official Records (1976).
276 ENMOD, Art. III.1.
277 *Id.*, Preamble.
278 *Id.*, Art. III.2.
279 *See* The UN Office at Geneva, Convention on the Prohibition of Military or Any Other Hostile Use of Environmental Modification Techniques (ENMOD), www.unog.ch/enmod.
280 Convention on International Civil Aviation [hereinafter Chicago Convention] (2006), Art. 1.
281 *Id.*, Art. 2.
282 *Id.*, Arts. 17–21, 29–36; Convention on Offences and Certain Other Acts Committed on Board Aircraft (1963), Art. 3.
283 Chicago Convention, Arts. 3, 5. The aims of the Chicago Convention are unstated beyond the Preamble's exhortations regarding "friendship and understanding among the nations and peoples of the world," the desire "to avoid friction and to promote … cooperation between nations and peoples," and the goal to develop civil aviation "in a safe and orderly manner." However, the aims of the International Civil Aviation Organization are enumerated, including ensuring "the safe and orderly growth of international civil aviation throughout the world" and meeting "the needs of the peoples of the world for safe, regular, efficient and economical air transport." Here, "safe" could include protection from environmental harm. *See id.*, Art. 44.

284 *Id.*, Arts. 3(c), 6, 8.
285 *Id.*, Art. 12.
286 *Id.*, Art. 11.
287 *Id.*, Arts. 12, 37, 43–79.
288 *Id.*, Art. 38.
289 International Civil Aviation Organization, *Environmental Protection, Volume II: Aircraft Engine Emissions*, Third edn (2008).
290 For example, the state could claim that the emissions are not safe. Such a state could make other claims, such as those concerning transboundary harm, if its environment were impacted by the emissions, and those concerning violations of sovereignty.
291 Chicago Convention, Art. 3 *bis*.
292 Convention on Offences and Certain Other Acrs Committed on Board Aircraft, Art. 4.
293 Chicago Convention, Arts. 3 *bis*, 4.
294 *Id.*, Arts. 84–88.
295 The L1 Lagrangian point is the location between the Earth and the Sun where their gravitational forces are counterbalanced by the centripetal (center-seeking) force required for an object to orbit the sun. An object there would consequently also be directly between the Earth and the Sun. This is approximately 1 percent of the distance from the Earth to the Sun, or about four times the distance from the Earth to the Moon.
296 Treaty on Principles Governing the Activities of States in the Exploration and Use of Outer Space, including the Moon and Other Celestial Bodies [hereinafter Outer Space Treaty] (1967), Arts. I, III, IX. Note that a subsequent UN General Assembly resolution clarified that the first cited phrase was intended to encourage consideration of developing countries' needs and to stimulate cooperation. *See* Declaration on International Cooperation in the Exploration and Use of Outer Space for the Benefit and in the Interests of All States Taking into Particular Account the Needs of Developing Countries, UNGA Res 122 (LI) (1996).
297 Outer Space Treaty, Art. XI.
298 *Id.*, Art. VI.
299 *Id.*, Art. I.
300 *Id.*, Art. VII; Convention on International Liability for Damage Caused by Space Objects (1972), Arts. I(c), II.
301 *Id.*, Art. V.1.
302 UN Economic and Social Council, Substantive Issues Arising in the Implementation of the International Covenant on Economic, Social and Cultural Rights: General Comment 12 (U E/C.12/1999/5) (1999), 4–5.
303 International Covenant on Economic, Social and Cultural Rights [hereinafter ICESCR] (1966), Art. 11. Note that the USA is not a party to this treaty.
304 *Id.*, Arts. 11–12. *See* Han Somsen, *Towards a Law of the Mammoth? Climate Engineering in Contemporary EU Environmental Law*, 7 EUROPEAN JOURNAL OF RISK REGULATION 109 (2016).
305 United Nations Universal Declaration on Human Rights, UNGA Res 217 A [hereinafter UDHR] (1948), Arts. 19, 26, 27; International Covenant on Civil and Political Rights [hereinafter ICCPR] (1966), Art. 19; ICESCR, Arts. 13, 15; Charter of Fundamental Rights of the European Union (2000), Art. 13. *See* Rhona K. M.

Smith, *Textbook on International Human Rights*, Seventh edn (Oxford: Oxford University Press, 2016), 339. *See also* Antarctic Treaty, Art. II; Outer Space Treaty, Art. I; UNCLOS, Art. 238.

306 ICESCR, Art. 15.1(c). *See* subsection 3.4.4.

307 UDHR, Art. 27(1); ICESCR, Arts. 15.1(b), 15.2; Farida Shaheed, Report of the Special Rapporteur in the Field of Cultural Rights, Farida Shaheed: The Right to Enjoy the Benefits of Scientific Progress and Its Applications (A/HRC/20/26) (2012).

308 Rio Declaration, A/RES/47/190, Principle 10. *See also* World Charter for Nature, UNGA Res 37/7 (1982), Princ. 23; Neil Craik, *International EIA Law and Geoengineering: Do Emerging Technologies Require Special Rules?*, 5 CLIMATE LAW 111 (2015).

309 UDHR, Art. 19; ICCPR, Art. 19.2. These include a "freedom to seek, receive and impart information and ideas." The precise contours of states' concomitant duties to fulfill this right are unclear.

310 *See* the discussion of the Aarhus Convention, subsection 3.3.3.2, and of customary international law, subsection 3.6.2, as well as UNFCCC, Art. 4.1(i).

311 *See* World Conference on Science, Declaration on Science and the Use of Scientific Knowledge (1999), para. 41; Anna-Maria Hubert *Code of Conduct for Responsible Geoengineering*: www.ucalgary.ca/grgproject/files/grgproject/revised-code-of-conduct-for-geoengineering-research-2017-hubert.pdf

312 *See* Chapter 6.

313 Jesse L. Reynolds, Jorge L. Contreras, and Joshua D. Sarnoff, *Solar Climate Engineering and Intellectual Property: Toward a Research Commons*, 18 *MINNESOTA JOURNAL OF LAW, SCIENCE & TECHNOLOGY* 1 (2017).

314 Agreement Establishing the World Trade Organization, Annex 1C, Agreement on Trade-Related Aspects of Intellectual Property Rights (1994), Art. 31.

315 *Id.*, Art. 27.2.

316 Charter of the United Nations, Chapter IV.

317 *Id.*, Chapters V, VII.

318 Statute of the International Court of Justice (1945).

319 Declaration of the United Nations Conference on the Human Environment [hereinafter Stockholm Declaration] (1972), Preamble, Proclamations 3, 6.

320 *Id.*, Principle 1.

321 *Id.*, Principle 18. *See also* Principle 20.

322 *Id.*, Principles 21, 24.

323 A/RES/47/190.

324 Rio Declaration, Principle 1.

325 *Id.*, Principles 2, 3.

326 *Id.*, Principle 9.

327 *Id.*, Principle 14.

328 *Id.*, Principles 1, 3, 7, 10, 15–17, 19, 27. *See* sections 3.6 and 3.7.

329 Johannesburg Declaration on Sustainable Development (2002), Annex.

330 The Future We Want (2012), para. 167.

331 The 2030 Agenda for Sustainable Development (2015).

332 *Id.*, Preamble para. 31.
333 *Id.*, target 13.3.
334 Provisions for Co-operation between States in Weather Modification (1980), footnote.
335 *Id.*, para. 1(a).
336 *Id.*, para. 1(h).
337 *Id.*, para. 1(b).
338 *Id.*, para. 1(c)–(d).
339 *Id.*, para. 1(e)–(f).
340 *See* Karen N. Scott, *International Law in the Anthropocene: Responding to the Geoengineering Challenge*, 34 MICHIGAN JOURNAL OF INTERNATIONAL LAW 309, 328 (2013).
341 UNESCO Expert Meeting, *Geoengineering: The Way Forward?* (November 2010), www.unesco.org/new/en/natural-sciences/about-us/single-view/news/geoengineering_the_way_forward/.
342 The World Climate Research Programme was established in 1980 by the International Council for Science and the WMO. Now with the additional sponsorship of the IOC of UNESCO, it is a leading international coordinator of climate research.
343 *See* UNESCO Expert Meeting, *supra* note 341.
344 *Engineering the Climate: Research Questions and Policy Implications*, UNESCO SCOPE UNEP Policy Briefs No. 14 (2011).
345 Doug Wallace et al., *Ocean Fertilization: A Scientific Summary for Policy Makers* (UNESCO 2010).
346 WMO, Report of the Expert Team on Weather Modification Meeting, Phitsanulok, Thailand, 17–19 March 2015, at 7, www.wmo.int/pages/prog/arep/wwrp/new/documents/WMO_expert_mtg_Phisanulok_2015_report_FINAL.pdf. The American Meteorological Society statement concludes: "The potential to help society cope with climate change and the risks of adverse consequences imply a need for adequate research, appropriate regulation, and transparent deliberation." AMS, Geoengineering the Climate System: A Policy Statement of the American Meteorological Society, www2.ametsoc.org/ams/index.cfm/about-ams/ams-statements/statements-of-the-ams-in-force/geoengineering-the-climate-system/. *See also* Alan Robock and Roelof Bruintjes, WMO Statement on Geoengineering: Draft, www.wmo.int/pages/prog/arep/wwrp/new/documents/WMO_Statement_Geoengineering.pdf.
347 *See* WMO, Report of the Expert Team on Weather Modification Meeting, *supra* note 346, at 14.
348 Rajendra K. Pachauri et al., *Climate Change 2014: Synthesis Report* (Cambridge: Cambridge University Press, 2014), 89. That page also contains references to where the Fifth Assessment Report discusses climate engineering.
349 Ottmar Edenhofer et al., eds., IPCC Expert Meeting on Geoengineering Lima, Peru 20–22 June 2011 Meeting Report (2012), www.ipcc.ch/pdf/supporting-material/EM_GeoE_Meeting_Report_final.pdf.
350 Resolution Adopted by the General Assembly 62/215: Oceans and the Law of the Sea (2007), para. 97.
351 *Id.*, paras. 97–98.

352 *See* Nico Schrijver, *Sovereignty over Natural Resources: Balancing Rights and Duties* (Cambridge: Cambridge University Press, 2008).
353 Right to Exploit Freely Natural Wealth and Resources (626 (VII)) (1952); Declaration on Permanent Sovereignty over Natural Resources (1803 (XVII)) (1962).
354 Stockholm Declaration, Principle 21; London Convention, Preamble para. 3; CLRTAP, Preamble para. 5; UNCLOS, Art. 193; Vienna Ozone Convention, Preamble para. 2; Rio Declaration, Principle 2; UNFCCC, Preamble para. 8; CBD, Art. 3.
355 Stockholm Declaration, Principle 21. *See, e.g., Trail Smelter Case. United States of America, Canada. April 16, 1938, and March 11, 1941*, 3 Reports of International Arbitral Awards 1905 (1941).
356 London Convention, Preamble para. 3; CLRTAP, Preamble para. 5; UNCLOS, Art. 194.2; Vienna Ozone Convention, Preamble para. 2; Rio Declaration, Principle 2; UNFCCC, Preamble para. 8; CBD, Art. 3.
357 International Court of Justice, *Legality of the Threat or Use of Nuclear Weapons, Advisory Opinion*, International Court of Justice Reports 226 (1996).
358 Rio Declaration, Principle 2.
359 *See* Kerryn Brent, Jeffrey McGee, and Amy Maguire, Does the "No-Harm" Rule Have a Role in Preventing Transboundary Harm and Harm to the Global Atmospheric Commons from Geoengineering?, 5 *Climate Law* 35 (2015).
360 International Law Commission, Prevention of Transboundary Harm from Hazardous Activities, in *Official Records of the General Assembly, Fifty-sixth Session*, Supplement No. 10 (A/56/10) (2001), Arts. 1, 2(a). Note that the ILC Articles are not themselves legally binding, but instead represent experts' best effort to capture current custom as practiced by states.
361 *Id.*, Arts. 3–4, 6–8, 13, 16.
362 *Id.*, Art. 9.
363 *Id.*, Art. 10(b), (e).
364 *See* David Reichwein et al., State Responsibility for Environmental Harm from Climate Engineering, 5 *Climate Law* 142 (2015).
365 International Law Commission, Prevention of Transboundary Harm, *supra* note 360, Art. 9.3.
366 Reichwein et al., *supra* note 364, at 169–70. *See also* Brent, McGee, and Maguire, *supra* note 359.
367 *See* Birnie, Boyle, and Redgwell, *supra* note 17, at 180–82, esp. footnote 440, citing the version found in the Sixth Report on International Liability for Injurious Consequences Arising out of Acts Not Prohibited by International Law (1990), Art. 20.
368 Bodansky, *supra* note 3, at 312; Reichwein et al., *supra* note 364, at 180.
369 Daniel Bodansky, *Governing Climate Engineering: Scenarios for Analysis*, 11–47 The Harvard Project on Climate Agreements Discussion Papers (2011), 15.
370 René Lefeber, *Climate Change and State Responsibility*, in INTERNATIONAL LAW IN THE ERA OF CLIMATE CHANGE (Rosemary Rayfuse and Shirley V. Scott Cheltenham, eds., Cheltenham: Edward Elgar, 2012), 333–40.
371 International Law Commission, *Draft Articles on Responsibility of States for Internationally Wrongful Acts, with Commentaries*, in REPORT OF THE

INTERNATIONAL LAW COMMISSION, 53RD SESSION, OFFICIAL RECORDS OF THE GENERAL ASSEMBLY (Geneva: United Nations, 2001), Art. 2.

372 Ralph Bodle, *Geoengineering and International Law: The Search for Common Legal Ground*, 46 TULSA LAW REVIEW 305, 306–307 (2010); Reichwein et al., *supra* note 364.

373 International Law Commission, *supra* note 371, Arts. 1–2, 30–31.

374 *Id.*, Arts. 34–37.

375 *See* Chapter 5 on liability.

376 International Law Commission, *supra* note 371, Arts. 31.2, 36.2.

377 *Id.*, Arts. 40–41.

378 *Id.*, Art. 40.2.

379 Report of the International Law Commission on the Work of its Twenty-eighth Session (1976), Art. 19, p. 75.

380 *See* Eva M. Kornicker Uhlmann, *State Community Interests, Jus Cogens and Protection of the Global Environment: Developing Criteria for Peremptory Norms*, 11 GEORGETOWN INTERNATIONAL ENVIRONMENTAL LAW REVIEW 101 (1998); Lefeber, *supra* note 370, at 342–33.

381 International Law Commission, *supra* note 371, Art. 25.

382 International Court of Justice, *Case Concerning the Gabčíkovo-Nagymaros Project* (1997).

383 International Law Commission, *supra* note 371, Art. 25. *See* Scott, *supra* note 340, at 348, fn. 271; Reichwein et al., *supra* note 364, at 174.

384 International Law Commission, *supra* note 371, Art. 49.

385 *Id.*, Art. 54.

386 UNFCCC, Preamble para. 1. *See* Dinah Shelton, *Common Concern of Humanity*, 5 *IUSTUM AEQUUM SALUTARE* 33, 39 (2009).

387 International Law Commission, *supra* note 371, Art. 52.

388 *Id.*, Arts. 50–51.

389 *Id.*, Art. 53.

390 *Id.*, Art. 49.

391 *Id.*, Art. 49, commentary (4).

392 *Draft Principles on the Allocation of Loss in the Case of Transboundary Harm Arising out of Hazardous Activities, with Commentaries*, in Report of the International Law Commission, 58th session, Official Records of the General Assembly (United Nations, 2006), Principles 1–2. Environmental damage to areas beyond national jurisdiction are not covered.

393 *Id.*, Principle 5.

394 *Id.*, Principle 4.1.

395 *Id.*, Principle 4, 6.

396 *Id.*, Principle 7. *See* Joshua B. Horton, Andrew Parker, and David Keith, Liability for Solar Geoengineering: Historical Precedents, Contemporary Innovations, and Governance Possibilities, 22 *NYU Environmental Law Journals* 225 (2015); Jesse Reynolds, An Economic Analysis of Liability and Compensation for Harm from Large-Scale Solar Climate Engineering Field Research, 5 *Climate Law* 182 (2015).

397 Gunther Handl, An International Legal Perspective on the Conduct of Abnormally Dangerous Activities in Frontier Areas: The Case of Nuclear Power Plant Siting, 7 *Ecology Law Quarterly* 1 (1978); Joni S. Charme, Transnational Injury and

Ultra-Hazardous Activity: An Emerging Norm of International Strict Liability, 4 *Journal of Law and Technology* 75 (1989).

398 International Law Commission, *Allocation of Loss*, Principle 4, commentary (13).

399 Report of the International Law Commission, *supra* note 20, Guideline 7.

400 *Id.*, at 294.

401 *Id.*

402 CBD, Preamble para. 3; UNFCCC, Preamble para. 1.

403 UNCLOS, Art. 194(1); Montreal Protocol, Art. 5; Rio Declaration, Principle 7; CBD, Preamble para. 3; UNFCCC, Arts. 3.1, 4.1; Kyoto Protocol.

404 UNFCCC, Art. 3.1.

405 Paris Agreement, Art. 1.2(a).

406 UNFCCC, Art. 4.1(g); Paris Agreement, Art 10.2.

407 Rio Declaration, Principle 15; CBD, Preamble para. 9; UNFCCC, Art. 3.3; Oslo Protocol, Preamble paras. 3–4; London Protocol, Article 3.1; Kiev Protocol, Art. 3.4.

408 UNFCCC, Art. 3.3.

409 Jesse L. Reynolds and Floor Fleurke, *Climate Engineering Research: A Precautionary Response to Climate Change?*, 7 CARBON AND CLIMATE LAW REVIEW 101 (2013). *See also* Elizabeth Tedsen and Gesa Homann, *Implementing the Precautionary Principle for Climate Engineering*, 7 CARBON AND CLIMATE LAW REVIEW 90 (2013); Reichwein et al., *supra* note 364, at 172–3.

410 Res. LC-LP.2; London Protocol.

411 This may be because, in some cases, following the polluter pays principle would be contrary to the public interest. In the international context, side payments to actual and potential polluters is inconsistent with the principle, yet encourages the participation of these often necessary states. In the case of climate engineering, requiring the implementing state or other actor of climate engineering to be responsible for damages may create disincentives for it to undertake an action that offers large net benefits, despite its harm.

412 Rio Declaration Principle 16; London Protocol, Art. 3.2.

413 *See* Chapter 5.

414 Charter of the United Nations (1945), Art. 74.

415 UNFCCC, Art. 3.1.

416 William C. G. Burns, *Climate Geoengineering: Solar Radiation Management and its Implications for Intergenerational Equity*, 4 STANFORD JOURNAL OF LAW, SCIENCE, AND POLICY 37 (2011).

417 Bodansky, *supra* note 3.

418 Jay Michaelson, *Geoengineering: A Climate Change Manhattan Project*, 17 STANFORD ENVIRONMENTAL LAW JOURNAL 73 (1998); William Daniel Davis, *What Does "Green" Mean?: Anthropogenic Climate Change, Geoengineering, and International Environmental Law*, 43 GEORGIA LAW REVIEW 901 (2009).

419 Bodansky, *supra* note 3; Bodle, *supra* note 372; Rex J. Zedalis, *Climate Change and the National Academy of Sciences' Idea of Geoengineering: One American Academic's Perspective on First Considering the Text of Existing International Agreements*, 19 EUROPEAN ENERGY AND ENVIRONMENTAL LAW REVIEW 18 (2010); Catherine Redgwell, *Geoengineering the Climate: Technological Solutions to Mitigation-Failure or Continuing Carbon Addiction?*, 5 CARBON

AND CLIMATE LAW REVIEW 178 (2011); Gerd Winter, *Climate Engineering and International Law: Last Resort or the End of Humanity?,"* 20 REVIEW OF EUROPEAN COMMUNITY AND INTL. ENV. LAW 277 (2011); Scott, *supra* note 340; David A. Wirth, *Engineering the Climate: Geoengineering as a Challenge to International Governance*, 40 BOSTON COLLEGE ENVIRONMENTAL AFFAIRS LAW REVIEW 413 (2013); Jesse Reynolds, *Climate Engineering Field Research: The Favorable Setting of International Environmental Law*, 5 WASHINGTON AND LEE JOURNAL OF ENERGY, CLIMATE, AND THE ENVT. 417 (2014); Wil Burns and Simon Nicholson, *Governing Climate Engineering*, in NEW EARTH POLITICS: ESSAYS FROM THE ANTHROPOCENE (Simon Nicholson and Sikina Jinnah, eds., Cambridge, MA: MIT Press, 2016).

420 Scott Barrett, *The Incredible Economics of Geoengineering*, 39 ENVTL. AND RESOURCE ECONOMICS 45 (2008); David G. Victor, *On the Regulation of Geoengineering*, 24 OXFORD REVIEW OF ECONOMIC POLICY 322 (2008); Albert Lin, *Geoengineering Governance*, 8 ISSUES IN LEGAL SCHOLARSHIP (2009); John Virgoe, *International Governance of a Possible Geoengineering Intervention to Combat Climate Change*, 95 CLIMATIC CHANGE 103 (2009); Adam D. K. Abelkop and Jonathan C. Carlson, *Reining in Phaëthon's Chariot: Principles for the Governance of Geoengineering*, 21 TRANSNATIONAL LAW AND CONTEMPORARY PROBLEMS 763 (2013); Honegger, Sugathapala, and Michaelowa, *supra* note 59; Tuomas Kuokkanen and Yulia Yamineva, *Regulating Geoengineering in International Environmental Law*, 7 CARBON AND CLIMATE LAW REVIEW 161 (2013); Michael Zürn and Stefan Schäfer, *The Paradox of Climate Engineering*, 4 GLOBAL POLICY 266 (2013); Vishal Garg, *Engineering a Solution to Climate Change: Suggestions for an International Treaty Regime Governing Geoengineering*, UNIVERSITY OF ILLINOIS JOURNAL OF LAW, TECHNOLOGY & POLICY 197 (2014); Ian D. Lloyd and Michael Oppenheimer, *On the Design of an International Governance Framework for Geoengineering*, 14 GLOBAL ENVIRONMENTAL POLITICS 45 (2014).

421 Tracy Hester, A *Matter of Scale: Regional Climate Engineering and the Shortfalls of Multinational Governance*, 7 CARBON AND CLIMATE LAW REVIEW 168 (2013); Proelss and Hong, *supra* note 158; K.-U. Schrogl and L. Summerer, *Climate Engineering and Space*, 129 ACTA ASTRONAUTICA 121 (2016).

422 Verlaan, *supra* note 94.

423 Jesse Reynolds, *The Regulation of Climate Engineering*, 3 LAW, INNOVATION AND TECHNOLOGY 113 (2011); Lisa Dilling and Rachel Hauser, *Governing Geoengineering Research: Why, When and How?*, 121 CLIMATIC CHANGE 553 (2013); Andy Parker, *Governing Solar Geoengineering Research as It Leaves the Laboratory*, 372 PHILOSOPHICAL TRANSACTIONS OF THE ROYAL SOCIETY A: MATHEMATICAL PHYSICAL AND ENGINEERING SCIENCES 20140173 (2014); Reynolds, *supra* note 396; Albert Lin, *The Missing Pieces of Geoengineering Research Governance*, 100 MINNESOTA LAW REVIEW 2509 (2016).

424 Peterson, *supra* note 93; Karen N. Scott, *The Day After Tomorrow: Ocean CO_2 Sequestration and the Future of Climate Change*, 18 GEORGETOWN INTL. ENVTL. LAW REVIEW 45 (2005); Freestone and Rayfuse, supra note 148; Rosemary Rayfuse, Mark G. Lawrence, and Kristina M. Gjerde, *Ocean Fertilisation and Climate Change: The Need to Regulate Emerging High Seas Uses*, 23 THE INTERNATIONAL

JOURNAL OF MARINE AND COASTAL LAW 297 (2008); Randall S. Abate and Andrew B. Greenlee, *Sowing Seeds Uncertain: Ocean Iron Fertilization, Climate Change, and the International Environmental Law Framework*, 27 PACE ENVTL. LAW REVIEW 555 (2010); Harald Ginzky, *Ocean Fertilization as Climate Change Mitigation Measure Consideration under International Law*, 7 JOURNAL FOR EUROPEAN ENVIRONMENTAL AND PLANNING LAW 57 (2010); Kerstin Güssow et al., *Ocean Iron Fertilization: Why Further Research is Needed*, 34 MARINE POLICY 911 (2010); Scott, *supra* note 148; Boschen, *supra* note 180; Ginzky and Frost, *supra* note 180; Matthew Hubbard, *Barometer Rising: The Cartagena Protocol on Biosafety as a Model For Holistic International Regulation of Ocean Fertilization Projects and Other Forms of Geoengineering*, 40 WILLIAM & MARY ENVTL. LAW AND POLICY REVIEW 591 (2016).

425 Randall S. Abate, A *Tale of Two Carbon Sinks: Can Forest Carbon Management Serve as a Framework to Implement Ocean Iron Fertilization as a Climate Change Treaty Compliance Mechanism?*, 1 SEATTLE JOURNAL OF ENVIRONMENTAL LAW 1 (2011); Mark A. Latham, *The BP Deepwater Horizon: A Cautionary Tale for CCS, Hydrofracking, Geoengineering and Other Emerging Technologies with Environmental and Human Health Risks*, 36 WILLIAM AND MARY ENVIRONMENTAL LAW AND POLICY REVIEW 31 (2011); William R. Travis, *Geo-Engineering the Climate: Lessons from Purposeful Weather and Climate Modification* (2012), www.colorado.edu/geography/class_homepages/geog_3402_s07/Travis_geoengineering.pdf; Rachel Hauser, *Using Twentieth-Century U.S. Weather Modification Policy to Gain Insight into Global Climate Remediation Governance Issues*, 5 WEATHER, CLIMATE, AND SOCIETY 180 (2013); Scott Barrett, *Solar Geoengineering's Brave New World: Thoughts on the Governance of an Unprecedented Technology*, 8 REVIEW OF ENVIRONMENTAL ECONOMICS AND POLICY 249 (2014); Jesse Reynolds, *The International Regulation of Climate Engineering: Lessons from Nuclear Power*, 26 JOURNAL OF ENVIRONMENTAL LAW 269 (2014); Edward J. Larson, *The Red Dawn of Geoengineering: First Step Toward an Effective Governance for Stratospheric Injections*, 14 DUKE LAW AND TECHNOLOGY REVIEW 157 (2016); Hubbard, *supra* note 424.

426 Till Markus and Harald Ginzky, *Regulating Climate Engineering: Paradigmatic Aspects of the Regulation of Ocean Fertilization*, 5 CARBON AND CLIMATE LAW REVIEW 477 (2011); Michael C. Branson, A *Green Herring: How Current Ocean Fertilization Regulation Distracts from Geoengineering Research*, 54 SANTA CLARA LAW REVIEW 163 (2014); Grant S. Wilson, *Murky Waters: Ambiguous International Law for Ocean Fertilization and Other Geoengineering*, 49 TEXAS INTERNATIONAL LAW JOURNAL 507 (2014).

427 Cinnamon P. Carlarne, *Arctic Dreams and Geoengineering Wishes: The Collateral Damage of Climate Change*, 49 COLUMBIA JOURNAL OF TRANSNATIONAL LAW 602 (2011); Daniel Bodansky, *The Who, What, and Wherefore of Geoengineering Governance*, 121 CLIMATIC CHANGE 539 (2013); Jane C. S. Long, *A Prognosis, and Perhaps a Plan, for Geoengineering Governance*, 7 CARBON AND CLIMATE LAW REVIEW 177 (2013); Edward A. Parson and Lia N. Ernst, *International Governance of Climate Engineering*, 14 THEORETICAL INQUIRIES IN LAW 307 (2013); Anthony E. Chavez, *Using Legal Principles to Guide Geoengineering Deployment*, 24 N.Y.U. ENVIRONMENTAL LAW JOURNAL 59 (2016).

428 Jung-Eun Kim, *Implications of Current Developments in International Liability for the Practice of Marine Geo-engineering Activities*, 4 ASIAN JOURNAL OF INTERNATIONAL LAW 235 (2014); Horton, Parker, and Keith, *supra* note 396; Reynolds, *supra* note 396; Barbara Saxler, Jule Siegfried, and Alexander Proelss, *International Liability for Transboundary Damage Arising from Stratospheric Aerosol Injections*, 7 LAW, INNOVATION AND TECHNOLOGY 112 (2015).

429 Anthony E. Chavez, *Exclusive Rights to Saving the Planet: The Patenting of Geoengineering Inventions*, 13 NORTHWESTERN JOURNAL OF TECHNOLOGY AND INTELLECTUAL PROPERTY 1 (2015); Reynolds, Contreras, and Sarnoff, *supra* note 313.

430 Doelle, *supra* note 58.

431 Edward A. Parson, *Climate Engineering in Global Climate Governance: Implications for Participation and Linkage*, 3 TRANSNATIONAL ENVIRONMENTAL LAW 89 (2013); Joshua B. Horton, *Geoengineering and the Myth of Unilateralism: Pressures and Prospects for International Cooperation*, 4 STANFORD JOURNAL OF LAW, SCIENCE, AND POLICY 56 (2011).

432 Albert Lin, *Does Geoengineering Present a Moral Hazard?*, 40 ECOLOGY LAW QUARTERLY 673 (2013); Parson, "*supra* note 431; Reynolds, *supra* note 26.

433 Burns, *supra* note 416; Reynolds and Fleurke, *supra* note 409; Tedsen and Homann, supra note 409; Brent, McGee, and Maguire, *supra* note 359; Craik, *supra* note 308; Reichwein et al., *supra* note 364.

434 Chiara Armeni, *Global Experimentalist Governance, International Law and Climate Change Technologies*, 64 INTERNATIONAL & COMPARATIVE LAW QUARTERLY 875 (2015); Janine Sargoni, *The Best of Both Worlds: Maximising the Legitimacy of the EU's Regulation of Geoengineering Research*, 7 EUROPEAN JOURNAL OF RISK REGULATION 87 (2016); Somsen, *supra* note 304.

435 See John D. Graham and Jonathan Baert Wiener, eds., *Risk vs. Risk: Tradeoffs in Protecting Health and the Environment* (Cambridge, MA: Harvard University Press, 2009).

436 As above, a "good" is used here in the descriptive sense of a product or service, not in any normative sense. *See* section 3.1.

437 Ben Kravitz et al., A *Multi-Model Assessment of Regional Climate Disparities Caused by Solar Geoengineering*, 9 ENVIRONMENTAL RESEARCH LETTERS ART. 074013 (2014). For a further discussion of SRM's problem structure, see Reynolds, *supra* note 396, at 186–90.

438 Christopher J. Preston, *Framing an Ethics of Climate Management for the Anthropocene*, 130 CLIMATIC CHANGE 359 (2015); Davor Vidas et al., *International Law for the Anthropocene? Shifting Perspectives in Regulation of the Oceans, Environment and Genetic Resources*, 9 ANTHROPOCENE 1 (2015).

4

US Law

Albert C. Lin

Both federal law and state law play a significant role in governing environmental matters in the United States. Some of these laws, such as the National Environmental Policy Act (NEPA), are generally applicable laws that could be implicated by any geoengineering technique. This chapter first considers these generally applicable laws. Other laws are likely to be relevant only to a limited subset of geoengineering techniques. This latter group of laws includes laws specific to an environmental medium, such as the Clean Air Act, as well as laws governing specific technologies, such as state and local ordinances governing the cultivation of genetically modified organisms.

4.1. GENERALLY APPLICABLE LAWS

Laws that could be triggered by almost any type of geoengineering project include statutes that impose procedural duties and those having substantive effects. Procedural statutes may create obligations to provide notice to the government or public of planned activities, analyze the environmental effects of such activities, allow public comment, and monitor and report on the activities when they are carried out. Substantive statutes include wildlife protection laws and laws providing for extraordinary government powers in emergency situations.

4.1.1. *Procedural Statutes*

NEPA, which was the first law to require environmental impact assessment, is a federal procedural statute that would apply to many geoengineering activities. Some states have adopted laws patterned after NEPA that would apply to geoengineering activities having state involvement or requiring state or local approval. In addition, federal and state weather modification statutes also impose procedural obligations that geoengineering projects could trigger.

4.1.1.1. National Environmental Policy Act

4.1.1.1.1. NEPA BASICS The federal government, the leading sponsor of basic research in the United States, is likely to be involved significantly in any major geoengineering research.[1] Furthermore, deployment of large-scale geoengineering in United States territory would likely involve the federal government as well because of the potential implications for international relations and global climate policy. Federal research, funding, or approval of geoengineering would implicate NEPA. Although NEPA's requirements may be time-consuming, expensive, and subject to litigation, they are ultimately procedural in nature. Provided that the necessary procedures are followed, the statutory requirements themselves are unlikely to serve as an insuperable barrier to geoengineering experimentation or deployment. Nonetheless, the NEPA process often generates public scrutiny and creates opportunities for those opposed to a planned project to seek to halt it, either politically or in court.

NEPA applies to "major federal actions significantly affecting the quality of the human environment."[2] Major federal actions include activities directly undertaken by the federal government as well as activities that "are potentially subject to federal control and responsibility."[3] Accordingly, NEPA applies not only to federal programs and projects, but also to "actions approved by [federal] permit or other regulatory decision" and actions partially financed or assisted by the federal government.[4] In a project not directly undertaken by the federal government, the critical factor determining the applicability of NEPA is the presence of federal decision-making power, such as the need for a federal permit.[5]

NEPA requires federal agencies to prepare an environmental impact statement (EIS) for "major federal actions significantly affecting the quality of the human environment."[6] The EIS must describe the environmental impacts of the proposed action, alternatives to the proposed action, and irreversible and irretrievable commitments of resources that the proposed action would involve.[7] NEPA's implementing regulations, promulgated by the Council on Environmental Quality (CEQ), specify the steps an agency must follow in preparing an EIS. These steps include scoping the issues, circulating a draft EIS to appropriate agencies and the public, responding to any comments received on the draft, and issuing a final EIS.[8]

The CEQ regulations also detail how an agency should proceed if it is not apparent at the outset that a federal action will have a significant environmental impact. The agency may proceed under a categorical exclusion if the contemplated action falls within categories of actions that the agency previously determined as not having significant impacts.[9] When a contemplated action falls under a categorical exclusion, the agency simply invokes the exclusion and proceeds without further analysis. Otherwise, the agency must prepare an environmental assessment (EA).[10] An EA briefly discusses the need for the proposal, alternatives to the proposed action, environmental impacts, and agencies and persons consulted.[11] Based on its analysis in the EA, the agency must either issue a finding of no significant impact (FONSI) or prepare a full-blown EIS.[12] An agency may prepare an EA even when a categorical exclusion might apply, such as where the agency wishes to document the absence of environmental impacts. Conversely, an agency may not rely on a categorical exclusion when an action corresponds to an existing category but may have a significant environmental effect in that instance.[13]

NEPA contains no citizen suit provisions. However, members of the public who would be affected by an action may sue federal agencies for NEPA violations under the Administrative Procedure Act's judicial review provisions.[14] Upon finding a NEPA violation, courts may order the agency to rectify any procedural shortcomings. Courts also may enjoin an agency from proceeding with its proposed action until it has corrected those shortcomings. In deciding whether to issue an injunction, courts must balance the equities and consider the public interest.[15]

Notwithstanding the general applicability of NEPA to federal agency actions, a statute may exempt an action from NEPA analysis. Statutory exemptions include: an exemption for actions taken by the Environmental Protection Agency (EPA) under the Clean Air Act (CAA), such as issuance of a permit;[16] an exemption for certain specified actions under the Clean Water Act;[17] and an exemption for federal emergency assistance provided under the Stafford Act.[18] Some geoengineering projects could trigger CAA permitting requirements, as this chapter later explains, and thus could be exempt from NEPA once a CAA permit is obtained.

In addition to these statutory exemptions, the CEQ regulations allow for special accommodations in emergencies. Specifically, "when emergency circumstances require taking actions with significant environmental impacts and there is not sufficient time to follow the regular [EIS] process," an agency may consult with the CEQ regarding "alternative arrangements."[19] Emergency circumstances generally involve a sudden, unanticipated event or unforeseeable conditions demanding unusual or immediate action.[20]

Alternative arrangements are intended to serve as a substitute means of NEPA compliance rather than as a complete exemption.[21] Such arrangements "are limited to actions necessary to control the immediate impacts of the emergency."[22] Whether climate change-related conditions would qualify as emergency circumstances is an open question that this chapter will consider in conjunction with government authority to act in emergencies.[23] For the moment, it is worth noting that some climate impacts will be foreseeable and gradual, where others will be more sudden and unexpected.

4.1.1.1.2. NEPA APPLIED TO GEOENGINEERING Geoengineering projects subject to federal control or having significant federal involvement would require NEPA analysis. Full-scale geoengineering deployment by the federal government would surely require preparation of an EIS. Even federally sponsored field research may require NEPA analysis.[24] Private interest in conducting geoengineering research could give rise to research projects lacking NEPA analysis if such projects are privately funded and involve no federal permitting or control. Some researchers, however, have expressed reluctance to proceed with field experiments without federal involvement or sanction.[25] NEPA analysis of geoengineering projects, public or private, would provide some assurance to the public and to environmental regulators that environmental concerns were analyzed and considered. The NEPA process also could offer a means of inviting and incorporating public input.

If the federal government were to embark on a geoengineering research program, that would raise the issue of whether the government would have to prepare a programmatic EIS in addition to the environmental analysis required for individual research projects.[26] A programmatic EIS could address general environmental issues relating to the establishment of a geoengineering research program. It could also help to frame subsequent project-specific federal actions.[27] However, unless federal geoengineering research activities constitute a single proposal or are systematically connected, no programmatic EIS would be required.[28]

The specific NEPA obligations triggered by a geoengineering project would depend in part on the federal agencies involved. In addition to complying with the CEQ NEPA regulations, each federal agency must follow its own regulations governing NEPA compliance. The Department of Energy (DOE) and the National Science Foundation (NSF) are active sponsors of climate change-related research, and their NEPA regulations may be especially relevant to geoengineering research activities. As it turns out, these regulations contain categorical exclusions that may apply to some field experiments, particularly those occurring on a smaller scale.

The DOE conducts and sponsors research on energy technologies and other matters relevant to climate change, and its technical expertise could be applied to geoengineering research.[29] The DOE's NEPA-implementing procedures include a categorical exclusion for "[s]iting, construction, modification, operation, and decommissioning of facilities for small-scale research and development projects; conventional laboratory operations ... and small-scale pilot projects (generally less than 2 years) frequently conducted to verify a concept before demonstration actions."[30] However, this categorical exclusion does not extend to "demonstration actions, meaning actions that are undertaken at a scale to show whether a technology would be viable on a larger scale and suitable for commercial deployment."[31] Other DOE categorical exclusions potentially applicable to geoengineering research projects include an exclusion for "[o]utdoor terrestrial ecological and environmental research in a small area (generally less than 5 acres)," and an exclusion for "[s]mall-scale, temporary surveying, site characterization, and research activities in aquatic environments."[32]

The NSF's NEPA regulations also could be relevant to geoengineering field experiments in light of that agency's significant sponsorship of federal research. Reasoning that the environmental effects of research are largely long term, "basically speculative[,] and unknowable in advance," the NSF regulations categorically exclude from NEPA analysis most research projects.[33] Geoengineering research projects may not qualify for this categorical exclusion, however. If a research project itself has potential environmental effects, the NSF regulations require preparation of at least an EA.[34] Indeed, the regulations explicitly require at least an EA for projects involving "weather modification, or other techniques that may alter a local environment," as well as "[c]ases where field work affecting the natural environment will be conducted."[35]

4.1.1.2. State Counterparts to NEPA

Fifteen states, the District of Columbia, and Puerto Rico have enacted analogs to NEPA.[36] Like NEPA, these state environmental policy acts (SEPAs) generally require government agencies to analyze the environmental impacts of specified actions.[37] The precise details of SEPAs vary from state to state. Some SEPAs apply only to state agencies, while others apply to local governments as well.[38] Where the scope of SEPAs extends to local government actions, a wide range of private activities, such as development projects necessitating local government permits, may be subject to SEPA requirements.[39] Courts generally may review whether a state or local agency has complied with a SEPA.[40] And in contrast to NEPA, some SEPAs, such as the California Environmental Quality Act (CEQA), have substantive requirements that may limit or prohibit projects whose environmental impacts exceed a certain threshold.[41]

CEQA is a fairly detailed statute that provides a useful comparison with NEPA.[42] CEQA requires preparation of an environmental impact report (EIR) for "any project which [state and local agencies] propose to carry out or approve that may have a significant effect on the environment."[43] Because the statute applies to zoning decisions and the granting of permits, licenses, leases, and other entitlements,[44] a wide range of otherwise private activities may come under its scope. The EIR mandated by CEQA must describe significant effects, including any growth-inducing effects, proposed measures to mitigate or minimize significant effects, and alternatives to the proposed project.[45] The threshold for preparing an EIR under CEQA is arguably lower than the threshold for preparing an EIS under NEPA; CEQA requires an EIR if an action "may have a significant effect on the environment," whereas NEPA requires an EIS for "actions significantly affecting the quality of the human environment."[46] If it is unclear whether an EIR is necessary, CEQA directs an agency to prepare an initial study to determine if the project may have a significant effect on the environment.[47]

In addition to its procedural requirements, CEQA also imposes a substantive obligation on an agency to mitigate or avoid significant environmental effects "whenever it is feasible to do so."[48] Mitigation measures are to be "fully enforceable through permit conditions, agreements, or other measures."[49] Nonetheless "[i]f economic, social, or other conditions make it infeasible to mitigate one or more significant effects," an agency has the discretion to carry out or approve a project.[50] In such circumstances, the agency must make a finding that "specific overriding economic, legal, social, technological, or other benefits of the project outweigh the significant effects on the environment."[51]

Geoengineering activities could implicate SEPAs to the extent that activities require state or local government approval. For example, projects involving new land uses, construction of new buildings, or the release of pollutants may require state or local permits and thus also may trigger SEPA obligations. Field research activities per se typically do not require state or local approval. Accordingly, so long as geoengineering research does not otherwise involve land use changes or other activities requiring state or local approval, it may not necessarily trigger SEPA review. However, as discussed below, some states do require a permit prior to conducting weather modification operations. Geoengineering activities that require a state weather modification permit would likely be subject to SEPA obligations as well.

4.1.1.3. Weather Modification Reporting Act

Whereas NEPA applies only to activities with non-trivial federal involvement, the federal Weather Modification Reporting Act of 1972 (WMRA) applies to federal and nonfederal activities, whether public or private. WMRA requires any person "engag[ing] or attempt[ing] to engage, in any weather modification activity" to submit a report of such activity in advance to the National Oceanic and Atmospheric Administration (NOAA).[52] The statute defines weather modification as "any activity performed with the intention of producing artificial changes in the composition, behavior, or dynamics of the atmosphere."[53] Operators are required to maintain records of their weather modification activities and to submit reports during and after completion of such activities.[54] Except for trade secrets, information submitted pursuant to WMRA is available to the public.[55]

WMRA is primarily a disclosure statute. However, "[w]hen consideration of a weather modification activity report and related information indicates that a proposed project may significantly depart from the practices or procedures generally employed in similar circumstances to avoid danger to persons, property, or the environment, or indicates that success of federal research projects may be adversely affected if the proposed project is carried out as described," NOAA may notify the operator and state officials of such possibility and make appropriate recommendations.[56]

WMRA was originally enacted to govern cloud-seeding projects and similar short-term weather modification techniques. The ultimate purpose of geoengineering, however, involves long-term modification of climate as opposed to short-term modification of weather. Nevertheless, most solar radiation management (SRM) activity and even some carbon dioxide removal (CDR) techniques would fall within WMRA's definition of weather modification. The statutory definition specifically encompasses "influenc[ing] in any way the natural development cycle of clouds or their environment," "modifying the solar radiation exchange of the earth or clouds, through the release of gases, dusts, liquids, or aerosols into the atmosphere," and "[m]odifying the characteristics of land or water surfaces by dusting or treating with powders, liquid sprays, dyes, or other materials."[57]

In sum, WMRA's requirements are fairly minimal yet broadly applicable. In contrast to NEPA, they could serve as an important mechanism to ensure that the government and general public have notice of private geoengineering activities. WMRA's recordkeeping and reporting requirements also could be pivotal in tracking geoengineering activities that are actually carried out.

4.1.1.4. State Weather Modification Laws

At the state level, a number of jurisdictions have laws governing weather modification.[58] Adopted primarily to address cloud-seeding efforts, these laws generally provide for the licensing of cloud seeders and the issuance of permits for cloud-seeding operations.[59]

Texas' regulatory scheme provides an illustrative example. Texas law defines weather modification activities as "changing or controlling, or attempting to change or control, by artificial methods the natural development of atmospheric cloud forms or precipitation forms that occur in the troposphere."[60] To engage in a weather modification project, an operator must be licensed and obtain a permit.[61] To obtain a license to conduct weather modification activities, a person must demonstrate "competence in the field of meteorology that is reasonably necessary to engage in weather modification and control activities."[62] To obtain a permit for a specific project, an applicant must show that the proposed weather modification "will not significantly dissipate the clouds and prevent their natural course of developing rain in the area in which the operation is to be conducted to the material detriment of persons or property in that area."[63] The applicant must also provide proof of financial responsibility, i.e., "the ability to respond in damages for liability that might reasonably result from the operation."[64] In addition, the applicant must give public notice of planned operations, and the state must hold a public hearing if at least 25 persons request it.[65] Notwithstanding these general permitting requirements, the state may exempt by rule "research, development, and experiments conducted by state and federal agencies, institutions of higher learning, and bona fide nonprofit research organizations," as well as "laboratory research and experiments."[66] Written reports are required for all weather modification activities, whether exempt or conducted under a permit.[67]

California law serves as a notable exception to the license-and-permit approach. California law defines "weather resources management" broadly to include "attempting to produce by physical means any of the following: cloud water conversion, light adjustment, or weather hazard suppression."[68] California previously required licenses and permits for weather modification, but it repealed those requirements in 1984.[69] Under current law, a weather modification operator need only provide advance notice to the state Department of Water Resources and to the public through a newspaper of general circulation.[70]

Many geoengineering activities could be subject to state weather modification laws because of broad statutory definitions of weather modification. Whereas geoengineering field experiments may be able to take advantage of exceptions to permitting requirements for specified research activities, full-scale deployment would not. Techniques most likely to be implicated include marine cloud brightening, cirrus cloud modification, and other efforts to manipulate clouds or change precipitation patterns. Even stratospheric aerosol release efforts would likely fall within the definition of weather modification in many jurisdictions. Furthermore, because geoengineering activities could modify the weather in more than one jurisdiction, multiple licenses and permits may be required.[71] Many states' definitions of weather modification generally refer to weather modification activities and their effects but contain no explicit geographic restrictions.[72] Thus, it is unclear whether geoengineering activities occurring outside a state's boundaries but having effects within that state would require a license or permit. Notably, some states prohibit the carrying out of weather modification operations within that state "for the purpose of affecting weather in any other state which prohibits such operations."[73]

4.1.1.5. Proposed State and Local Initiatives

Thus far, no government in the United States, whether at the federal, state, or local level, has enacted legislation explicitly addressing geoengineering. Such legislation may be forthcoming, however. In 2015, for example, a bill was introduced in the Rhode Island legislature to regulate geoengineering. The bill defined geoengineering broadly as activities "designed to effect a change in the area climate," including "attempts to remove carbon dioxide from the atmosphere" and "solar radiation management or cloud whitening, or similar process[es]."[74] The bill would have required any proposed geoengineering activity within the state to obtain a permit and be subject to at least two public hearings.[75] Re-drafted versions of the bill, modified to focus on SRM and oversight of potential research, were introduced in 2016 and 2017.[76]

4.1.2. *Wildlife Laws*

Geoengineering activities could affect wildlife and thereby trigger various federal and state laws that protect wildlife. This section focuses on the most prominent of these laws: the federal Endangered Species Act, state endangered species acts, the Migratory Bird Treaty Act, and the Marine Mammal Protection Act.

4.1.2.1. The Endangered Species Act and State Analogs

4.1.2.1.1. ENDANGERED SPECIES ACT The federal Endangered Species Act (ESA) provides for the protection of federally designated threatened and endangered species. Section 4 of the ESA authorizes the federal government to list protected species and to designate critical habitat for such species.[77] Two key provisions of the ESA protect listed species: Section 7, which applies only to federal agencies; and Section 9, which applies to all persons.

If an action is authorized, funded, or carried out by a federal agency, Section 7 requires the agency to engage in a consultation process with the US Fish and Wildlife Service or, in the case of marine and anadromous species, NOAA Fisheries (collectively referred to as the "wildlife agencies"). The purpose of consultation is to "insure" that the proposed agency action "is not likely to jeopardize the continued existence of any endangered species or threatened species or result in the destruction or adverse modification" of critical habitat.[78] The action agency initiates consultation by asking the wildlife agencies if any listed species may be present in the project area.[79] If listed species may be present and the action agency determines the project is likely to affect any such species, the agency must engage in formal consultation with the wildlife agency. Formal consultation generates a determination regarding whether the action is likely to jeopardize a protected species or adversely modify its critical habitat.[80] A jeopardy determination is made if an action "reasonably would be expected, directly or indirectly, to reduce appreciably the likelihood of both the survival and recovery of a listed species in the wild by reducing the reproduction, numbers, or distribution of that species."[81] Upon a finding of no jeopardy and no adverse modification, the action agency may proceed. However, if there is a finding of jeopardy or adverse modification, the wildlife agency issues a biological opinion proposing reasonable and prudent action alternatives that are not expected to result in jeopardy or adverse modification.[82] The action agency may then proceed under one of the reasonable and prudent action alternatives.

Section 9 of the ESA makes it unlawful for any person to take any endangered species within the United States, the territorial sea of the United States, or the high seas.[83] This prohibition applies to all actions, federal and nonfederal, and by regulation protects threatened species in addition to endangered species.[84] The term "take" means "to harass, harm, pursue, hunt, shoot, wound, kill, trap, capture, or collect, or to attempt to engage in any such conduct."[85] It includes not only acts that directly harm protected species, but also "significant habitat modification that actually kills or injures wildlife."[86] The government may bring enforcement actions seeking injunctive relief, civil penalties, and even criminal sanctions for violations of the ESA or its regulations.[87] The ESA also authorizes citizen suits against any private or public "person" to enforce the statute.[88]

Two important exceptions soften the seemingly absolute take prohibition of Section 9. First, a federal agency that undergoes the Section 7 consultation process may obtain authorization for an "incidental take," meaning that the harm to the animals is a side effect of the action and not its purpose, provided that the take will not lead to jeopardy or adverse modification and that the agency adopts reasonable and prudent measures to minimize the impact of its action.[89] Second, a nonfederal actor may obtain an incidental take permit under Section 10 of the ESA if it prepares a habitat conservation plan that specifies the impacts likely to result from the taking, the steps the applicant will take to minimize and mitigate impacts on protected species, the funding to implement such steps, alternative actions the applicant considered and reasons why they were not taken, and other necessary measures.[90] A Section 10 incidental take permit shall issue if, inter alia, the taking will be incidental and "will not appreciably reduce the likelihood of the survival and recovery of the species in the wild."[91]

Any geoengineering project that threatens to harm individual members of a protected species or significantly modify the habitat of a protected species could trigger Section 9's take prohibition. In addition, federal involvement in a geoengineering project could trigger Section 7's consultation requirements as well as its prohibition on jeopardy and adverse modification. In ESA litigation over a geoengineering project, at least two issues would likely be disputed: (1) whether a project caused or would cause harm to a species or its habitat, and (2) whether any of a project's effects rise to the level of a take (under Section 9) or jeopardy (under Section 7). If a project directly harms a protected species, causation and take elements under Section 9 would likely be satisfied, and perhaps the jeopardy element of Section 7 would be satisfied as well. Each of these issues may be subject to dispute, however, if plaintiffs allege that future harm would be caused indirectly by changed habitat or climatic conditions following a geoengineering project. Courts largely interpret Section 9 to prohibit only habitat modification that foreseeably causes actual death or injury to identifiable animals.[92] Some effects of geoengineering projects on protected species may not be foreseeable, at least initially, given the unprecedented nature and scale of such projects. Even for foreseeable effects, plaintiffs may still face a difficult task of linking geoengineering activity specifically to observed detrimental effects on a species.

However, the fact that geoengineering has the avowed purpose of causing detectable changes in climate would strengthen the case for causation. Furthermore, as harmful effects on protected species become more obvious, continuation of an ongoing geoengineering project could be subject to the prohibitions of Section 7 and/or Section 9. Under such circumstances, courts may order a halt to geoengineering activity until the necessary procedures are followed and any required incidental take authorizations or permits are issued.

4.1.2.1.2. STATE ENDANGERED SPECIES ACTS States have traditionally been responsible for managing wildlife within their borders, and most states have laws providing some degree of protection to endangered species.[93] Generally, state endangered species acts authorize the listing of endangered species within the state, and state lists may include species that are not federally endangered.[94] State endangered species acts nonetheless vary widely: Some laws only prohibit trafficking or intentionally harming protected species, whereas others establish more comprehensive schemes patterned on the federal ESA.[95] Several states, for instance, have enacted provisions comparable to ESA Section 7, requiring agencies to consult with wildlife agencies on proposed state projects to ensure that those projects do not jeopardize a protected species.[96]

Exemplifying the more comprehensive approach, the California Endangered Species Act (CESA) provides for the listing of protected species, prohibits the take of such species, and establishes a process for issuing incidental take permits.[97] The protections of CESA are somewhat broader than those of the federal ESA in that they extend to candidate species – species proposed for listing – as well as threatened and endangered species.[98] However, CESA defines take more narrowly than the federal ESA to mean "hunt, pursue, catch, capture, or kill, or attempt to hunt, pursue, catch, capture, or kill."[99] Courts have ruled that the statute prohibits the killing of protected species in the course of lawful activity,[100] but that it does not prohibit the adverse modification of habitat in and of itself.[101] The precise scope of the term "take" nonetheless remains somewhat uncertain as to whether it requires mortality.[102]

If an action will result in take that is incidental to otherwise lawful activity, a person may apply for a state incidental take permit similar to that available under the federal ESA. The California Department of Fish and Game may issue a permit for incidental take if impacts will be minimized and fully mitigated, the applicant ensures adequate funding to implement mitigation measures, and if the take would not jeopardize the continued existence of the species.[103] In addition, a person who has obtained a federal incidental take permit (under ESA § 10) or incidental take authorization (under ESA § 7) may seek a consistency determination from the state in lieu of obtaining a state incidental take permit.[104]

4.1.2.2. Other Wildlife Statutes

Geoengineering projects that harm wildlife may implicate other statutes in addition to ESA. Statutes such as the Migratory Bird Treaty Act (MBTA) and the Marine Mammal Protection Act (MMPA) protect specific types of wildlife.[105]

4.1.2.2.1. MIGRATORY BIRD TREATY ACT MBTA imposes strict liability for taking a migratory bird.[106] While the statute clearly forbids the intentional killing of a migratory bird, case law is divided on whether the term "take" extends to the inadvertent killing of a migratory bird.[107] In contrast to the ESA, whose definition of "take" includes "harm," neither MBTA nor its regulations define take to include harm. The text of MBTA contains no definition of take, and MBTA regulations' definition of take is limited to "pursue, hunt, shoot, wound, kill, trap, capture or collect," or to attempt to undertake these actions.[108] Accordingly, some courts require that a defendant "take an affirmative action to cause migratory bird deaths" before imposing liability.[109] Also in contrast to the ESA, MBTA's implementing regulations do not presently authorize the issuance of permits for incidental takes or deaths of migratory birds.[110] The absence of a permitting system further suggests that the statute may not govern actions that incidentally harm migratory birds.

MBTA is a criminal statute.[111] Although the statute does not authorize citizen suits,[112] private parties sometimes sue federal agencies for MBTA violations under the Administrative Procedure Act's judicial review provisions. Courts are divided on whether MBTA's take prohibition applies to the federal government and whether that prohibition may be enforced through an action for injunctive relief.[113]

Assuming that an MBTA cause of action is available, any challenge to a geoengineering project under MBTA likely would encounter issues of causation and harm similar to those encountered under the ESA. MBTA's narrow definition of take could also present problems for plaintiffs. In many instances, harms to migratory birds caused by geoengineering projects may be inadvertent and thus potentially outside the scope of MBTA and its regulations.[114] Although MBTA is a strict liability statute,[115] courts may be reluctant to construe its take prohibition broadly because it is a criminal statute.[116] At a minimum, courts would likely require that any harm to migratory birds be foreseeable before imposing liability.[117]

4.1.2.2.2. MARINE MAMMAL PROTECTION ACT MMPA generally prohibits the take of marine mammals (such as whales, dolphins, and seals) in US waters or lands, as well as the take of marine mammals on the high seas by any person or vessel subject to the jurisdiction of the United States.[118] The statute defines take as "to harass, hunt, capture, or kill," or to attempt to do so. Although MMPA's definition of take may initially appear similar to the take definition found in MBTA, MMPA's inclusion of harassment suggests a wider scope. MMPA defines harassment as "any act of pursuit, torment, or annoyance which has the potential to injure a marine mammal … or has the potential to disturb a marine mammal or marine mammal stock in the wild by causing disruption of behavioral patterns."[119] Pertinent regulations further specify that take includes "the doing of any negligent or intentional act which results in disturbing or molesting a marine mammal."[120] Also in contrast to MBTA, MMPA establishes a permitting process for unintentional take. NOAA Fisheries may issue an incidental take authorization if the take would involve a small number of individuals, have no more than a negligible impact on the species at issue, and have no more than an unmitigable adverse impact on the availability of species for subsistence uses.[121]

Civil or criminal penalties may be imposed for violations of MMPA.[122] Although the statute does not explicitly authorize citizen suits, a party with conservationist, economic, or other interests in marine mammal protection may bring suit to challenge the issuance of a permit or the failure to apply for a required permit.[123]

Operational activities that incidentally harm marine mammals, such as Navy sonar operations, fall within the scope of MMPA.[124] At first glance, it may seem unlikely that marine geoengineering activities would affect marine mammals as directly or severely as Navy sonar operations. Nonetheless, efforts to brighten marine clouds, enhance ocean upwelling or downwelling, or fertilize the oceans all could affect marine mammals. Because MMPA's protections extend to individual mammals and prohibit the disruption of their behavioral patterns, some of these geoengineering projects could require an MMPA incidental take authorization.

4.1.3. *Emergency Authorities*

Various authorities grant federal, state, and local governments extraordinary powers to act in emergency situations generally or in emergencies involving environmental hazards. Such powers could enable government actors to regulate or halt geoengineering projects that give rise to emergency circumstances. Conversely, such powers also could be invoked by governments to initiate some types of geoengineering in response to a climate-related emergency.

4.1.3.1. Federal

Federal emergency authority may be based on statutory grants of specified authorities or on the US Constitution. Some statutory grants are restricted, authorizing fairly limited responses in narrow factual circumstances. Other sources of emergency powers are far broader. This section canvasses the sources of federal emergency authority potentially most relevant to geoengineering and considers how they might apply to geoengineering activities.

4.1.3.1.1. STAFFORD ACT The Robert T. Stafford Disaster Relief and Emergency Assistance Act ("Stafford Act") provides an important statutory basis for declaring an emergency.[125] The Stafford Act authorizes the president of the United States, upon the request of the governor of an affected state, to issue a declaration of a major disaster or emergency exceeding a state's capacity to respond.[126] A "major disaster" is defined as "any natural catastrophe … or, regardless of cause, any fire, flood, or explosion … which … causes damage of sufficient severity and magnitude to warrant major disaster assistance."[127] An emergency is defined as "any occasion or instance for which … Federal assistance is needed to supplement State and local efforts and capabilities to save lives and to protect property and public health and safety, or to lessen or avert the threat of a catastrophe."[128] An emergency declaration is typically made in response to and in advance of a recognized threat, whereas a major disaster declaration typically follows a catastrophic event and provides broader authority to recover from damage caused by the event.[129]

A declaration of emergency or major disaster under the Stafford Act makes federal assistance – technical, financial, and otherwise – available to state and local governments. Under such a declaration, the president or the Federal Emergency Management Agency (FEMA) may coordinate disaster relief assistance and provide federal assistance and support to save human lives, prevent suffering, or mitigate severe damage.[130] In addition, the president or FEMA may direct any federal agency to use its authorities and resources to support state and local responses and recovery efforts. Typically, FEMA issues a mission assignment to another federal agency to undertake such efforts, and that agency is subsequently reimbursed through the Disaster Relief Fund.[131] Most actions taken under the Stafford Act – including the provision of general federal assistance, provision of assistance essential to meeting immediate threats to life and property, removal of debris, and repair of damaged facilities – is statutorily exempt from NEPA compliance.[132]

4.1.3.1.2. ENVIRONMENTAL STATUTES CONTAINING EMERGENCY AUTHORITIES Several environmental laws contain grants of federal emergency authority to respond to environmentally hazardous situations. The Comprehensive Environmental Response, Compensation, and Liability Act authorizes the president to respond to any release or threat of release of a hazardous substance or pollutant that "constitutes a public health or environmental emergency" if no other person will respond in a timely manner.[133] The Clean Air Act authorizes the EPA to seek a court order or issue an administrative order against any person causing or contributing to air pollution that presents "an imminent and substantial endangerment to public health or welfare, or the environment."[134] Similarly, the Clean Water Act authorizes the EPA to bring suit to restrain any person causing or contributing to water pollution that presents "an imminent and substantial endangerment to the health of persons or to the welfare of persons where such endangerment is to the livelihood of such persons."[135] And the Resource Conservation and Recovery Act authorizes the EPA to seek a court order or issue an administrative order against any person who has contributed or is contributing to the "handling, storage, treatment, transportation or disposal of solid waste or hazardous waste [that] may present an imminent and substantial endangerment to health or the environment."[136] Each of these statutes could empower the United States to halt or control an environmental emergency created by geoengineering activities.

Other emergency statutes, in contrast, might be invoked in support of government geoengineering efforts. The Flood Control and Coastal Emergency Act, for instance, establishes an emergency fund to be expended by the Army Corps of Engineers for various purposes, including "preparation for emergency response to any natural disaster," "flood fighting and rescue operations," and "emergency protection of federally authorized hurricane or shore protection."[137] Activities authorized under the statute typically include flood fighting and management projects such as levee protection or dam maintenance. However, one can imagine an argument that geoengineering projects aimed at decreasing the severity or scope of climate-related disasters also qualify as appropriate emergency response measures. Such projects might include cloud brightening intended to provide relief from a heatwave, or cloud seeding intended to supply water in a drought.

4.1.3.1.3. NATIONAL EMERGENCIES ACT The Stafford Act and environmental emergency authorities address emergencies or disasters that are not necessarily national in scope. The National Emergencies Act (NEA), in contrast, governs "national emergencies" declared by the president. The NEA is a procedural statute that requires the president to transmit to Congress and publish in the *Federal Register* a proclamation specifying the legal basis for such an emergency declaration.[138] The statute also requires that Congress meet every six months "to consider a vote on a joint resolution to determine whether [an] emergency shall be terminated."[139] Notwithstanding this latter requirement, Congress rarely votes on whether a declared emergency should continue, and courts essentially have held the requirement unenforceable.[140]

A declaration of national emergency under the NEA may lead to the exercise of various emergency powers, including the conduct of military operations, imposition of trade restrictions, freezing of assets, suspension of minimum wage requirements in public contracts, and seizure of radio and TV stations and telecommunications networks.[141] In recent years, national emergencies have been declared almost exclusively to respond to terrorist threats or to advance specific foreign policy objectives.[142]

Although the NEA governs the procedures to be followed in declaring a national emergency, the statute itself provides no authority to declare a national emergency.[143] Rather, an emergency declaration must have an independent basis in some other statute or the US Constitution.[144] One statute frequently cited as the basis for a national emergency declaration is the International Emergency Economic Powers Act (IEEPA). The IEEPA defines a national emergency as "any unusual and extraordinary threat, which has its source in whole or substantial part outside the United States, to the national security, foreign policy or economy of the United States," and it authorizes the freezing of assets and prohibition of financial transactions with persons designated by executive order.[145]

4.1.3.1.4. CONSTITUTIONAL SOURCES OF EMERGENCY FEDERAL EXECUTIVE AUTHORITY Presidential assertions of authority in an emergency may rest not only on statutory powers, but also on powers expressed or implied by the US Constitution.[146] Constitutional sources of emergency authority may include specific clauses in the Constitution such as the Commander-in-Chief Clause and the Executive Power Clause. Emergency authority may also rest on broader notions of presidential power to conduct foreign affairs and a presidential duty to defend the nation.

Emergency authority derived from the Constitution largely involves presidential power over the military and foreign affairs. Most prominently, the Commander-in-Chief Clause states: "The President shall be Commander in Chief of the Army and Navy of the United States."[147] Though the precise scope of the president's powers as commander in chief is contested, the clause is commonly understood to confer broad substantive war powers on the president.[148] Further, the clause implies presidential authority over foreign affairs, a power also reflected in his or her constitutionally enumerated powers to make treaties and receive foreign representatives.[149] Presidential authority over foreign affairs is broad but somewhat overlapping with congressional authority.[150] A related – and somewhat contested – notion is the asserted presidential duty to defend the nation.[151] The argument for the existence of this duty is rooted in the presidential oath to preserve, protect, and defend the Constitution.[152] Defending the Constitution, proponents contend, necessarily implies a broader obligation to defend the nation. However, even if one assumes the existence of this obligation, it could justify only those measures necessary to preserve the nation and the Constitution.[153] It is uncertain how grave an environmental threat would need to be in order to implicate this demanding standard.

Presidential power over the military and foreign affairs, whether grounded in the Commander-in-Chief Clause or otherwise, could support US efforts to block geoengineering activities by a foreign state or entity. At present, it may seem unlikely that foreign geoengineering efforts would be so contrary to US interests as to warrant a military response. US scientists have played a leading role in geoengineering research to date, and the United States is sometimes perceived as more receptive to geoengineering than other nations because of the country's technological bent and its reluctance to address climate change by mitigating greenhouse gas (GHG) emissions with sufficient vigor.[154] Furthermore, direct use of geoengineering as a weapon is unlikely, given the existence of more effective, more reliable, and cheaper weapons.[155] Nonetheless, international conflict over whether, when, and how geoengineering should be deployed is a real possibility, and the potential for unilateral deployment only heightens this conflict.[156]

Constitutionally grounded presidential authority to act in an emergency is not limited to matters involving the military and foreign affairs. The Executive Power Clause, found in Article II, Section 1 of the Constitution, provides that "[t]he executive Power shall be vested in a President of the United States of America."[157] Much debate surrounds the specific powers, if any, that this provision grants.[158] Under a narrow reading, the clause "merely designates the President as the holder of a certain office" and itself confers no authority.[159] In contrast, a broader interpretation of the clause suggests not only the power to execute and enforce laws enacted by Congress, but also the power to address national emergencies and other situations.[160] Indeed, one commentator describes the clause as "perhaps … the most important constitutional power for government crisis management."[161] The text of the clause sets out no obvious limits to such power, but there is historical support for subjecting any power granted by the clause to a requirement of reasonableness.[162] Simply put, a reasonably defined general executive power should be exercised in a manner that reflects efficacy, proportionality, and respect for background constraints.[163]

4.1.3.1.5. GEOENGINEERING AND FEDERAL EMERGENCY AUTHORITIES Could an emergency declaration under the Stafford Act, National Emergencies Act, or other authority serve as a legal foundation for geoengineering deployment? On the one hand, various weather-related disasters, such as a heatwave, flood, or drought, could trigger a presidential emergency or disaster declaration. Such a declaration in turn might lead to proposals to deploy certain SRM techniques, which would presumably be faster-acting than CDR techniques.[164]

On the other hand, any relief provided by SRM schemes – assuming that they work – may be of limited value in addressing the circumstances of a specific emergency. Stratospheric aerosols, for example, would take months to have any cooling effect, may not prevent or directly counter specific weather disasters, and may be unable to halt processes that have passed critical tipping points.[165] Marine cloud brightening offers the prospect of rapid cooling in some coastal regions, but with more geographically limited effects.[166] It is debatable, moreover, whether these sorts of measures would qualify as properly authorized emergency assistance. Under the Stafford Act, for example, "assistance authorized by an emergency declaration is limited to short-term assistance, essential to save lives, to protect property and public health and safety, or to lessen or avert the threat of a catastrophe."[167] In some circumstances – such as a heatwave – marine cloud brightening might lessen or avert a catastrophe. Stratospheric aerosol deployment, in contrast, is generally framed as a long-term rather than a short-term response to climate change, and thus would be difficult to justify as an emergency response.

More generally, there are various practical and philosophical difficulties involved in defining a climate emergency and in considering geoengineering as an emergency response. Climate change itself does not fall squarely within some conceptions of emergency because of the long time scales on which it operates, though New York courts have construed "emergencies" under that state's Environmental Quality Review Act to include slow-acting and long-anticipated events, such as homelessness and prison overcrowding.[168] Catastrophic weather events more obviously represent emergencies, but the absence of clear causal links between particular weather disasters and climate change weakens the case for using geoengineering to respond to weather disasters (though the science of attributing the magnitude of particular weather events to climate change is rapidly improving). Furthermore, the framing of geoengineering as a potential response to future climate emergencies has been widely criticized.[169] For one, it makes heroic assumptions regarding decision-makers' ability to define, recognize, and respond to a climate emergency.[170] For another, it discounts or distracts from other policy options for combating climate change, particularly mitigation of GHG emissions.[171] As geoengineering commentators widely concede, achieving the collective cooperation necessary to decrease GHG emissions may be difficult but is ultimately essential.[172] Finally, the framing of geoengineering as an emergency response utilizes emotionally charged rhetoric that may encourage irrational and undemocratic decision-making.[173] The resulting concern is that an emergency declaration could be used to justify the suspension of ordinary processes and protections of civil society in favor of pressing forward with an inadequately vetted geoengineering project.[174]

4.1.3.2. State and Local Emergency Authorities

State and local governments, like the federal government, possess emergency authorities. In general, state executives have the authority to declare emergencies and to exercise broad emergency powers.[175] Although state law governs the authority and role of state institutions in emergencies, state institutions often must coordinate their actions with federal actions and institutions.[176]

The discussion here considers California law regarding emergency authorities as an example. The California Emergency Services Act authorizes the governor to declare an emergency when conditions of disaster or extreme peril are likely to exceed the resources of any local jurisdiction, thereby requiring mutual aid.[177] Such a declaration authorizes the governor to expend emergency funds and use state resources, such as the National Guard, and releases financial assistance to local agencies.[178] An emergency declaration also has the effect of legally immunizing state and local officials responding to the emergency.[179] During a state of emergency, the governor may suspend any regulatory statute or state regulation if strict compliance with the statute or regulation would prevent, hinder, or delay mitigation of the emergency's effects.[180] In addition, the governor may also "commandeer or utilize any private property or personnel deemed by him necessary" to carry out his responsibilities.[181]

The California Emergency Services Act also authorizes local governments to declare a local emergency. Such a declaration enables local officials to issue "orders and regulations necessary for the protection of life and property."[182] In a local emergency, state agencies may provide aid to local governments.[183]

Like their federal counterparts, state and local emergency authorities could be invoked in response to geoengineering-related hazards. In addition, climate change-related disasters or threats also might prompt state and local officials to consider geoengineering deployment as an emergency response. In either circumstance, federal involvement is likely because of the relatively limited resources available to state and local governments and because of the potential national or international implications of any government actions initiating or responding to geoengineering.

4.1.4. *Public Trust Doctrine*

One additional body of law that could be relevant to almost any type of geoengineering is the public trust doctrine. The public trust doctrine imposes a duty on states to protect public rights to natural resources. Rooted in state constitutions, statutes, and common law, the doctrine "represents a fundamental, inherent attribute of state sovereignty."[184] Resources subject to the doctrine include tidal and submerged lands, inland navigable waters, and fish and wildlife.[185] In a few states, the doctrine's protection extends to additional resources such as groundwater.[186]

States "ha[ve] an affirmative duty to take the public trust into account" when administering public trust resources and a further duty to exercise continuous supervision over the trust.[187] Objectives to be promoted through the public trust doctrine include navigation, commerce, fisheries, recreation, and environmental protection.[188] Though states have the authority to convey private rights to trust resources, they also have an accompanying responsibility not to substantially impair public use and enjoyment of those resources.[189] Ultimately, the role of the state as trustee "is to look forward to the future and take actions in the present that will first, maintain and, second, enhance the trust property for the benefit of future generations."[190] The public trust duty may arise in various legal contexts: The government can invoke the public trust against private holders of rights over trust resources, the government can assert the public trust as a defense against takings claims by property owners, and the public may invoke the doctrine as a source of affirmative government obligations.[191]

A number of geoengineering activities could affect public trust resources. Marine cloud brightening projects would presumably use seawater and could impact submerged lands or tidal waters. State jurisdiction over – and public trust responsibility for – these resources generally extends up to 3 miles from the coastline.[192] Stratospheric aerosol release could affect the hydrological cycle, with implications for a state's surface water and groundwater resources, as well as fish and wildlife. Furthermore, air capture devices, accelerated weathering projects, and efforts to modify land albedo all could affect existing ecosystems and have detrimental effects on wildlife. Finally, some courts' recognition that public trust doctrine extends to the atmosphere[193] raises the possibility that almost any geoengineering project could implicate the doctrine.

Due to the public trust doctrine's flexible, evolving, and context-specific nature, application of the doctrine to geoengineering is difficult to predict with precision. Nonetheless, several of the doctrine's key features are of particular relevance to geoengineering. First, the doctrine imposes duties and limits primarily on states, not the federal government.[194] Though future decisions on geoengineering may occur primarily at the federal level, state input into those decisions should reflect state responsibility for public trust resources. Second, the public trust doctrine has both a substantive component, checking legislative or agency discretion at the margins, and a procedural component.[195] The procedural aspect of public trust doctrine obliges states to take public trust values into account in their decision-making processes, and courts may be asked to step in if those values are ignored. Third, the doctrine embraces a broad and future-oriented consideration of public trust values. This broad perspective may complicate application of the doctrine to geoengineering. Geoengineering projects may protect some public trust resources by combating the effects of climate change while at the same time harming other public trust resources. Reviewing courts may allow the political branches broad discretion in balancing these potential conflicts.[196] Indeed, despite numerous legal efforts to apply the public trust doctrine to the atmosphere, none so far has resulted in an order mandating regulators to take action on climate change.[197]

4.2. FEDERAL ENVIRONMENTAL LAWS APPLICABLE TO SPECIFIC TECHNIQUES

As Chapter 2 explains, the term geoengineering encompasses a wide range of possible techniques for countering elevated GHG concentrations in the atmosphere. The details involved in implementing any of these techniques have yet to be worked out, and the applicability of domestic laws will depend on the particulars of a specific geoengineering project. Relevant details may include location, type of technology, scale, and expected effects.[198] This section canvasses the primary federal statutes that could be triggered by specific geoengineering techniques. Most of these statutes provide for delegation of administrative and enforcement authority to states, and individual state programs may differ substantively and procedurally in important ways.

4.2.1. *Clean Air Act*

Geoengineering projects involving the dispersal of substances into the atmosphere could be subject to regulation under the CAA. To illustrate how the CAA's various provisions might apply to some geoengineering techniques, the discussion below focuses on the release of sulfate aerosols into the stratosphere.[199]

As an initial matter, stratospheric aerosol schemes raise the question of whether the deliberate release of substances into the atmosphere would constitute air pollution.[200] The term air pollution typically describes the unintentional by-products of industrial, commercial, or individual activity, not air emissions released deliberately to achieve a societal goal.[201] However, the CAA defines air pollutant broadly to mean "any air pollution agent or combination of such agents, including any physical, chemical, biological … substance or matter which is emitted into or otherwise enters the ambient air," as well as "any precursors to the formation of any air pollutant."[202] This definition does not refer in any way to the intent behind an air pollutant release. Moreover, the Supreme Court has noted that the "capacious" statutory definition "embraces all airborne compounds of whatever stripe … without distinguishing between atmospheric layers."[203] The broad statutory language and Supreme Court precedent argue in favor of classifying stratospheric aerosols as air pollutants, regardless of why or where they were released. In other words, the geoengineered release of stratospheric aerosols would constitute air pollution under the CAA.

Several parts of the CAA could be relevant to the geoengineered release of stratospheric aerosols. Title I regulates air emissions from stationary sources, which include "any building, structure, facility, or installation which emits or may emit any air pollutant."[204] Title II governs air emissions from mobile sources. Title IV regulates emissions of pollutants that contribute to acid deposition, including sulfur dioxide, from fossil fuel-fired combustion devices. Title VI implements international obligations to regulate ozone-depleting substances. Each of these regulatory programs applies to public as well as private sources, although the president may exempt federal emission sources from CAA requirements if doing so would be "in the paramount interest of the United States."[205] Violations of the CAA may generally be enforced by the federal government or by members of the affected public through a citizen suit.[206]

4.2.1.1. Title I/Stationary Source Regulation

Title I of the CAA directs the EPA to identify air pollutants that may reasonably be anticipated to endanger public health or welfare and to establish national ambient standards for these so-called "criteria pollutants."[207] Under the cooperative federalism approach incorporated into Title I, states play a central role in achieving and determining how to achieve the ambient standards established by the EPA.[208] Each state must develop a state implementation plan (SIP) that provides for the attainment, maintenance, and enforcement of these ambient standards.[209] The SIP, which is subject to EPA approval, establishes several types of limits on pollutant emissions from individual stationary sources. These include limits determined by the state to be necessary to achieve the ambient standards,[210] limits under the prevention of significant deterioration (PSD) program (if the area is in attainment of the national ambient standards),[211] limits under the nonattainment new source review program (if the area is not in attainment of the national ambient standards),[212] and limits under the visibility protection program (if emissions may impair visibility in mandatory Class I federal areas, which include most national parks and wilderness areas).[213] Pollution sources whose emissions of regulated pollutants exceed specified threshold amounts must obtain permits and abide by permit conditions. Violations of SIP requirements may be enforced by the EPA or the state, or through a citizen suit.[214] Title I also requires the EPA to impose emission standards on sources of hazardous air pollutants.[215] All of the emission limits and monitoring and reporting requirements applicable to an individual stationary source of air pollution are incorporated into an operating permit, as mandated under Title V of the CAA.[216]

Two of the criteria pollutants for which the EPA has set national ambient standards are sulfur dioxide and particulate matter (coarse particulate matter and fine particulate matter are subject to separate standards).[217] Stratospheric aerosol releases could generate both sulfur dioxide and particulate matter. Specifically, stratospheric aerosol projects contemplated by would-be geoengineers include the release of sulfuric acid, hydrogen sulfide, or sulfur dioxide via aircraft, long tubes supported by high-altitude balloons, artillery shells, or other means.[218] In the atmosphere, these various sulfur species can react to form not only sulfate aerosols – as project proponents intend – but also sulfur dioxide.[219] Sulfate aerosols themselves in turn can serve as precursors to the formation of particulate matter.[220] Thus, a geoengineering project involving the release of stratospheric aerosols from a stationary source – such as an artillery piece or a hose attached to the ground – could trigger CAA Title I permitting requirements. The applicability of specific SIP requirements would depend on whether the project would release or generate sulfur dioxide or particulates in quantities exceeding regulatory thresholds for individual sources, or so as to impair visibility in mandatory Class I federal areas.[221]

Geoengineering projects other than stratospheric aerosol release also may trigger Title I requirements to the extent they emit air pollutants from stationary sources. For example, biochar facilities would potentially be subject to regulation as stationary sources because they may emit particulates, volatile organic compounds, carbon monoxide, and hazardous air pollutants.[222] Likewise, particulate emissions from the mining and processing of rock for enhanced weathering projects could lead to regulation under Title I.

4.2.1.2. Title II/Mobile Source Regulation

Title II of the CAA addresses air pollution from mobile sources through such measures as limits on motor vehicle tailpipe emissions and regulation of fuel and fuel additives. Although Title II generally focuses on pollution from surface vehicles, it also authorizes the EPA to promulgate "emission standards applicable to the emission of any air pollutant from any class or classes of aircraft engines."[223] Such standards, which have generally followed the recommendations of the International Civil Aviation Organization, are set by the EPA but implemented and enforced by the Federal Aviation Administration (FAA).[224] These standards presently govern emissions of nitrogen oxides, hydrocarbons, and carbon monoxide.[225] No aircraft engine standard currently governs sulfur emissions, but the EPA could use its Title II authority to develop standards to govern the deliberate release of stratospheric aerosols from aircraft engines.[226] At this time, however, it seems unlikely that the geoengineered release of stratospheric aerosols would involve the emission of sulfur directly from an aircraft engine. To avoid corrosion and interference with combustion processes, a geoengineering scheme would more likely be designed to release sulfur through a nozzle from a dedicated tank.[227] Title II, which focuses on engine emission standards, does not explicitly extend to this latter sort of sulfur release.[228]

If a geoengineering project were to incorporate sulfur into aircraft fuel, other regulations also could be relevant. The FAA regulates jet fuel composition to help achieve the engine emission standards set by the EPA. Specifically, the Federal Aviation Act requires the FAA to prescribe "standards for the composition or chemical or physical properties of an aircraft fuel or fuel additive to control or eliminate aircraft emissions" that the EPA has determined endanger the public health or welfare.[229] The FAA does not directly regulate aircraft fuel, however; rather, it indirectly regulates fuel composition by specifying that aircraft operators use fuel that meets specifications identified by the aircraft engine manufacturer.[230] Internationally accepted standards for jet fuel limit its sulfur content to 0.30 percent by weight.[231]

4.2.1.3. Title IV/Regulation of Acid Deposition

Enacted to address the problem of acid deposition, Title IV of the CAA regulates emissions of sulfur dioxide and nitrogen oxides from fossil fuel-fired electric utility units.[232] Under this program, the EPA allocates a finite number of emission allowances each year to pollution sources based on overall target pollution levels.[233] Each regulated source must have sufficient allowances to cover the pollution it emits, and it may trade allowances with other sources in order to achieve this mandate.[234] In its current form, Title IV applies only to fossil fuel-fired electric utility units. Other sources of sulfur dioxide, including possible geoengineering projects that could release sulfate aerosols, lie outside the scope of Title IV. The estimated quantities of sulfur that geoengineering might involve nonetheless are of such magnitude that they could contribute measurably to acid deposition.[235] Releases of sulfate aerosols in sufficient amounts could cause concern and stimulate support for an amendment to Title IV that would govern such releases.

4.2.1.4. Title VI/Regulation of Ozone Depleting Substances

Stratospheric sulfur aerosols serve as a catalyst for chemical reactions that deplete ozone and therefore could reduce ozone levels in the stratosphere.[236] Depending on the magnitude of these reductions, the geoengineered release of stratospheric aerosols could come under the oversight of the international regime for protecting the ozone layer – the Montreal Protocol on Substances that Deplete the Ozone Layer – and its domestic counterparts. The Montreal Protocol, discussed in more detail in Chapter 3, phases out or reduces the production and consumption of stratospheric ozone-depleting substances (ODS).[237] Title VI of the CAA translates the United States' obligations under the Montreal Protocol into domestic law.[238] In addition to listing specific ODS, Title VI authorizes the EPA to regulate ODS emissions and identify additional substances that cause or contribute significantly to harmful effects on stratospheric ozone.[239] Stratospheric sulfur aerosols are not currently listed under Title VI as an ODS.[240] However, if the EPA finds that such substances cause or contribute significantly to harmful effects on stratospheric ozone, the agency could add them to the list of regulated substances and establish controls on their production and use.

4.2.2. *Clean Water Act*

Geoengineering projects involving the dispersal of substances into waters of the United States could be subject to regulation under the Clean Water Act (CWA). Violations of the CWA are subject to civil or criminal prosecution, and citizen suits may be brought against any person for violating an effluent standard or limitation established under the statute.[241] Depending on their design specifics, a range of techniques might implicate the CWA, including ocean fertilization, stratospheric aerosol release, marine cloud brightening, and enhanced weathering.

4.2.2.1. NPDES Permitting

At the heart of the CWA is a prohibition on the discharge of any pollutant into navigable waters without a National Pollutant Discharge Elimination System (NPDES) permit.[242] The statute defines "discharge of a pollutant" as "any addition of any pollutant to navigable waters from any point source" and "any addition of any pollutant to the waters of the contiguous zone or the ocean from any point source *other than a vessel or other floating craft*."[243] This definition extends permitting requirements to point sources discharging to internal waters as well as non-vessel point sources up to 200 miles from the US shoreline.[244] However, it exempts from NPDES permit regulation any discharges from vessels occurring more than 3 miles from shore.[245] (Such discharges are potentially subject to regulation under the Ocean Dumping Act, discussed below.) An NPDES permit incorporates various pollution control requirements and may be issued by the EPA or a state authorized by the EPA to issue such permits.[246]

To understand the reach of the CWA, one must consider the statute's key terms. The statutory prohibition on discharge is limited to point sources, which are defined as discernible and discrete conveyances. The focus on point sources reflects an assumption that such sources are easier to identify and control than other sources of water pollution.[247] The statute prohibits discharge of pollutants, which are defined broadly to include "dredged spoil, solid waste, incinerator residue, sewage, garbage, sewage sludge, munitions, chemical wastes, biological materials, radioactive materials, heat, wrecked or discarded equipment, rock, sand, cellar dirt and industrial, municipal, and agricultural waste discharged into water."[248] This definition encompasses a wide range of substances that geoengineering projects might discharge into protected waters.

The CWA defines navigable waters as "waters of the United States, including the territorial seas" – i.e., those portions of the ocean lying within three miles of the US coast.[249] By regulation, the term includes waters used or susceptible to use in interstate commerce; interstate waters, including interstate wetlands; tributaries to traditionally navigable waters; waters with a significant nexus to traditionally navigable waters; and waters and wetlands adjacent to waters that are otherwise determined to be waters of the United States.[250] Litigation has centered on – and will likely continue to center on – whether the regulatory definition extends to a particular non-navigable body of water or wetland and whether the regulatory definition is consistent with the statute.[251]

Several types of geoengineering techniques could require an NPDES permit. However, ocean fertilization efforts, which as described in Chapter 2 would probably involve the release of a fertilizing agent such as iron filings on the high seas,[252] would generally lie outside the CWA's jurisdiction. While it is possible that the design parameters of a specific ocean fertilization project could subject it to NPDES permitting requirements, such projects are more likely to be subject to the Ocean Dumping Act than the CWA. The release of a fertilizing agent into the ocean from a vessel or floating craft would not require an NPDES permit, provided that the release occurs beyond the 3-mile limit to US territorial seas. Likewise, the release of a fertilizing agent into the ocean from any non-vessel source would not require an NPDES permit if it occurs more than 200 miles from the US coastline.

Moreover, even if ocean fertilization activity were to occur within the geographic limits of the CWA, it would require an NPDES permit only if the discharged fertilizer constitutes a pollutant. As noted above, the statute defines pollutant to include various wastes, such as chemical wastes. It is not obvious, however, that a fertilizing agent would constitute a chemical waste or any other waste within the statutory definition of pollutant, as opposed to a useful product. One might draw a rough analogy to pesticide spraying, an activity that the Sixth Circuit has held to constitute a discharge of pollutants under the CWA when excess chemical pesticides and pesticide residues contaminate waters.[253] Similarly, any waste products generated by the ocean fertilization process,[254] or any excess fertilizer itself, would arguably constitute a pollutant under the statute's broad definition of the term. If a CWA discharge permit is ultimately required, it may not be issued if a discharge "will cause unreasonable degradation of the marine environment."[255]

Depending on their design and effects, other geoengineering techniques could also implicate NPDES permitting requirements. As the CWA requires a permit only for the discharge of a pollutant from a point source, the specific manner by which a pollutant is added to waters of the United States could be critical. For example, stratospheric aerosol release and marine cloud brightening could eventually disperse pollutants into waters of the United States, but would probably not require an NPDES permit because of the indirect route by which such pollutants reach the water. In contrast, a land-based enhanced weathering project might require a permit if its design and layout were such that it qualified as a point source dispersing pollutants into waters of the United States. Enhanced weathering to increase seawater alkalinity or store carbon in the ocean could also require a NPDES permit. Such projects would presumably release ground-up calcium carbonate, calcium silicate, or other materials directly into the ocean or territorial seas.[256] These releases would raise issues similar to those raised by ocean iron fertilization with respect to application of the CWA (as well as the Ocean Dumping Act). Some enhanced weathering techniques also could require the intake of large volumes of seawater.[257]

The CWA imposes technology-based standards on cooling water intake structures to protect against potentially harmful effects on aquatic ecosystems, but does not otherwise regulate open seawater intakes.[258] The intake of water for industrial facilities may be subject to state regulatory requirements, however.[259]

4.2.2.2. Other CWA and Related Requirements

The alteration of geographic features or the development of infrastructure in conjunction with a geoengineering project could trigger further permitting requirements under the CWA or related statutes. These requirements include: (1) Section 404 permits for the discharge of dredged or fill material, (2) Section 401 state certification of activities subject to a federal permit, (3) Section 303(d) pollution limits established to achieve water quality standards, and (4) Rivers and Harbors Act permits for structures constructed over navigable waters.

The discharge of dredged or fill material into waters of the United States, including wetlands, requires a CWA Section 404 permit from the US Army Corps of Engineers (Corps).[260] In deciding whether to issue a Section 404 permit, the Corps evaluates the probable impacts (including cumulative impacts) of the proposed activity and its intended use on the public interest.[261] Section 404 permits may not be issued if the discharge will cause or contribute to significant degradation of waters of the United States or if there exists a practicable alternative with less adverse impacts on the aquatic ecosystem.[262] Even if these conditions are not present, the EPA may prohibit the Corps from issuing a Section 404 permit if the discharge will have an unacceptable effect on municipal water supplies, shellfish beds and fishery areas, wildlife, or recreational areas.[263]

Construction activity associated with a geoengineering project could discharge dredged or fill material into waters of the United States. In addition, specific geoengineering projects may discharge materials into US waters as a carbon dioxide removal mechanism. For example, spreading olivine (a very common silicate mineral) in the marine environment has been suggested as a means of capturing carbon, combating ocean acidification, and replenishing beaches.[264] Such activity would require a Section 404 permit if done within 3 miles of shore.[265]

If an activity discharging into navigable waters is required to obtain a federal permit (such as a Section 404 permit), the permit applicant must also obtain a Section 401 certification from the state determining that the activity will comply with state clean water standards.[266] This certification allows the state to protect its interests when federal permitting decisions are made by essentially giving it a veto power over the proposed activity. Under Section 401, a state may also impose effluent limitations or other permit conditions to ensure compliance with state clean water standards.[267]

CWA provisions governing water quality standards also could become relevant to geoengineering projects. CWA Section 303(d) requires states to identify water bodies within their boundaries that fail to meet established water quality standards.[268] For these water bodies, states must establish pollution budgets, known as total maximum daily loads (TMDLs), and allocate the allowable amount of pollution among individual pollution sources.[269] Through this process, states can impose pollution limits that are additional to those already governing individual pollution sources, in order to achieve desired ambient water quality.

Water quality standards could be used someday to manage rising carbon dioxide concentrations in the oceans. Though increased carbon dioxide in the ocean is causing ocean acidification, the EPA historically has not sought to regulate oceanic carbon dioxide levels through Section 303(d), nor has it established water quality standards for ocean acidity. However, in 2010 the EPA issued a memorandum encouraging states to evaluate the impacts of ocean acidification on their coastal waters.[270] The memorandum instructed states to identify waters impaired by ocean acidification if supporting data is available, while also acknowledging that such data "is largely absent or limited at this point in time."[271] Depending on future policy and legal developments,[272] Section 303(d) could become relevant to some marine-based geoengineering techniques in the future. Geoengineering projects that store carbon in the oceans, for example, could be subject to limitations imposed through the TMDL process if the projects alter ocean acidity or dissolved carbon dioxide levels.

In addition to the just-discussed CWA requirements, Section 10 of the Rivers and Harbors Act mandates Corps approval prior to the construction of any structure in or over any navigable water of the United States.[273] The Corps considers various factors, including effects on navigability and the environment, in deciding whether to issue a Section 10 permit.[274] Notably, the Rivers and Harbors Act defines the term "navigable waters" more narrowly than the CWA. Under the Rivers and Harbors Act, navigable waters include only "those waters that are subject to the ebb and flow of the tide and/or are presently used, or have been used in the past, or may be susceptible for use to transport interstate or foreign commerce,"[275] and ocean waters extending to 3 miles from the shore.[276] A geoengineering project involving the construction of a structure over such waters would require a Section 10 permit.

4.2.3. *Ocean Dumping Act*

The dumping of material into the ocean as part of an ocean iron fertilization project or other geoengineering activities may be subject to the Ocean Dumping Act. This statute, which comprises Titles I and II of the Marine Protection, Research, and Sanctuaries Act, implements the United States' obligations under the London Convention. That treaty, discussed in Chapter 3, generally forbids the "deliberate disposal at sea of wastes or other matter."[277]

Except as authorized by permit, the Ocean Dumping Act prohibits three classes of activities: (1) the transport of any material from the United States for the purpose of dumping it into ocean waters; (2) the transport of any material from any location by any vessel or aircraft registered in the United States for the purpose of dumping it into ocean waters; and (3) the dumping of any material into the territorial sea of the United States or into a zone contiguous to the territorial sea.[278] Dumping is broadly defined as "a disposition of material,"[279] and material as "matter of any kind or description, including, but not limited to, dredged material, solid waste, incinerator residue, garbage, sewage, sewage sludge, munitions, radiological, chemical, and biological warfare agents, radioactive materials, chemicals, biological and laboratory waste, wreck or discarded equipment, rock, sand, excavation debris, and industrial, municipal, agricultural, and other waste."[280]

Notwithstanding the Act's prohibitions, the EPA may issue a permit for ocean dumping if it "will not unreasonably degrade or endanger human health, welfare, or amenities, or the marine environment, ecological systems, or economic potentialities."[281] In evaluating permit applications, the EPA considers the need for the proposed dumping, alternatives to ocean dumping, effects on human health and welfare, effects on fisheries resources and marine ecosystems, and persistence and permanence of effects.[282] The EPA's regulations provide for the issuance of various types of permits, including general permits, special permits, and research permits.[283]

The Ocean Dumping Ban Act, a 1988 amendment to the Ocean Dumping Act, added a more stringent prohibition on dumping sewage sludge and industrial waste into ocean waters.[284] Of particular relevance to geoengineering is the definition of industrial waste, which "means any solid, semisolid, or liquid waste generated by a manufacturing or processing plant."[285] Permits for the dumping of such waste are available only in case of an emergency "which poses an unacceptable risk relating to human health and admits of no other feasible solution."[286]

Violations of the Ocean Dumping Act are subject to civil and criminal penalties, and the federal government may seek a court injunction against imminent or continuing violations.[287] The US Coast Guard is responsible for conducting surveillance and other enforcement activities to prevent unlawful dumping.[288] In addition to government prosecutions, any person with standing may bring a civil suit to enjoin any other person from violating the act.[289]

Although the Ocean Dumping Act was originally aimed at preventing the disposal of sewage waste and toxic materials into the ocean, the statute may apply to some marine-based geoengineering activities.[290] First, the prohibition on the dumping of industrial waste would preclude any US-based effort to dispose of carbon from industrial processes in solid, semisolid, or liquid form at sea. The statute presumably does not bar the disposal of carbon in gaseous form or disposal in geological formations under the sea bottom. Second, the requirement to obtain an ocean dumping permit likely would apply to ocean fertilization activities that involve US-flagged vessels, transport of materials from the United States, or which take place in the territorial sea or contiguous zone of the United States. The broad statutory definition of dumping to include any disposition of material would probably extend to ocean fertilization activities even though they would not involve the direct disposal of waste material.

Ocean fertilization field experiments may qualify for an Ocean Dumping Act research permit. Such permits may issue if "the scientific merit of the proposed project outweighs the potential environmental or other damage that may result."[291] As discussed in Chapter 6, recent efforts to carry out ocean fertilization research suggest the limitations of applying the Ocean Dumping Act's permitting requirements to geoengineering.[292] In one such effort, Planktos, Inc. planned to release 100 tons of iron dust from a US-flagged ship into the Pacific Ocean near the Galapagos Islands.[293] Although the experiment would have taken place far outside of US waters, the EPA notified Planktos that an Ocean Dumping Act permit would be required if the project involved a US-flagged vessel.[294] Planktos responded that it needed no permit because it would not be employing a US-flagged vessel.[295] Ultimately, however, the company abandoned the experiment in 2008 in the wake of strong opposition and weak financial support.[296] Four years later, the same American businessman involved in the Planktos effort conducted an iron fertilization experiment off the Canadian coast without having obtained a permit from the United States or any other national authorities.[297] The experiment was framed as an effort to restore salmon runs by boosting phytoplankton populations, but its financing was dependent in part on assumptions that the project would generate marketable carbon credits.[298] The experiment was arguably within the jurisdiction of the United States in that it was undertaken by a US citizen in international waters. Nevertheless, the experiment was beyond the reach of the Ocean Dumping Act because it did not involve US vessels or transport of materials from the United States, nor did it take place in US waters.

4.2.4. *Statutes Governing Hazardous Substances and Hazardous Waste*

Geoengineering projects that use or generate hazardous substances may trigger the application of statutes governing hazardous substances or hazardous waste, including the Toxic Substances Control Act (TSCA), the Resource Conservation and Recovery Act (RCRA), and the Comprehensive Environmental Response, Compensation, and Liability Act (CERCLA).

4.2.4.1. TSCA

TSCA authorizes the EPA to regulate a chemical substance at any point in its life cycle. Under Section 6 of TSCA, the EPA can regulate the manufacture, processing, distribution, use, or disposal of any chemical substance if it determines that such an activity "presents an unreasonable risk of injury to health or the environment."[299] Possible regulatory measures include restrictions on use, regulations governing disposal, and prohibitions on manufacture or distribution.[300] Although narrow judicial interpretations of the original version of Section 6 left the EPA reluctant to invoke its authority,[301] recent statutory amendments are expected to address some of the provision's weaknesses. These amendments make clear that the EPA's risk determinations are to be based solely on health and environmental factors, that its risk management decisions must factor in costs and feasibility, and that the agency is not required to choose the least burdensome method of regulating risk.[302]

Another important provision of TSCA, Section 5' requires manufacturers of new chemical substances to submit a premanufacture notice and any available health and safety data to the EPA.[303] Section 5 also authorizes the EPA to evaluate significant new uses of existing chemicals and to require manufacturers to notify the EPA of such uses.[304] The EPA may take action to regulate unreasonable risks presented by a new chemical or a significant new use of existing chemicals.[305] Such actions may include prohibiting or limiting the manufacture or distribution of a chemical.[306]

The EPA could rely on its TSCA authority to regulate chemicals that geoengineers might use in the release of stratospheric aerosols. Current proposals for aerosol release largely assume the use of various sulfur compounds to generate sulfur aerosols in the stratosphere. These compounds include hydrogen sulfide, a corrosive and "very dangerous" gas; sulfur dioxide, which is "dangerous to life and health";[307] and carbonyl sulfide, a carcinogen and greenhouse gas.[308] Each of these compounds poses some risk of injury to health or the environment and thus could be subject to regulation under TSCA. However, the historical difficulties faced by the EPA in applying Section 6 make it uncertain whether the EPA actually would use TSCA to regulate a geoengineering project. One of the several difficulties the EPA would face in relying on Section 6 would be to establish that the hazards involved in deploying these substances outweigh their potential benefits in cooling the Earth.

TSCA Section 5 offers an alternative source of regulatory authority, particularly if a geoengineering project were to use a novel chemical substance. For instance, to avoid some of the drawbacks of using sulfur, some researchers have suggested the release of nanoparticles specifically engineered to optimize particle distribution.[309] Such a scheme would trigger Section 5's notice requirement and could lead to EPA oversight. Indeed, even geoengineering projects that employ non-novel chemicals could be subject to Section 5 as a significant new use. For Section 5's notice requirement to apply to non-novel chemicals, the EPA first would have to determine through a rulemaking that geoengineering constitutes a significant new use.[310] Based on the information submitted in the Section 5 notice, the EPA could regulate to protect against unreasonable risks or issue a temporary order to allow the development of sufficient risk information.[311] Given the long delays the EPA has encountered in developing its nanomaterials regulatory policy,[312] however, it seems unlikely that it would be able to use Section 5 – or TSCA generally – to regulate geoengineering.

4.2.4.2. RCRA

RCRA, the primary federal law dealing with hazardous waste management, sets up a "cradle-to-grave" system designed to regulate hazardous wastes from the point of their generation up until their ultimate disposal.[313] Hazardous wastes include solid, liquid, or contained gaseous wastes that the EPA has listed as hazardous or that exhibit certain hazardous characteristics.[314] Persons who generate, transport, treat, store, or dispose of hazardous waste are subject to strict requirements governing the handling and disposal of such waste.[315] Generators must keep records of their waste and prepare a hazardous waste manifest to keep track of the waste.[316] Transporters must keep records of their activities and deliver waste to the facility designated by the generator in the manifest.[317] And treatment, storage, and disposal facilities must obtain a permit and satisfy various standards governing their design and operation.[318] Violations of RCRA may give rise to civil or criminal penalties, and may also be the subject of citizen suits.[319]

At first glance, it may seem implausible that any geoengineering proposal would proceed if it is likely to generate substantial quantities of hazardous waste. Nonetheless, a number of geoengineering techniques could fall within RCRA's purview, including bioenergy with carbon capture and sequestration (BECCS), biochar, and air capture. Combustion processes in bioenergy facilities would generate ash and other waste, some of which could be hazardous.[320] Biochar production similarly could generate hazardous material. Air capture schemes ideally would make complete reuse of the caustic chemicals that they may require, but may nonetheless generate unintended by-products.[321] Whether material generated by such processes would trigger RCRA depends in part on what is done with the material. If used legitimately as a fertilizer or construction material, for example, such material might not be deemed a waste at all, let alone a hazardous waste. RCRA regulates only discarded material, not useful products or secondary materials that contribute usefully to an ongoing production process.[322] Distinguishing between ongoing production of a useful product or materials and the generation and treatment of discarded material has proven to be a complex and difficult endeavor for the EPA and the regulated community.[323] Assuming that the material generated by a geoengineering process ultimately qualifies as discarded material, the applicability of RCRA would then depend on whether this material is hazardous in nature.[324]

Geoengineering might implicate RCRA in another way. Air capture and BECCS would generate large volumes of carbon dioxide that would likely require geologic sequestration. Captured CO_2 typically is not considered hazardous waste. Nonetheless, it may contain hazardous constituents in some instances.[325] At present, the EPA has conditionally excluded from RCRA regulation CO_2 captured from industrial sources and injected into underground wells.[326] CO_2 disposed of in this manner is instead regulated under the Safe Drinking Water Act underground injection control program.[327]

4.2.4.3. CERCLA

CERCLA generally addresses the release of hazardous substances into the environment from improperly disposed waste and other sources. Section 104 of the statute authorizes federal cleanups of hazardous waste sites, and Section 106 authorizes the federal government to compel private cleanups of such sites. The discussion below focuses on Section 107, which imposes strict liability for clean-up costs and damage to natural resource on persons responsible for the release or threat of release of hazardous substances.[328]

Liability attaches under Section 107 if (1) there is a release or threat of release of a hazardous substance, (2) the release or threatened release occurs from a facility, (3) the release or threatened release causes the incurrence of response costs, and (4) the defendant is a "potentially responsible party" (PRP). Persons who may be held liable as PRPs include "the owner and operator of a vessel or a facility"; "any person who at the time of disposal of any hazardous substance owned or operated any facility at which such hazardous substances were disposed of"; "any person who by contract, agreement, or otherwise arranged for disposal or treatment … of hazardous substances"; and "any person who accepts or accepted any hazardous substances for transport to disposal or treatment facilities, incineration vessels or sites selected by such person."[329] The statute defines a facility as:

> (A) any building, structure, installation, equipment, pipe or pipeline … well, pit, pond, lagoon, impoundment, ditch, landfill, storage container, motor vehicle, rolling stock, or aircraft, or (B) any site or area where a hazardous substance has been deposited, stored, disposed of, or placed, or otherwise come to be located; but does not include any consumer product in consumer use or any vessel.[330]

Hazardous substances include substances specifically listed by the EPA under CERCLA as well as substances deemed hazardous under other specified environmental statutes.[331] Liability under CERCLA is presumed to be joint and several, though responsible parties may avoid joint and several liability if they can prove a reasonable basis for apportioning harms.[332]

CERCLA imposes liability for natural resource damages as well as clean-up costs. Natural resource damages may be recovered for harm to "land, fish, wildlife, biota, air, water, ground water, drinking water supplies, and other such resources belonging to, managed by, held in trust by, appertaining to, or otherwise controlled" by federal, state, or local governments or any Indian tribe.[333] Notably, CERCLA defines public trust resources somewhat more broadly than common law public trust doctrine to include any natural resource held or controlled by a government, and not just tidal and submerged lands, waterways, and fish and wildlife.[334] Damages to purely private resources are not recoverable under CERCLA, although they may be the subject of private tort claims.

Geoengineering activities involving the release or threat of release of a hazardous substance could trigger CERCLA liability. Such activities might include the release of stratospheric aerosols using hydrogen sulfide (a hazardous substance under CERCLA) or the dispersal of minerals containing a hazardous substance into the oceans or land.[335] Such releases would likely qualify as releases from a facility under CERCLA's broad definition of "facility." In determining liability, an interesting question would arise regarding whether a geoengineering project operator qualifies as a PRP. In some instances, the geoengineering project operator would be deemed a PRP as a current owner or operator of a facility. But two additional PRP categories might be relevant in other instances: past owners and operators – i.e., persons "who at the time of disposal of any hazardous substance owned or operated any facility at which such hazardous substances were disposed of"; and arrangers – i.e., persons "who by contract, agreement, or otherwise arranged for disposal or treatment … of hazardous substances."

Both of these latter PRP categories require that there be a "disposal" of a hazardous substance. Geoengineering project operators may contend, however, that the deliberate release of materials containing hazardous substances as part of a geoengineering project does not constitute disposal. CERCLA defines disposal as "the discharge, deposit, injection, dumping, spilling, leaking, or placing of any solid waste or hazardous waste into or on any land or water so that such solid waste or hazardous waste or any constituent thereof may enter the environment or be emitted into the air or discharged into any waters, including ground waters."[336] This definition suggests that disposal involves the handling or mishandling of waste materials, but perhaps not the deliberate deployment of useful products.[337] The case law suggests that such a distinction between useful products and waste materials is relevant to – though not necessarily dispositive of – whether a disposal has occurred.[338] Namely, to be liable as an arranger, a defendant must have entered into a transaction with the intention that at least a portion of the product be disposed of.[339] Accordingly, the Supreme Court has held that a chemical manufacturer cannot be held liable as an arranger merely for selling a useful product that the purchaser later disposed of in a way that led to contamination.[340] Similarly, a court of appeals concluded that the use of asbestos as a building material did not constitute a disposal under CERCLA.[341] In each instance, the courts considered the usefulness of a product as one factor in determining whether a transaction's primary purpose was disposal.[342]

Applying the above line of reasoning to geoengineering projects, one might draw an analogy between pesticide spraying and the release of hazardous substances by a stratospheric aerosol project. Both activities involve the deliberate deployment of potentially useful yet hazardous substances. The proper application of pesticide in suitable quantities constitutes the use of a product, but the discharge of excess pesticide into the environment arguably constitutes a disposal. Likewise, one might reason that any excessive use of hazardous substances for geoengineering purposes constitutes a disposal. In the case of pesticides, CERCLA moots the issue by explicitly exempting pesticide application activities from liability.[343] In the case of geoengineering, however, no parallel exemption exists to protect project operators from potential liability.

CERCLA nonetheless does contain one exemption that could be relevant to some geoengineering project operators. Releases of hazardous substances that are undertaken in compliance with specified federal permits (such as a CWA NPDES permit) are exempt from CERCLA liability.[344] Geoengineering project operators who obtain such a permit would be protected from CERCLA liability for releases complying with such a permit.

Finally, one other issue specific to geoengineering relates to CERCLA Section 104, which authorizes the federal government to respond to the release or threat of release of a hazardous substance or "of any *pollutant or contaminant* which may present an imminent and substantial danger to the public health or welfare."[345] In contrast to Section 107 liability, the scope of which is limited to hazardous substances, Section 104 authorizes a government response to pollutants or contaminants as well as hazardous substances. CERCLA defines "pollutant or contaminant" to include any substance that may cause death, disease, or abnormalities in any organism.[346] Thus, a federal response action may not necessarily result in clean-up cost liability for geoengineering project operators, but "could nonetheless … impose substantial restrictions on an ongoing climate engineering project that arguably created a threat to human health or the environment."[347]

4.3. OTHER FEDERAL LAWS APPLICABLE TO SPECIFIC TECHNIQUES

Federal laws other than environmental laws also may be relevant to specific geoengineering techniques. Of particular importance are the laws and regulations governing airspace and the regulatory framework governing genetically modified organisms.

4.3.1. *Airspace Regulation*

Proposals for delivering aerosols into the stratosphere include the use of airplanes, rockets, artillery shells, or balloons.[348] All of these proposals would in some way implicate the oversight of the FAA, which has jurisdiction over US airspace.[349] The construction of large structures on the ground for the purposes of capturing carbon dioxide from the atmosphere also could affect airspace and thus might be subject to FAA oversight as well.

4.3.1.1. Amateur Rockets, Balloons, and Kites

The FAA regulations at 14 C.F.R. Part 101 govern the operation of moored balloons, kites, amateur rockets, and unmanned free balloons. Specifically, these regulations apply to: (1) amateur rockets, defined as unmanned rockets that are "propelled by a motor or motors having a combined total impulse of 889,600 Newton-seconds … or less" and that "[c]annot reach an altitude greater than 150 kilometers above the earth's surface";[350] (2) moored balloons having a diameter greater than 6 feet or a gas capacity exceeding 115 cubic feet; (3) kites weighing more than 5 pounds; and (4) unmanned free balloons carrying a payload exceeding specified criteria.[351] Generally, these devices may not be operated in a manner that creates a hazard to other persons or their property, nor may they be operated in restricted areas without permission.[352]

Further regulations govern the operation of each of these categories of devices. The regulations governing amateur rockets classify amateur rockets into three subcategories: Class 1 model rockets, which weigh no more than 1,500 grams and use no more than 125 grams of propellant; Class 2 high-power rockets, which include amateur rockets other than model rockets that are propelled by motors having a combined impulse of 40,960 Newton-seconds or less; and Class 3 advanced high-power rockets, which include any other amateur rockets.[353]

In general, amateur rockets must be launched on a suborbital trajectory and may not cross into foreign territory.[354] In addition, Class 2 and Class 3 rockets may not be operated into any cloud, at any altitude where horizontal visibility is less than 5 miles or where clouds or obscuring phenomena of more than five-tenths coverage prevails, and at less than a specified distance from other persons or property.[355] Further restrictions prohibit operations between sunset and sunrise, in controlled airspace, and near airports without prior FAA authorization.[356] Class 2 and Class 3 rockets also are subject to a requirement that advance notice be given to the nearest FAA air traffic control facility of planned activity, including location, altitude, and duration.[357] Notice is not required for the operation of Class 1 rockets.[358]

Moored balloons and kites are subject to operating restrictions somewhat more restrictive than those governing amateur rockets: They may not be operated less than 500 feet from the base of any cloud, more than 500 feet above the surface of the Earth, from any area where the ground visibility is less than 3 miles, or within 5 miles from an airport boundary.[359] These devices also are subject to lighting and marking requirements designed to ensure visibility,[360] and 24-hours' advance notice is required for any operations more than 150 feet above the Earth's surface.[361]

Finally, unmanned free balloons are subject to similar altitude and safety restrictions. Such balloons may not be operated at any altitude where clouds or obscuring phenomena of more than five-tenths coverage prevails, at any altitude below 60,000 feet standard pressure altitude where horizontal visibility is less than 5 miles, over congested populated areas during the first 1,000 feet of ascent, and below 2,000 feet within designated airspaces near airports.[362] Such balloons must be equipped with devices to ensure visibility and a way to terminate the flight safely.[363] Operators must provide notice to the nearest FAA air traffic control facility 6 to 24 hours in advance of operations, as well as notice after a balloon is launched.[364] Operators must also monitor and record the course of the balloon and provide balloon position reports to the air traffic control facility if requested.[365]

The design and operation of stratospheric aerosol experiments or projects may be limited substantially by the various restrictions governing the operation of amateur rockets, moored balloons, kites, and unmanned free balloons. A person intending to conduct operations that deviate from these restrictions must request a waiver or authorization from the FAA to ensure the operations are safely conducted.[366] Such requests must be submitted at least 45 days in advance and provide detailed information regarding planned operations.[367]

4.3.1.2. Other Rockets

Persons operating unmanned rockets other than amateur rockets – i.e., rockets propelled by a motor or motors having a combined total impulse exceeding 889,600 Newton-seconds or that can reach an altitude greater than 150 kilometers above the Earth's surface – must comply with the more extensive regulatory requirements found in 14 C.F.R. Chapter III.[368] These regulations, which implement the Commercial Space Launch Act (CSLA), govern space transportation activities conducted in the United States or by a US citizen, with the exception of space transportation activities carried out by the federal government.[369] Enacted to facilitate commercial space launches by the private sector, the CSLA requires a license or permit in order to deploy a launch vehicle, operate a launch site or reentry site, or reenter a reentry vehicle into the Earth's atmosphere.[370] FAA analysis of license applications includes a safety review, policy review, payload review, financial responsibility determination, and environmental determination.[371] As part of the safety review, the applicant must show that it has assessed the hazards of launch operations and has proposed ways to mitigate them, and the FAA conducts a risk analysis of each launch.[372]

Experimental permits are available under the CSLA for "[r]esearch and development to test new design concepts, new equipment, or new operating techniques."[373] Such permits are processed more quickly than commercial space licenses and allow an unlimited number of launches for a particular vehicle design.[374] Applicants for experimental permits must provide a program description, flight test plan, operational safety documentation, and sufficient information to allow the FAA to analyze environmental impacts.[375]

While nonfederal operators could become involved in the launching of rockets for geoengineering research or deployment purposes, the federal government is at least equally likely to engage in these activities. Launches by the federal government would be subject not to the CSLA, but rather to regulatory requirements specific to the government agency conducting the launches. For example, the Department of Defense has established launch safety standards intended to protect the general public against falling debris, exposure to toxic substances and radiation, and other potential hazards.[376] These standards provide for, inter alia, FAA issuance of warnings or restrictions regarding military operations hazardous to civilian aircraft, or the halting of such operations if necessary to protect aircraft.[377]

4.3.1.3. Airplanes

The Federal Aviation Act gives the FAA general authority over aviation safety in the United States and charges the agency with overseeing the use of airspace by civilian and military operations.[378] The FAA regulates the design of aircraft and aircraft parts, establishes and enforces airworthiness standards, licenses pilots, and regulates flight operations.[379]

The use of aircraft to deliver stratospheric aerosols may require the development of planes specially designed to carry and distribute large quantities of material at high altitudes.[380] The testing of new aircraft design concepts or new aircraft equipment would require an experimental certificate from the FAA.[381] Subsequent design and production of a specific model of aircraft would require further FAA certification.[382] In addition, individual aircraft would have to be registered and certified as airworthy before they could be legally operated, and they would have to be inspected for airworthiness on an annual basis.[383] The Federal Aviation Act defines "aircraft" as "any contrivance invented, used, or designed to navigate, or fly in, the air."[384] Accordingly, the FAA has interpreted the term to apply to any aircraft, manned or unmanned, large or small, and has stated that all such aircraft are potentially subject to FAA regulation, including registration requirements.[385] The use of drones would be subject to the FAA's new regulations governing small unmanned aircraft systems (UAS).[386] These rules, which apply to UAS over 0.55 pounds and under 55 pounds, provide for a streamlined web-based registration process and simplified method for marking such aircraft.[387]

FAA rules also govern the location and manner of flights. For example, FAA regulations generally require fixed-wing aircraft to be operated at an altitude that allows the aircraft to conduct an emergency landing without undue hazard to persons or property on the surface.[388] Except when necessary for takeoff or landing, aircraft generally must operate at least 500 feet above uncongested areas and at least 1,000 feet above the highest obstacle within congested areas.[389] Further FAA regulations divide airspace into various categories, each of which is subject to different rules of operation.[390] An air traffic control service is provided in the airspace categories defined as controlled airspace.[391] Particularly relevant to the proposed use of aircraft to deliver aerosols to the stratosphere is Class A airspace, a controlled airspace which extends over the entire United States from 18,000 feet above mean sea level upward to the undefined boundary between airspace and outer space.[392] In Class A airspace, prior clearance from air traffic control is required for aircraft operations, and pilots must operate under instrument flight rules and maintain two-way radio communications with air traffic control.[393]

4.3.1.4. Large Ground Structures

In addition to its jurisdiction over aircraft, the FAA also shares regulatory authority with local governments over structures that could interfere with navigable airspace or the safe and efficient use of airspace, facilities, or equipment for aviation.[394] Persons proposing construction or alteration of structures exceeding 200 feet in height, as well as some shorter structures located near airport runways, must submit a Notice of Proposed Construction or Alteration to the FAA.[395] Notice is not required for a new structure if it will be shielded by existing structures or natural terrain, is located in an already congested area, and will not adversely affect safety in air navigation.[396] The FAA presumes structures exceeding 2,000 feet in height to be "a hazard to air navigation that results in inefficient use of airspace," and a person proposing such a structure has the burden of demonstrating to the contrary.[397] Based on the Notice of Proposed Construction or Alteration, the FAA evaluates potential effects on air safety and navigation, determines whether the structure will create a hazard to navigable airspace, and decides whether to require the adoption of any safety measures.[398] If the FAA issues a Determination of Hazard, a proponent may negotiate with the FAA to change the project so that it qualifies for a "no hazard" determination, or it may pursue an appeal.[399] The FAA's determination of whether a hazard to air navigation exists does not automatically decide whether construction may proceed; rather, that decision is left to local governments, which have jurisdiction over land use.[400]

4.3.2. *Federal Regulation of Genetically Modified Organisms*

The term geoengineering is sometimes defined to include proposals to genetically modify crops to increase their albedo or their ability to absorb carbon dioxide.[401] Such activities could trigger the application of federal law governing genetically modified organisms (GMOs). No federal statute specifically aims at governing GMOs.[402] Rather, federal GMO regulatory policy relies on a patchwork of pre-existing laws, as described in a 1986 policy document, the Coordinated Framework for Regulation of Biotechnology ("Coordinated Framework"). The Coordinated Framework calls on three federal agencies – the Animal and Plant Health Inspection Service (APHIS), the EPA, and the Food and Drug Administration (FDA) – to use these laws to regulate GMO risks to health and the environment.[403]

The APHIS is responsible for overseeing plant pest risks that GMOs might create. The Plant Protection Act gives the APHIS the authority to regulate the movement of potential plant pests, referred to as "regulated articles."[404] "Regulated articles" include:

> any organism which has been altered or produced through genetic engineering, if the donor organism, recipient organism, or vector or vector agent belongs to any [designated] genera or taxa ... and meets the definition of plant pest ... or any other organism or product altered or produced through genetic engineering which [the APHIS] determines is a plant pest or has reason to believe is a plant pest.[405]

Any person intending to introduce regulated articles into the environment must either provide a notification to APHIS or obtain a permit prior to introduction.[406] A person may use the notification process, provided that various precautions are followed, if a species is not a noxious weed and meets specified criteria designed to reduce the risk that the species will be a plant pest.[407] Most GMO field tests are conducted under the notification process, which involves relatively prompt and cursory review by the APHIS and little subsequent oversight.[408]

Species not meeting the criteria for using the notification process, including plants engineered to produce toxic substances, require a permit prior to field testing.[409] A permit application must provide detailed information regarding the GMO and planned introduction, and the APHIS may impose various conditions to prevent the establishment or spread of plant pests.[410] Persons seeking to undertake GMO releases beyond field testing typically file a petition for deregulation.[411] The APHIS grants the petition if it determines that the GMO "does not pose a potential for plant pest risk,"[412] and a GMO thereafter may be planted or moved without further regulatory supervision.

It is worth emphasizing that the APHIS oversight focuses on plant pest risks to agriculture, rather than broader human health and environmental risks.[413] Historically, most genetically modified plants received at least some APHIS review because the genetic engineering process used material from a plant pest such as a bacterium or virus.[414] However, the development of new genetic engineering techniques has enabled a growing number of genetically modified plants to be created without using DNA sequences from plant pests.[415] Unless these GMOs are themselves plant pests, they may lie outside the scope of the Plant Protection Act and escape APHIS scrutiny.[416]

EPA oversight of GMOs, which is based primarily on the Federal Insecticide, Fungicide, and Rodenticide Act (FIFRA),[417] is even narrower in scope than that of the APHIS. Under the Coordinated Framework, the EPA regulates only plants that are genetically engineered to express pesticidal substances. With minor modifications, the EPA applies FIFRA's requirements, which typically govern chemical pesticides, to crops genetically engineered to produce pesticides (which the EPA refers to as "plant-incorporated protectants" (PIPs)).[418] These requirements include obtaining an experimental use permit to conduct field trials of plants containing PIPs on ten or more acres[419] and registering a PIP prior to commercialization.[420] The EPA grants a pesticide registration if, inter alia, the pesticide "will perform its intended function without unreasonable adverse effects on the environment" and "when used in accordance with widespread and commonly recognized practice it will not generally cause unreasonable adverse effects on the environment."[421]

FDA oversight of GMOs, which derives from the FDA's jurisdiction over adulterated foods, covers only genetically modified plants to the extent they are found in foods or food products.[422] Most foods containing GMOs are deemed by the FDA to be "generally recognized as safe" – which means that they receive no additional scrutiny because of their GMO components.[423] FDA policy will not be discussed further here because of its limited relevance to geoengineering.

In general, federal oversight of GMOs under the Coordinated Framework is minimal, and oversight would likely be minimal for plants genetically engineered to have increased albedo. Commodity crops such as corn and soybeans, because of the vast acreage devoted to their cultivation, would be the most likely targets for such engineering. Most commodity crops currently grown in the United States are already genetically modified[424] and have undergone the process of APHIS notification and deregulation. Although GMOs with increased albedo would potentially be subject to APHIS oversight as well, the agency would probably have little reason ab initio to expect such plants to pose serious threats as plant pests or noxious weeds. Plant developers might even escape APHIS oversight by avoiding the use of genetic material from plant pests to modify a plant variety. Finally, the EPA is unlikely to exercise much oversight of plants engineered for increased albedo. Standing alone, greater reflectivity probably does not constitute a PIP that would trigger EPA review.

4.4. STATE AND LOCAL LAWS APPLICABLE TO SPECIFIC TECHNIQUES

Geoengineering activities could be subject to state common law as well as state regulatory schemes. State common law tort claims, including public and private nuisance, trespass, negligence, and strict liability for abnormally dangerous activities, are addressed in Chapter 5. The discussion below considers how state or local laws other than tort law might govern specific geoengineering techniques.

4.4.1. *White Roofs: Building Standards and Other Land Use Regulation*

Land use planning has traditionally been a matter of state and local rather than federal control.[425] Through zoning enabling acts, most states have granted local governments the power to regulate building height, size, and design; population density; and the location of varied land uses.[426] Standards in some jurisdictions may support the use of white roofs and other albedo modification techniques, whereas standards in other jurisdictions may discourage the use of such techniques. State and local standards also may govern the placement or design of structures deployed by such techniques. Such techniques would ordinarily not be considered as climate engineering unless carried out on a very large scale, but they will be discussed briefly here for the sake of comprehensiveness.

One example of supportive standards at the state level is California's "cool roof" requirements, adopted in 2005 with the objectives of reflecting heat, reducing energy use, and reducing local temperatures.[427] California's requirements set minimum solar reflectance values for new or reroofed buildings.[428] A small but increasing number of other jurisdictions have enacted similar requirements.[429]

Some state and local laws, in contrast, may hinder or prevent the installation of white roofs. White roofs can cause glare or other visual impacts that could be subject to local ordinances as well as nuisance law. Ordinances may specify the maximum level of glare from a building or prohibit the use of specific materials.[430] Other standards that may be relevant to white roofs include historic preservation requirements[431] and mandates that the architectural design of a building be consistent with the existing terrain or surrounding community.[432] The latter mandates have sometimes been challenged as vague or overly subjective, but typically are upheld as permissible exercises of local authority to promote the general welfare.[433]

Obstacles to the installation of white roofs also may arise from the enforcement of restrictive covenants by homeowners' associations (HOAs).

HOAs are governing bodies created to manage planned communities and enforce restrictive covenants.[434] Planned communities frequently use restrictive covenants to ensure a common design theme and prevent undesired activities or land uses within a neighborhood.[435] Some communities have used restrictive covenants to block the installation of solar panels[436] and could similarly try to obstruct the installation of white roofs. Restrictions that may be pertinent to white roof installations include architectural style requirements and building materials specifications.[437] Under an HOA, additions or renovations to existing residences typically must be approved by an architectural review committee.[438] Although HOA decisions are sometimes challenged, courts generally defer to HOA actions and uphold the enforcement of restrictive covenants.[439]

HOAs generally have the power to amend restrictive covenants and thus could accommodate white roof adoption, if so desired.[440] Changes to restrictive covenants typically require the approval of a majority of members, however, and are sometimes difficult to accomplish. Homeowners seeking to overcome restrictive covenants and unable to do so through the amendment process may request that state legislatures enact laws invalidating covenants that serve as barriers to public policy goals.[441]

4.4.2. *Air Capture Devices: Zoning Requirements*

Zoning requirements also may be relevant to the siting of air capture devices or other large geoengineering equipment. Although the size and design of any air capture device are yet to be determined, for discussion purposes one might assume that such a device could fold up into a standard shipping container.[442] An air capture device of that size could raise issues similar to those raised in siting other large structures such as wind turbines or cell phone towers.[443] Smaller air capture devices that do not raise such issues might be treated as utility structures or industrial facilities and deemed permissible in zoning districts that allow such uses.

Typically, a zoning designation lists specific uses that are permitted and allows other similar uses.[444] If an air capture device does not qualify as a permitted use, such as an industrial use in an industrial zone, it would require a special use permit or a use variance.[445] A special use permit allows a use that has been specifically authorized by the local legislative body. The applicant for a special use permit must meet specific predetermined conditions as well as more general standards concerning effects on the environment, visual resources, neighborhood character, property values, and historic values.[446] A variance, like a special use permit, allows the use of property in a manner otherwise prohibited by a zoning ordinance. However, a variance typically requires that an applicant demonstrate unnecessary hardship such as the lack of other economically viable uses for the parcel. Whether a person proposing to install an air capture device would be able to establish such hardship is uncertain.[447] Widespread proliferation of air capture devices could lead to the establishment of zoning designations specifically authorizing the erection of such devices. Air capture devices might even be treated as public utilities, and statutes might be enacted to facilitate their siting and construction.[448]

A zoning ordinance also may establish height limitations and setback requirements that are specific to individual use zones.[449] Limits on the height or placement of a structure could serve as a barrier to the erection of large air capture devices.[450] To qualify for a dimensional variance from height and setback requirements, one typically must show practical difficulties in conforming to existing standards.[451]

Finally, in addition to the special use permit, use variance, or dimensional variance that may be required to install an air capture device, building permits and other local or state approvals may be necessary to construct the device.

4.4.3. *State and Local Regulation of GMOs*

The genetic modification of crops to increase albedo could trigger state and local regulation in addition to the federal oversight discussed above. Although GMO regulation began at the federal level, there also have been numerous state and local efforts to regulate the cultivation of GMOs, sometimes through popular initiatives. Various jurisdictions, particularly in the western United States, have enacted bans, moratoria, or limits on the testing or cultivation of genetically engineered crops.

In the United States local efforts to restrict GMO cultivation first appeared in California, where several counties and cities enacted bans on genetically modified crops.[452] Several local jurisdictions in other states then adopted similar measures.[453] In response, some states have enacted legislation blocking local governments from putting such restrictions in place or limiting their ability to do so.[454] Local GMO ordinances have also faced judicial challenges.[455]

Three local measures governing GMOs in Hawaii, whose year-round warm weather attracts much GMO field research activity, have proven particularly contentious.[456] Kauai County enacted an ordinance requiring that growers annually report the cultivation of GMOs.[457] Hawaii County adopted measures generally banning the cultivation of GMOs, with an exception for crops located where GMOs were previously grown.[458] Maui County voters approved a moratorium on GMO operations and mandated as a condition of lifting the moratorium that the county complete a health and environmental study and find that proposed cultivation practices would be safe.[459] All three local measures were challenged in the federal court and invalidated. In each instance, state law, which charges the Hawaii Department of Agriculture with the authority to oversee the introduction, propagation, and control of plants, was held to preempt the local measure.[460] The legal challenges to these measures also raised the issue of federal preemption. Both the Hawaii County regulation of GMOs and the Maui County moratorium were deemed federally preempted to the extent that the federal government had regulated GMOs as plant pests.[461]

Finally, although no ban on the cultivation of GMOs has been enacted at the state level, some states have enacted regulations specific to GMOs.[462] These state regulations typically require growers to obtain a permit before planting GMOs, though such requirements may be lifted once a particular GMO is found not to pose a threat to agriculture.[463]

4.5. CONCLUSION

Geoengineering activity, whether in the form of field research or deployment, could be subject to various environmental laws that are not directly aimed at regulating geoengineering. Procedural obligations to report on weather modification activities and to conduct environmental impact assessments likely would apply to most geoengineering projects. Substantive laws governing pollution, wildlife protection, land use, and other matters could also be relevant, depending on the design, location, and effects of a specific geoengineering project.

NOTES

1 *See* BIPARTISAN POLICY CENTER (BPC), TASK FORCE ON CLIMATE REMEDIATION RESEARCH 13 (2011).
2 42 U.S.C. § 4332(C) (2012).
3 40 C.F.R. § 1508.18 (2014).
4 *Id.*
5 *See Ka Makani 'O Kohala Ohana Inc. v. County of Hawai'i Dept. of Water Supply*, 295 F.3d 955, 960–61 (9th Cir. 2002).
6 42 U.S.C. § 4332(C) (2012).
7 *Id.*
8 40 C.F.R. Pts. 1502, 1503 (2014).
9 *Id.* §§ 1507.3, 1508.4 (2014).
10 *Id.* §§ 1501.3, 1508.9 (2014).
11 *Id.* § 1508.9 (2014).
12 *Id.* § 1501.4(c) (2014).

13 *Id.* § 1508.4 (2014).
14 5 U.S.C. §§ 701–706; *see Public Citizen v. U.S. Trade Representative*, 5 F.3d 549, 551 (D.C. Cir. 1993).
15 *See Winter v. Natural Resources Defense Council*, 555 U.S. 7, 32 (2008).
16 15 U.S.C. § 793(c)(1) (2012).
17 33 U.S.C. § 1371(c) (2012).
18 42 U.S.C. § 5159; *see infra* text accompanying note 132 for further discussion of this exemption.
19 40 C.F.R. § 1506.11 (2014).
20 *See Natural Resources Defense Council v. Winter*, 518 F.3d 658, 680–82 (9th Cir. 2008), *rev'd on other grounds*, 555 U.S. 7 (2008).
21 *See* Department of Homeland Security, NEPA Alternative Arrangements for Critical Physical Infrastructure in New Orleans, 71 Fed. Reg. 14712, 14712 (Mar. 23, 2006).
22 *Id.* at 14712.
23 *See infra* section 4.1.3.
24 Geoengineering research is discussed in Chapter 6 of this volume.
25 *See, e.g.*, David Rotman, *A Cheap and Easy Plan to Stop Global Warming*, 116 *MIT Technology Rev.* 52 (Feb. 2013), www.technologyreview.com/featuredstory/511016/a-cheap-and-easy-plan-to-stop-global-warming/ (noting that David Keith, a leading proponent of stratospheric aerosol field experimentation, "is adamant that work will not go forward unless he and his colleagues receive public funding and approval from established scientific agencies"). *See* Chapter 6 for a detailed discussion of the legal and policy issues involved in climate engineering research and experimentation.
26 *See* Council on Environmental Quality, Final Guidance for Effective Use of Programmatic NEPA Reviews, 79 Fed. Reg. 76986 (Dec. 23, 2014); Michael Boots, Council on Environmental Quality, Memorandum: Effective Use of Programmatic NEPA Reviews (Dec. 18, 2014), https://energy.gov/sites/prod/files/2016/05/f31/effective_use_of_programmatic_nepa_reviews_18dec2014.pdf.
27 *See* Boots, *supra* note 26, at 10.
28 *See Found. on Econ. Trends v. Lyng*, 817 F.2d 882, 884–86 (D.C. Cir. 1987) (rejecting contention that programmatic EIS was required for agency's animal productivity research program as a whole); *see also Kleppe v. Sierra Club*, 427 U.S. 390, 414–15 (1976) (holding that regional programmatic EIS was not required absent a proposal for regionwide action).
29 *See, e.g.*, US GLOBAL CHANGE RESEARCH PROGRAM, OUR CHANGING PLANET: THE U.S. GLOBAL CHANGE RESEARCH PROGRAM FOR FISCAL YEAR 2016, 51–52 (2015).
30 10 C.F.R. Pt. 1021, Subpt. D, App. B, § B3.6 (2015).
31 *Id.*
32 10 C.F.R. Pt. 1021, Subpt. D, App. B, §§ B3.8, 3.16 (2015).
33 45 C.F.R. § 640.3(b) (2014).
34 *Id.*
35 *Id.* § 640.3(b)(2), (4) (2014).
36 *See* Paul Weiland et al., *Environmental Impact Review*, in GLOBAL CLIMATE CHANGE AND U.S. LAW 153, 165 (Michael B. Gerrard & Jody Freeman eds.,

American Bar Association, 2nd edn 2014). For a list of these statutes, see Daniel R. Mandelker, NEPA Law and Litigation § 12:2 (2012).

37 *See* Mandelker, *supra* note 36, at § 12:1.

38 *See id.* at § 12:1.

39 *See id.* at § 12:1.

40 *See id.* at § 12:6; *see, e.g.*, CAL. PUB. RES. CODE §§ 21168, 21168.5 (West 2014).

41 CAL. PUB. RES. CODE §§ 21002, 21002.1 (West 2014); *see* Mandelker, *supra* note 36, at § 12:1. Washington's State Environmental Policy Act also provides substantive authority to state and local governments to require mitigation measures and to condition or deny a proposal based on environmental impacts. WASH. REV. CODE § 43.21C.060 (West 2014); *see also* WASH. ADMIN. CODE § 197-11-660 (West 2014).

42 CAL. PUB. RES. CODE §§ 21000–21177 (West 2014).

43 *Id.* § 21100 (West 2014).

44 *Id.* § 21080 (West 2014); *Friends of Mammoth v. Bd. of Supervisors of Mono Cnty.*, 502 P.2d 1049, 1054–56 (Cal. 1972). New York and Washington similarly subject permits and land use approvals to their respective SEPAs. *See* Mandelker, *supra* note 36, at § 12:13.

45 CAL. PUB. RES. CODE § 21100 (West 2014).

46 *Id.* § 21100 (West 2014); 42 U.S.C. § 4332(2)(C) (2014); *see* Mandelker, *supra* note 36, at § 12:14.

47 CAL. PUB. RES. CODE § 21080(c)(2) (West 2014); CAL. CODE REG. § 15063 (West 2014).

48 *Id.* § 21002.1(b) (West 2014).

49 *Id.* § 21081.6(b) (West 2014).

50 *Id.* § 21002.1(c) (West 2014).

51 *Id.* § 21081(b) (West 2014).

52 15 U.S.C. §§ 330a, 330b (2012); 15 C.F.R. § 908.4 (2015).

53 15 U.S.C. § 330 (2012).

54 15 C.F.R. §§ 908.5, 908.6, 908.8 (2015).

55 *Id.* § 908.12 (2015).

56 *Id.* § 908.12 (2015).

57 *Id.* § 908.3(a) (2015) (listing specific activities subject to reporting requirement).

58 *See* Ray Jay Davis, *Law and Urban-Induced Weather Change*, 25 U. TOL. L. REV. 379, 391 (1994) (noting that "[a]t one time or another, two-thirds of the states … have had legislation focusing upon intended weather modification through cloud seeding"). *See, e.g.*, WASH. REV. CODE § 43.37 (West 2014); OR. REV. STAT. CH. 558 (2013); TEX. AGRIC. CODE ANN. TITLE 9, CH. 301 (WEST 2014).

59 *See* George W. Bomar, *Weather Modification and the Law*, SOUTHWEST HYDROLOGY, 22, 22 (Mar./Apr. 2007).

60 TEX. AGRIC. CODE ANN. § 301.001(5) (West 2011).

61 *Id.* §§ 301.101, 301.102.

62 *Id.* §§ 301.101, 301.103.

63 *Id.* § 301.107.

64 *Id.* §§ 301.107, 301.114.

65 *Id.* §§ 301.107, 301.110–301.112. Some states have their own administrative procedures and hearing requirements for issuance of these permits.

66 *Id.* § 301.102.

67 *Id.* § 301.117 (requiring also that license holders keep records of operations conducted under a permit).
68 CAL. WATER CODE § 402(a) (West 2012).
69 *See* Maurice Roos, *Status of Weather Modification Regulation in California*, 17 J. WEATHER MODIFICATION 71, 71 (1985).
70 CAL. WATER CODE §§ 410, 411 (West 2012).
71 *See* SPECIAL COMMISSION ON WEATHER MODIFICATION, NATIONAL SCIENCE FOUNDATION, WEATHER AND CLIMATE MODIFICATION 108 (1965) (noting that state licensing requirements "could place a serious burden on any [weather modification] experiment crossing State lines, and most experiments of any size would be likely to do so").
72 *See, e.g.*, TEX. AGRIC. CODE ANN. § 301.001(5); KS. STAT. § 82a-1402(g) (defining weather modification as "any operation or experimental process which has as its objective inducing change, by artificial means, in the composition, behavior, or dynamics of clouds and the atmosphere"); COLO. REV. STAT. § 36-20-104(10) (defining weather modification as "any program, operation, or experiment intended to induce changes in the composition, behavior, or dynamics of the atmosphere by artificial means").
73 N.M. STAT. § 75-3-12; *see also* COLO. REV. STAT. § 36-20-118.
74 H. 5480, 2015 Gen Assem., Reg. Sess. (R.I. 2015), http://webserver.rilin.state.ri.us/BillText/BillText15/HouseText15/H5480.pdf.
75 *Id.*
76 H. 7578, 2016 Gen Assem., Reg. Sess. (R.I. 2016), http://webserver.rilin.state.ri.us/BillText16/HouseText16/H7578.pdf; *see also* Forum for Climate Engineering Assessment, Forum Discussion – Rhode Island H 7578: The Climate Geoengineering Act of 2016, http://ceassessment.org/forum-discussion-rhode-island-h-7578-the-climate-geoengineering-act-of-2016/; H. 6011, 2017 Gen Assem., Reg. Sess. (R.I. 2017), http://webserver.rilin.state.ri.us/BillText/BillText17/HouseText17/H6011.pdf.
77 16 U.S.C. § 1533 (2012). Critical habitat includes "the specific areas within the geographical area occupied by the species … on which are found those physical or biological features (I) essential to the conservation of the species and (II) which may require special management considerations or protection" as well as "specific areas outside the geographical area occupied by the species … upon a determination by the Secretary that such areas are essential for the conservation of the species." *Id.* § 1532(5).
78 16 U.S.C. § 1536(a)(2) (2012). The Fish and Wildlife Service generally has taken the position that the elements of jeopardy and adverse modification are completely overlapping. *See* Fish & Wildlife Service, Endangered and Threatened Wildlife and Plants; Notice of Intent to Clarify the Role of Habitat in Endangered Species Conservation, 64 Fed. Reg. 31871, 31872 (June 14, 1999).
79 16 U.S.C. § 1536(c) (2012).
80 *Id.* § 1536(b)(3) (2012).
81 50 C.F.R. § 402.02 (2014).
82 16 U.S.C. § 1536(b)(3) (2012).
83 *Id.* § 1538(a)(1) (2012). The take prohibition applies to listed animal species; a somewhat narrower prohibition applies to listed plant species. *Id.* § 1538(a)(2)(B).
84 50 C.F.R. § 17.31 (2014). A threatened species is "any species which is likely to become an endangered species within the foreseeable future throughout all or a significant portion of its range." 16 U.S.C. § 1532(20) (2012).

85 16 U.S.C. § 1532(19) (2012).
86 *Babbitt v. Sweet Home Chapter of Communities for a Great Oregon*, 515 U.S. 687, 708 (1995) (upholding regulatory definition of "take" found in 50 C.F.R § 17.3).
87 16 U.S.C. § 1540 (2012).
88 *Id.* § 1540(g) (2012).
89 *Id.* § 1536(b)(4) (2012).
90 *Id.* § 1539(a)(2)(A) (2012).
91 *Id.* § 1539(a)(2)(B) (2012).
92 *Cf. Sweet Home*, 515 U.S. at 713 (O'Connor, J., concurring) ("In my view, then, the 'harm' regulation applies where significant habitat modification … proximately (foreseeably) causes actual death or injury to identifiable animals."); Tracy Hester, *Remaking the World to Save It: Applying U.S. Environmental Laws to Climate Engineering Projects*, 38 ECOLOGY L.Q. 851, 883–84 (2011); *see also* Alan M. Glen & Craig M. Douglas, *Taking Species: Difficult Questions of Proximity and Degree*, NATURAL RESOURCES & 65, 68 (Fall 2001) ("The challenge facing FWS and private plaintiffs in the post-*Sweet Home* era is producing sufficient evidence to demonstrate that death or actual injury was incurred by identifiable members of the species, and that the activity in question was the foreseeable and proximate cause of said death or actual injury"). Issues of causation and harm have complicated the application of the ESA to climate change generally. *See* Dave Owen, *Endangered Species Act*, in *Global Climate Change and U.S. Law*, *supra* note 36, 183, 191–96; Matthew Gerhart, *Climate Change and the Endangered Species Act: The Difficulty of Proving Causation*, 36 ECOLOGY L.Q. 167 (2009).
93 *See* Susan George & William J. Snape III, State Endangered Species Acts, in ENDANGERED SPECIES ACT: LAW, POLICY, AND PERSPECTIVES 344, 345 (Donald S. Baur & William Robert Irvin eds., American Bar Association, 2nd edn 2010).
94 *See id.* at 347.
95 *See id.* at 345–6.
96 *See id.* at 352; Dhananjay Manthripragada, *Species Protection Versus State Agency Autonomy: Who Wins under the California Endangered Species Act?*, 36 ENVTL. L. REP. 10644, 10646–48 (2006).
97 CAL. FISH & GAME CODE §§ 2070–2079 (listing), § 2080 (prohibiting take), § 2081(incidental take permits) (West 2014); *see* CAL. CODE REGS. tit. 14, § 783.1 (take prohibition) (West 2014).
98 CAL. FISH & GAME CODE § 2085 (West 2014).
99 *Id.* § 86 (West 2014). Few states explicitly include habitat modification as a take. *See* George & Snape, *supra* note 93, at 349.
100 *Dep't of Fish & Game v. Anderson-Cottonwood Irrigation Dist.*, 11 Cal. Rptr. 2d 222, 226–27 (Ct. App. 3d Dist. 1992).
101 *Envtl. Council of Sacramento v. City of Sacramento*, 142 Cal. App. 4th 1018, 1025 (Cal. App. 3d Dist. 2006).
102 The California Supreme Court recently granted review of a court of appeal decision rejecting the argument that a take can occur only when there is mortality. *See Center for Biological Diversity v. Dep't of Fish & Wildlife*, 169 Cal. Rptr. 3d 413, 443 (Cal. App. 2d Dist. 2014), *review granted and op. superseded*, 328 P.3d 67 (Cal. 2014).
103 CAL. FISH & GAME CODE § 2081 (West 2014). While § 2081 clearly applies to private parties, it is uncertain whether state agencies whose actions result in incidental

take are obligated to obtain an incidental take permit under § 2081. *See* Manthripragada, *supra* note 96, at 10646.

104 CAL. FISH & GAME CODE § 2080.1 (West 2014).

105 Another potentially relevant statute, not discussed further in this chapter, is the Bald and Golden Eagle Protection Act, 16 U.S.C. §§ 668-668c.

106 16 U.S.C. § 703(a) (2012).

107 *Compare United States v. Brigham Oil & Gas, L.P.*, 840 F. Supp. 2d 1202, 1208–14 (D.N.D. 2012) (discussing cases declining to apply MBTA to activities that unintentionally caused bird deaths), *with U.S. v. Apollo Energies Inc.*, 611 F.3d 679, 686 (10th Cir. 2010) (applying MBTA to bird deaths caused by unprotected oil field equipment) and *U.S. v. CITGO Petro. Corp.*, 893 F. Supp. 2d 841, 843–48 (S.D. Tex. 2012) (citing cases applying MBTA to unintentional killings of migratory birds and finding liability where defendant was aware that migratory birds were being trapped in oil tanks), *rev'd*, 801 F.3d 477 (5th Cir. 2015). *See generally* Conrad A. Fjetland, *Possibilities for Expansion of the Migratory Bird Treaty Act for the Protection of Migratory Birds*, 40 NAT. RESOURCES J. 47, 62 (2000) (stating that courts apply MBTA to bird deaths "associated with hunting, trapping, and poaching," and to bird deaths "directly resulting from the careless application or handling of dangerous chemicals," but not "to incidental or innocent bird deaths occurring from otherwise legal activities").

108 *See* 16 U.S.C. §§ 703–712 (2012) (statute); 50 C.F.R. § 10.12 (2014) (regulation defining take).

109 *CITGO Petro.*, 801 F.3d at 492.

110 *See Center for Biological Diversity v. Pirie*, 191 F. Supp. 2d 161, 167 (D.D.C. 2002) (quoting Fish & Wildlife Service letter denying an MBTA permit request on grounds that such permits are not available for unintended conduct). The Fish & Wildlife Service is currently considering various regulatory approaches for authorizing incidental take of migratory birds under MBTA. *See* Department of the Interior, Fish & Wildlife Service, Migratory Bird Permits; Programmatic Environmental Impact Statement, 80 Fed. Reg. 30032 (May 26, 2015).

111 *See* 16 U.S.C. § 707 (2012).

112 *See id.* §§ 703–712 (2012); *see also Newton County Wildlife Ass'n v. U.S. Forest Serv.*, 113 F.3d 110, 114 (8th Cir. 1997) (holding that MBTA does not create private rights of action, nor does the Administrative Procedure Act confer jurisdiction to allege a direct violation of MBTA).

113 *Compare Humane Soc'y of the U.S. v. Glicksman*, 217 F.3d 882, 887–888 (D.C. Cir. 2000) (holding that the federal government may be enjoined for violating MBTA), *with Sierra Club v. Martin*, 110 F.3d 1551, 1555 (11th Cir. 1997) (holding that MBTA does not apply to the federal government).

114 *See supra* note 107.

115 *See United States v. Morgan*, 311 F.3d 611, 615–16 (5th Cir. 2002) (holding that strict liability applies to misdemeanor offenses under MBTA).

116 *See Brigham Oil & Gas*, 840 F. Supp. 2d at 1211–12.

117 *See Apollo Energies*, 611 F.3d at 690.

118 16 U.S.C. § 1372(a) (2012).

119 *Id.* § 1362(13), (18) (2012).

120 50 C.F.R. § 216.3 (2014).

121 16 U.S.C. § 1371(a)(5) (2012); 50 C.F.R. Pt. 216 Subpt. I (2014).

122 16 U.S.C. § 1375 (2012).
123 *Id.* § 1374(d)(6) (2012) (authorizing party opposed to permit to obtain judicial review of terms and conditions or permit); *City of Sausalito v. O'Neill*, 386 F.3d 1186, 1202–03 (9th Cir. 2004).
124 *See Natural Resources Defense Council, Inc. v. Evans*, 279 F. Supp. 2d 1129 (N.D. Cal. 2003) (considering MMPA challenge to Navy operations using sonar).
125 42 U.S.C. §§ 5121–5207 (2012). *See generally* Michael B. Gerrard, *Emergency Exemptions from Environmental Laws After Disasters*, 20 NATURAL RESOURCES & ENVIRONMENT 10, 10 (Spring 2006).
126 *Id.* §§ 5170(a), 5191(a) (2012). The president may declare an emergency without a gubernatorial request if an emergency involves "federal primary responsibility." *Id.* § 5191(b).
127 *Id.* § 5122(2) (2012).
128 *Id.* § 5122(1) (2012).
129 *See* FRANCIS X. MCCARTHY & JARED T. BROWN, CONGRESSIONAL RESEARCH SERVICE, CONGRESSIONAL PRIMER ON RESPONDING TO MAJOR DISASTERS AND EMERGENCIES 1 (2015).
130 42 U.S.C. §§ 5170a, 5192 (2012).
131 *See* MCCARTHY & BROWN, *supra* note 129, at 8.
132 42 U.S.C. § 5159 (exempting actions taken under 42 U.S.C. §§ 5170a, 5170b, 5172, 5173, and 5192); *see also* 44 C.F.R. § 10.8(c).
133 42 U.S.C. § 9604(a)(4) (2012).
134 *Id.* § 7603 (2012).
135 33 U.S.C. § 1364 (2012).
136 42 U.S.C. § 6973(a) (2012).
137 33 U.S.C. § 701n (2012).
138 50 U.S.C. § 1621 (2012).
139 *Id.* § 1622(b) (2012).
140 *See* Patrick Thronson, *Toward Comprehensive Reform of America's Emergency Law Regime*, 46 737, 752–53 (2013); Jules Lobel, *Emergency Power and the Decline of Liberalism*, 98 YALE L.J. 1385, 1415–17 (1989).
141 *See* Thronson, *supra* note 140, at 753–61, 769–77. For a compilation of emergency powers activated by a presidential declaration of national emergency, see PATRICK THRONSON, COMPENDIUM OF EMERGENCY POWERS STATUTES (2013), https://prospectusmjlr.files.wordpress.com/2013/10/national_emergency_powers.pdf.
142 *See* HAROLD C. RELYEA, CONGRESSIONAL RESEARCH SERVICE, NATIONAL EMERGENCY POWERS 13–16 (2007) (listing national emergencies declared between 1976 and 2007).
143 50 U.S.C. §§ 1621, 1631 (2012); *see* Thronson, *supra* note 140, at 749–51; Lobel, *supra* note 140, at 1413–16.
144 The Federal Emergency Management Agency's National Response Framework includes a list of various authorities relevant to emergency action. *See* FEDERAL EMERGENCY MANAGEMENT AGENCY, NATIONAL RESPONSE FRAMEWORK: LIST OF AUTHORITIES AND REFERENCES, www.fema.gov/pdf/emergency/nrf/nrf-authorities.pdf.
145 50 U.S.C. § 1701(a) (2012); *see* Thronson, *supra* note 140, at 758.
146 *See* Saikrishna Bangalore Prakash, *A Taxonomy of Presidential Powers*, 88 *B.U. L. REV.* 327, 329–33 (2008) (discussing sources of presidential powers); Lobel, *supra* note 140, at 1404.

147 U.S. CONST. art. II, § 2, cl. 1.
148 *See Youngstown Sheet & Tube v. Sawyer*, 343 U.S. 579, 587 (1952); *id.* at 635–38 (Jackson, J., concurring) ("Presidential powers are not fixed but fluctuate, depending upon their disjunction or conjunction with those of Congress."); David J. Barron & Martin S. Lederman, *The Commander in Chief at the Lowest Ebb – Framing the Problem, Doctrine, and Original Understanding*, 121 HARV. L. REV. 689, 770–72 (2008).
149 U.S. CONST. art. II, § 2, cl. 2; *id.* § 3, cl. 4.
150 *See* Barron & Lederman, *supra* note 148, at 743–44.
151 *See id.* at 745.
152 *See id.* at 745.
153 *See id.* at 746.
154 *See* CLIVE HAMILTON, EARTHMASTERS: THE DAWN OF THE AGE OF CLIMATE ENGINEERING 135–36 (2013).
155 *See* Climate Geoengineering Governance, *What Are the Security Implications of Solar Geoengineering? (Briefing Note 3)* (2015), geoengineering-governance-research.org/.../cgg-briefing-note-3securitysolarge-1.pdf.
156 *See* Albert Lin, *Geoengineering's Thermostat Dilemma*, in THE LAW OF THE FUTURE AND THE FUTURE OF LAW: VOLUME II (Sam Muller et al. eds., Hague Institute for the Internationalisation of Law 2012); David G. Victor et al., *The Geoengineering Option: A Last Resort Against Global Warming?*, 88 FOREIGN AFFAIRS 64, 71–72 (March/April 2009); *see also* Climate Geoengineering Governance, *supra* note 155 (noting likely involvement of the military in deploying solar geoengineering and the likelihood that nation-states would be blamed for extreme weather events).
157 U.S. CONST. art. II, § 1.
158 *See* Prakash, *supra* note 146, at 332; Gary Lawson, *Ordinary Powers in Extraordinary Times: Common Sense in Times of Crisis*, 87 B.U. L. REV. 289, 304–306 (2007); *see also* Saikrishna Bangalore Prakash, *The Imbecilic Executive*, 99 VA. L. REV. 1361, 1370–74 (2013) (discussing various theories of emergency executive power).
159 *See* Lawson, *supra* note 158, at 305.
160 *See id.* at 306; Prakash, *supra* note 158, at 1363–64.
161 *See* Lawson, *supra* note 158, at 304.
162 *See id.* at 306.
163 *See id.* at 306–08.
164 *See* JASON BLACKSTOCK ET AL., CLIMATE ENGINEERING RESPONSES TO CLIMATE EMERGENCIES 2 (2009), http://arxiv.org/ftp/arxiv/papers/0907/0907.5140.pdf (noting that geoengineering techniques aimed at reducing GHG concentrations would take several decades to have substantial climatic impacts).
165 *See id.* at 4, 6–7.
166 *See* NATIONAL RESEARCH COUNCIL (NRC), CLIMATE INTERVENTION: REFLECTING SUNLIGHT TO COOL EARTH 95 (2015).
167 44 C.F.R. § 206.63 (2014).
168 *See Spring-Gar Community Civic Ass'n, Inc. v. Homes for the Homeless, Inc.*, 540 N.Y.S.2d 453 (N.Y. App. Div. 1989) (acknowledging trial court's "recognition that an emergency situation currently exists in New York City with regard to the defendant city's legal and moral obligation to shelter a growing number of homeless

families"); *Hart Island Comm. v. Koch*, 520 N.Y.S.2d 977, 980–81 (N.Y. Sup. Ct. 1987) (upholding a claim that prison overcrowding constituted an emergency).

169 *See, e.g.*, Joshua B. Horton, *The Emergency Framing of Solar Geoengineering: Time for a Different Approach*, 2 THE ANTHROPOCENE REVIEW 147 (2015); Mike Hulme, *CAN SCIENCE FIX CLIMATE CHANGE?: A CASE AGAINST CLIMATE ENGINEERING* 23–26 (John Wiley 2014); Nils Markusson et al., *"In Case of Emergency Press Here": Framing Geoengineering as a Response to Dangerous Climate Change*, 5 WILEY INTERDISCIPLINARY REVIEWS: CLIMATE CHANGE 281, 281–90 (2014).

170 *See* Markusson et al., *supra* note 169, at 284 ("Defining any aspect of climate change as an 'emergency' is fraught with both scientific subjectivity and uncertainty"); Horton, *supra* note 169, at 2; Hulme, *supra* note 169, at 23.

171 *See* STEPHEN M. GARDINER, A PERFECT MORAL STORM: THE ETHICAL TRAGEDY OF CLIMATE CHANGE 357–58 (2011).

172 *See, e.g.*, NRC, *supra* note 166, at 12; Asilomar International Conference on Climate Intervention Technologies, Statement from the Conference's Scientific Organizing Committee, Mar 26, 2010, www.whoi.edu/fileserver.do?id=62483&pt=10&p=39472; ROYAL SOCIETY, GEOENGINEERING THE CLIMATE: SCIENCE, GOVERNANCE, AND UNCERTAINTY ix (2009) ("No geoengineering method can provide an easy or readily acceptable alternative solution to the problem of climate change"); Martin Bunzl, *Researching Geoengineering: Should Not or Could Not?*, 4 ENVIRON. RES. LETT., 045104, 2 (2009), doi: 10.1088/1748–9326/4/4/045104.

173 *See* Hulme, *supra* note 169, at 25.

174 *See* Gardiner, *supra* note 171, at 354–55 (warning that arguments from emergency are "open to manipulation" and potentially involve suspension of normal political processes and rights).

175 *See* Jim Rossi, *State Executive Lawmaking in Crisis*, 56 DUKE L.J. 237, 242 (2006).

176 *See id.* at 241–42.

177 CAL. GOVT. CODE §§ 8558, 8625 (West 2014).

178 *Id.* §§ 8628, 8655 (West 2014).

179 *Id.* § 8655 (West 2014).

180 *Id.* § 8571 (West 2014). The authority to waive existing laws and regulations varies from state to state. *See* Daniel G. Orenstein, *When Law Is Not Law: Setting Aside Legal Provisions During Declared Emergencies*, 41 J.L. MED. & ETHICS 73, 73 (2013).

181 CAL. GOVT. CODE § 8572 (West 2014).

182 *Id.* §§ 8630, 8634 (West 2014).

183 *Id.* § 8632 (West 2014).

184 Richard M. Frank, *The Public Trust Doctrine: Assessing Its Recent Past and Charting Its Future*, 45 U.C. DAVIS L. REV. 665, 686 (2012); *see National Audubon Soc'y v. Superior Court*, 658 P.2d 709, 712 (Cal. 1983) ("[T]he core of the public trust doctrine is the state's authority as sovereign to exercise a continuous supervision and control over the navigable waters of the state and the lands underlying those waters").

185 *See* Frank, *supra* note 184, at 677–79. Some legal scholars and advocates have argued for extending the public trust to air resources as a means of addressing

climate change. *See, e.g.*, MARY CHRISTINA WOOD, NATURE'S TRUST: ENVIRONMENTAL LAW FOR A NEW ECOLOGICAL AGE (2014).

186 *See* Frank, *supra* note 184, at 675–77.

187 *See National Audubon*, 658 P.2d at 727–28.

188 *See id.* at 719.

189 *See* Stephen E. Roady, *The Public Trust Doctrine*, in OCEAN AND COASTAL LAW AND POLICY 39, 41 (Donald C. Baur et al. eds., American Bar Association, 2008).

190 Alexandra B. Klass, *Renewable Energy and the Public Trust Doctrine*, 45 U.C. DAVIS L. REV. 1021, 1029 (2012).

191 *See* Albert C. Lin, *Public Trust and Public Nuisance: Common Law Peas in a Pod?*, 45 U.C. DAVIS L. REV. 1075, 1091–92 (2012).

192 43 U.S.C. §§ 1301(a), 1311 (2012); *U.S. v. California*, 447 U.S. 1, 3 (1980).

193 *See Sanders-Reed ex rel. Sanders-Reed v. Martinez*, 350 P.3d 1221, 1225 (N.M. Ct. App. 2015) (recognizing that public trust duty under New Mexico law extends to the atmosphere, but holding that arguments regarding that duty must be raised within the existing constitutional and statutory framework and not through a separate common law cause of action); *Foster v. Washington Dept. of Ecology*, No. 14-2-25295-1, slip op. at 8 (Wash. Super. Ct. Nov. 19, 2015) (noting that under Washington public trust doctrine, "[t[he navigable waters and the atmosphere are intertwined and to argue a separation of the two, or to argue that GHG emissions do not affect navigable waters, is nonsensical"); *see also Juliana v. US*, 217 F. Supp.3d 1224 (D. Or. 2016) (denying a motion to dismiss with respect to climate change-related public trust claims without determining whether the atmosphere is a public trust asset because the plaintiffs alleged public trust violations in connection with territorial sea as well). *But see, e.g., Filippone ex rel. Filippone v. Iowa Dept. of Natural Resources*, 2013 WL 988627, at *2-*3 (Iowa Ct. App. 2013) (affirming trial court's refusal to expand public trust to include the atmosphere).

194 *See PPL Montana, LLC v. Montana*, 132 S. Ct. 1215, 1235 (2012) ("the public trust doctrine remains a matter of state law"); *Alec L. ex rel. Loorz v. McCarthy*, 561 Fed. Appx. 7 (D.C. Cir. 2014) (mem. op.) (affirming dismissal of claim alleging abdication of federal public trust duty with respect to atmosphere); *see also* Klass, *supra* note 190, at 1037. *But see Juliana*, 217 F. Supp.3d at 1258–59 (holding that public trust doctrine applies to federal government, at least with respect to territorial seas). The courts have not resolved whether the federal government has a public trust duty over federal lands or waters, and some commentators have argued for recognition of such a duty particularly with respect to the exclusive economic zone. *See* Roady, *supra* note 189, at 58–59.

195 *See* Lin, *supra* note 191, at 1086; Klass, *supra* note 190, at 1071–72.

196 *Cf.* Ronald B. Robie, *Effective Implementation of the Public Trust Doctrine in California Water Resources Decision-Making: A View From the Bench*, 45 U.C. DAVIS L. REV. 1155, 1167–69 (2012) (explaining the deferential review that California courts apply to legislative and adjudicative actions by the State Water Resources Control Board).

197 *See, e.g., Foster*, slip op. at 9–10 (denying a petition for review where the state had already commenced rulemaking to establish greenhouse gas emission standards); *Sanders-Reed*, 350 P.3d at 1227 (concluding that "the courts cannot independently

regulate greenhouse gas emissions in the atmosphere" and affirming grant of summary judgment in favor of the state).

198 *See* Hester, *supra* note 92, at 873.

199 The following discussion comes with the caveat that the evolution of aerosols in the stratosphere and the ultimate effects of aerosol release involve substantial uncertainties. *See* NRC, *supra* note 166, at 62–63, 75.

200 *See* Hester, *supra* note 92, at 877 & n.113 (noting that the intentional discharge of chemicals into the air to fight forest fires has not triggered a need for CAA permits, and that historical attempts to modify weather "have fallen under separate state regulatory regimes rather than the Clean Air Act").

201 *See* Mark Sagoff, *The Principles of Federal Pollution Control Law*, 71 MINN. L. REV. 19, 22 (1986) (describing the economic approach to pollution, which "regard[s] pollution as the by-product of ordinary economic activity"); John Nagle, *The Idea of Pollution*, 43 U.C. DAVIS L. REV. 1, 41–49 (2009) (discussing the contextual nature of pollution). Generally, deliberate releases of chemical substances – such as aerial pesticides, forest fire suppressants, and rain-making chemicals – have not required CAA permits.

202 42 U.S.C. § 7602(g) (2012).

203 *Massachusetts v. EPA*, 549 U.S. 497, 500 & n.26 (2007).

204 42 U.S.C. § 7411(a)(3) (2012) (defining the stationary source for the purposes of CAA Section 111). The EPA has generally applied the same definition of stationary source to other portions of the CAA. *See* ARNOLD W. REITZE, STATIONARY SOURCE AIR POLLUTION LAW 164 (2005).

205 42 U.S.C. § 7418(b) (2012) (allowing for exemptions, except with respect to CAA §§ 111 and 112).

206 *Id.* §§ 7413, 7604 (2012).

207 *Id.* §§ 7408(a), 7409 (2012).

208 In cooperative federalism schemes, the federal government sets minimum regulatory standards, which may be implemented by either the states or the federal government, and states retain the ability to adopt more stringent standards. *See* David B. Spence, *Federalism, Regulatory Lags, and the Political Economy of Energy Production*, 161 U. PA. L. REV. 431, 470–71 (2013).

209 42 U.S.C. § 7410(a) (2012).

210 *Id.* § 7410(a)(2)(A) (2012).

211 *Id.* §§ 7475–7479 (2012).

212 *Id.* §§ 7501–7503 (2012).

213 *Id.* § 7491 (2012). *See* M. Lea Anderson, *The Visibility Protection Program*, in THE CLEAN AIR ACT HANDBOOK 201–30 (Julie R. Domike & Alec C. Zacaroli eds., American Bar Association, Chicago, 3rd edn, 2011).

214 42 U.S.C. §§ 7413, 7604 (2012).

215 *Id.* § 7412 (2012).

216 *Id.* §§ 7661-7661f (2012).

217 *See* 40 C.F.R. §§ 50.4, 50.5, 50.6, 50.7, 50.13, 50.17 (2014). In using SO_2 as the indicator for its regulatory standard, the EPA intends to address effects associated with exposure to gaseous sulfur oxides (SOx) generally. *See* Environmental Protection Agency, Primary National Ambient Air Quality Standard for Sulfur Dioxide, 75 Fed. Reg. 35519, 35536–37 (June 22, 2010).

218 *See* Jeffrey R. Pierce et al., *Efficient Formation of Stratospheric Aerosol for Climate Engineering by Emission of Condensible Vapor from Aircraft*, 37 *Geophysical Research Letters*, L18805 (2010), doi:10.1029/2010GL043975; Philip Rasch et al., *An Overview of Geoengineering of Climate Using Stratospheric Sulphate Aerosols*, 366 PHIL. TRANSACTIONS OF THE ROYAL SOC'Y 4007, 4013 (2008); Alan Robock et al., *Benefits, Risks, and Costs of Stratospheric Geoengineering*, 36 GEOPHYSICAL RESEARCH LETTERS L19703, 5 (2009), doi:10.1029/2009GL039209.

219 *See* Rasch et al., *supra* note 218, at 4010–11 (discussing atmospheric chemical reactions involving sulfur species).

220 *See* Hester, *supra* note 92, at 878; *see also* Primary National Ambient Air Quality Standard for Sulfur Dioxide, 75 Fed. Reg. at 35523 (noting that SOx includes particulate species that are regulated under the NAAQS for particulate matter).

221 *See* Hester, *supra* note 92, at 878.

222 *See* Adriana Downie et al., *Biochar as a Geoengineering Climate Solution: Hazard Identification and Risk Management*, 42 CRITICAL REVIEWS IN ENVTL. SCI. & TECH. 225, 238 (2012).

223 42 U.S.C. § 7571 (2012); *see* Michael J. Horowitz, Regulation of Mobile Sources: Motor Vehicles, Nonroad Engines, and Aircraft, in *THE CLEAN AIR ACT HANDBOOK*, *supra* note 213, at 351, 371 (noting EPA's issuance of such standards).

224 42 U.S.C. § 7572(a) (2012); *see* Michael Gerard Green, *Control of Air Pollutant Emissions from Aircraft Engines: Local Impacts of National Concern*, 5 ENVTL. LAWYER 513, 535–37 (1999).

225 40 C.F.R. Pt. 87 (2014); *see* FEDERAL AVIATION ADMINISTRATION, AVIATION EMISSIONS, IMPACTS & MITIGATION: A PRIMER 17 (2015) (noting that aircraft engines have certification requirements for CO, HC, NOx, and smoke emissions); Nathan Richardson, *Aviation, Carbon, and the Clean Air Act*, 38 COLUM. J. ENVTL. L. 67, 75–76 (2013) (discussing the history of EPA regulation of aviation emissions and noting that EPA standards generally are no more stringent than international standards).

226 *See, e.g.*, Anton Laakso et al., *Stratospheric Passenger Flights Are Likely an Inefficient Geoengineering Strategy*, 7 ENVTL. RES. LETT., 034021(2012) (investigating the use of commercial aircraft flying at higher altitudes and using sulfur-enhanced jet fuel as a method of delivering sulfur to the stratosphere).

227 *See* Alan Robock et al., *Regional Climate Responses to Geoengineering with Tropical and Arctic SO_2 Injections*, 113 J. GEOPHYS. RES., D16101, 5 (2008), doi:10.1029/2008JD010050.

228 For an argument that the statute authorizes the EPA to reduce emissions through operational changes independent of engine design, see Richardson, *supra* note 225, at 88–91.

229 49 U.S.C. § 44714 (2012).

230 *See* Environmental Protection Agency, Finding that Greenhouse Gas Emissions from Aircraft Cause or Contribute to Air Pollution that May Reasonably Be Anticipated to Endanger Public Health and Welfare, 81 Fed. Reg. 54422, 54427–28 (2016); *see also* Federal Aviation Administration, Advisory Circular: Approval of Propulsion Fuels, Additives, and Lubricating Oils, June 30, 2014, www.faa.gov/documentLibrary/media/Advisory_Circular/AC-20-24D.pdf.

231 *See* FEDERAL AVIATION ADMINISTRATION, *supra* note 225, at 19.

232 42 U.S.C. § 7651(b) (2012) (setting out the purpose of Title IV); 42 U.S.C. § 7651a(15) (2012) (defining "unit" as "a fossil fuel-fired combustion device").

233 *See* Debra J. Jezouit, *The Acid Rain Program*, in THE CLEAN AIR ACT HANDBOOK, *supra* note 213, at 449, 449.

234 *See id.* at 450.

235 *See* Alan Robock, *20 Reasons Why Geoengineering May Be a Bad Idea*, 64 Bull. of the Atomic Scientists 14, 16 (May/June 2008), http://thebulletin.org/2008/may/20-reasons-why-geoengineering-may-be-bad-idea. Noting that the 1991 eruption of Mt. Pinatubo injected 20 megatonnes (Mt) SO_2 into the stratosphere, one computer modeling analysis considered the injection of 3-5Mt of SO_2 per year. *See* Robock et al., *Regional Climate Responses*, *supra* note 227, at 4. Title IV was enacted with the goal of reducing annual SO_2 emissions from 23 million short tons to 13 million short tons. *See* Jezouit, *supra* note 233, at 449. *But cf.* NRC, REFLECTING SUNLIGHT, *supra* note 166, at 75 ("Introduction of stratospheric aerosols is likely to slightly increase the acidity of the snow and rain reaching the surface").

236 *See* NRC, REFLECTING SUNLIGHT, *supra* note 166, at 61, 75. The ozone-depleting effect of stratospheric aerosols may decrease with time as atmospheric concentrations of halogens and other ozone-depleting substances (ODS) decline. *See id.* at 75.

237 Montreal Protocol on Substances that Deplete the Ozone Layer, Sept. 16, 1987, S. Treaty Doc. No. 100–10, 1522 U.N.T.S. 29 [hereinafter Montreal Protocol].

238 42 U.S.C. §§ 7671-7671q (2012). *See* Diane E. McConkey & Peter H. Wyckoff, *Stratospheric Ozone Protection*, in THE CLEAN AIR ACT HANDBOOK, *supra* note 213, at 473–519.

239 42 U.S.C. § 7671a (2012).

240 40 C.F.R. Pt. 82, Subpt. A, Apps. A, B, & F (lists of controlled ODS) (2014).

241 33 U.S.C. §§ 1319, 1365(a) (2012).

242 *Id.* § 1311(a) (2012).

243 *Id.* § 1362(12) (2012) (emphasis added).

244 *See* Aaron C. Courtney et al., *Ocean Dumping and Marine Pollution*, in OCEAN AND COASTAL LAW AND POLICY, *supra* note 189, at 243, 244.

245 *See* Robin Kundis Craig, *Coastal Water Quality Protection*, in OCEAN AND COASTAL LAW AND POLICY, *supra* note 189, at 205, 218.

246 33 U.S.C. § 1342 (2012).

247 *Id.* § 1362(14) (2012).

248 *Id.* § 1362(6) (2012).

249 *Id.* § 1362(7), (8) (2012) (defining navigable waters and territorial seas).

250 33 C.F.R. § 328.3(a) (2014); 40 C.F.R. § 122.2 (2014). *See* Department of Defense and Environmental Protection Agency, Clean Water Rule: Definition of "Waters of the United States," 80 Fed. Reg. 37054 (June 29, 2015). On February 18, 2017, President Trump issued an executive order instructing the EPA and the Corps of Engineers to begin the process of withdrawing this regulation and narrow the definition of regulated waters. These agencies have begun the process. Department of Defense and Environmental Protection Agency, Intention to Review and Rescind or Revise the Clean Water Rule, 82 Fed. Reg. 12532 (March 6, 2017).

251 *See U.S. v. Riverside Bayview Homes, Inc.*, 474 U.S. 121 (1985) (holding the CWA applicable to wetland adjacent to navigable water); *Solid Waste Agency of Northern*

Cook County v. U.S. Army Corps of Eng'rs, 531 U.S. 159 (2001) (holding the CWA inapplicable to an isolated body of water); *Rapanos v. U.S.*, 547 U.S. 715, 759 (2006) (Kennedy, J., concurring) (stating that water or wetland must possess a significant nexus to waters that are or were navigable in fact in order to constitute "navigable waters" under the CWA).

252 *See* NATIONAL RESEARCH COUNCIL (NRC), CLIMATE INTERVENTION: CARBON DIOXIDE REMOVAL AND RELIABLE SEQUESTRATION 58 (2015) (noting the focus of ocean iron fertilization experiments on high-latitude surface waters of the Southern Hemisphere and other iron-limited waters in the sub-polar north Pacific and eastern equatorial Pacific).

253 *Nat'l Cotton Council of Am. v. U.S. Envtl. Protection Agency*, 553 F.3d 927, 935–37 (6th Cir. 2009); *see also* 33 U.S.C. § 1362(6) (2012) (defining pollutant to include "dredged spoil, solid waste, incinerator residue, sewage, garbage, sewage sludge, munitions, chemical wastes, biological materials, radioactive materials, heat, wrecked or discarded equipment, rock, sand, cellar dirt and industrial, municipal, and agricultural waste discharged into water").

254 The parties to the London Convention and London Protocol have characterized ocean fertilization activities other than legitimate scientific research as a form of dumping subject to the restrictions found in those respective treaties. Resolution LC-LP.1 (2008) on the Regulation of Ocean Fertilization, *available at* www.whoi.edu/fileserver.do?id=56339&pt=10&p=39373. This characterization rests on the premise that ocean fertilization would function as a means of dumping waste carbon in the oceans. *See* Aaron L. Strong et al., *Ocean Fertilization: Science, Policy, and Commerce*, 22 OCEANOGRAPHY 236, 251 (2009). The treatment of ocean fertilization under the London Convention and London Protocol is discussed in Chapter 3 of this book, Sec. III.

255 40 C.F.R. § 125.123 (2014); *see also* 33 U.S.C. § 1343 (2012) (ocean discharge criteria). Factors used to determine whether a discharge will cause unreasonable degradation include: the quantities, composition, and potential for bioaccumulation of pollutants; the potential transport of pollutants; and the composition and vulnerability of biological communities that may be exposed to pollutants. 40 C.F.R. § 125.122(a) (2014).

256 *See* NRC, CARBON DIOXIDE REMOVAL, *supra* note 252, at 49.

257 *See id.* at 42–43; *see also* Angela Haren Kelley, *A Call for Consistency: Open Seawater Intakes, Desalination, and the California Water Code*, 4 GOLDEN GATE U. ENVTL. L.J. 277, 283–86 (2011) (discussing the adverse effects of open seawater intakes).

258 33 U.S.C. § 1326(b) (2012); *see also* Environmental Protection Agency, National Pollutant Discharge Elimination System – Final Regulations to Establish Requirements for Cooling Water Intake Structures at Existing Facilities and Amend Requirements at Phase I Facilities, 79 Fed. Reg. 48300 (Aug. 15, 2014) (promulgating the final rule applying CWA § 316(b) to cooling water intake structures for power-generating and industrial facilities).

259 *See, e.g.*, CAL. WATER CODE § 13142.5(b) (West 2014) (requiring "best available site, design, technology, and mitigation measures feasible … be used to minimize the intake and mortality of all forms of marine life" for intake of seawater by industrial facilities).

260 33 U.S.C. § 1344 (2012).
261 33 C.F.R. § 320.4(a)(1) (2014).
262 40 C.F.R. § 230.10 (2014).
263 33 U.S.C. § 1344(c) (2012); *see Mingo Logan Coal Co. v. EPA*, 714 F.3d 608 (D.C. Cir. 2013).
264 *See* Roelof Dirk Schuiling, *A Natural Strategy Against Climate Change*, 1 J. CHEM. ENG'G & CHEM. RESEARCH 413, 415–16 (2014).
265 *See* 33 U.S.C. § 1362(7), (8) (2012) (defining navigable waters to include territorial seas and defining territorial seas to extend seaward from the coast a distance of 3 miles).
266 *Id.* § 1341(a) (2012).
267 *Id.* § 1341(d) (2012).
268 *Id.* § 1313(d)(1)(A) (2012).
269 *Id.* § 1313(d)(1)(C) (2012).
270 Environmental Protection Agency, Memo from Denise Keehner to Water Division Directors, Integrated Reporting and Listing Decision Related to Ocean Acidification, Nov. 15, 2010, www.epa.gov/sites/production/files/2016-01/documents/memo_integrated_reporting_and_listing_decisions_related_to_ocean_acidfication.pdf.
271 *Id.* at 1, 4.
272 A court recently rejected the claim that the EPA should have required Oregon and Washington to list waters impaired by ocean acidification as impaired water bodies. *See Center for Biol. Diversity v. EPA*, 90 F. Supp. 3d 1177 (W.D. Wash. 2015).
273 33 U.S.C. § 403 (2012). Furthermore, Section 13 of the Rivers and Harbors Act prohibits the discharge of refuse into navigable waters in the absence of an NPDES permit. *Id.* § 407 (2012); *see also id.* § 1342(a)(4), (5) (2012) (subsuming Rivers and Harbors Act Section 13 permitting under the NPDES permitting program).
274 33 C.F.R. § 322.5 (2014).
275 *Id.* § 329.4 (2014).
276 *Id.* § 329.12 (2014).
277 Convention on the Prevention of Marine Pollution by Dumping of Wastes and Other Matter, Arts. IV.1; III.1(a)(i), Dec. 29, 1972, 1046 U.N.T.S. 120.
278 33 U.S.C. § 1411 (2012).
279 *Id.* § 1402(f) (2012) (setting out various exceptions not applicable to ocean fertilization).
280 *Id.* § 1402(c) (2012).
281 *Id.* § 1412(a) (2012).
282 *Id.* § 1412(a) (2012); 40 C.F.R. Pt. 227 (2014).
283 40 C.F.R. § 220.3 (2014).
284 33 U.S.C. § 1414b(a) (2012).
285 *Id.* § 1412a(b) (2012).
286 *Id.* § 1412a(a) (2012).
287 *Id.* § 1415 (2012).
288 *Id.* § 1417(c) (2012).
289 *Id.* § 1415(g) (2012).
290 *See* Michael C. Branson, *A Green Herring: How Current Ocean Fertilization Regulation Distracts from Geoengineering Research*, 54 SANTA CLARA L. REV. 163, 181 (2014).

291 40 C.F.R. § 220.3(e) (2014).
292 *See* U.S. ENVIRONMENTAL PROTECTION AGENCY, MEASURING THE EFFECTIVENESS OF THE OCEAN DUMPING MANAGEMENT PROGRAM 4–5 to 4–6 (2012) (noting potential application of the Ocean Dumping Act to ocean fertilization).
293 *See* Hester, *supra* note 92, at 861.
294 *See id.* at 862 & n.44.
295 *See id.* at 862 & n.44; Sean Low et al., *Geoengineering Policy and Governance Issues*, in GEOENGINEERING RESPONSES TO CLIMATE CHANGE: SELECTED ENTRIES FROM THE ENCYCLOPEDIA OF SUSTAINABILITY SCIENCE AND TECHNOLOGY 169, 182 (Tim Lenton & Naomi Vaughan eds., Springer, 2012) (suggesting that Planktos "attempted to flout [the Ocean Dumping Act] regulation" of its iron fertilization experiment "by using a vessel flying under a different national flag").
296 *See* Hester, *supra* note 92, at 862; Low et al., *supra* note 295, at 182.
297 *See* Jeff Tollefson, *Ocean-Fertilization Project off Canada Sparks Furore*, 490 NATURE 458 (2012); Martin Lukacs, *World's Biggest Geoengineering Experiment 'Violates' UN Rules*, GUARDIAN, Oct. 15, 2012, www.theguardian.com/environment/2012/oct/15/pacific-iron-fertilisation-geoengineering.
298 *See* Tollefson, *supra* note 297, at 458–59.
299 15 U.S.C. § 2605(a) (2016).
300 *See id.*
301 *See Corrosion Proof Fittings v. EPA*, 947 F.2d 1201, 1215–23 (5th Cir. 1991).
302 *See* Frank R. Lautenberg, Chemical Safety for the 21st Century Act, Pub. L. No. 114–182, § 6 (2016).
303 15 U.S.C. § 2604(a)(1)(A) (2016).
304 *Id.*
305 *Id.* § 2604(f) (2016).
306 *Id.*
307 Robock et al., *supra* note 218, at 4.
308 *See* Philip J. Rasch et al., supra note 218, at 4014.
309 *See* Shigenao Maruyama et al., *Possibility for Controlling Global Warming by Launching Nanoparticles into the Stratosphere*, 10 J. THERMAL SCI. & TECH., No. 2, Paper No. 15-00081 (2015); David W. Keith, *Photophoretic Levitation of Engineered Aerosols for Geoengineering*, 107 PROCEEDINGS OF THE NAT'L ACAD. OF SCI. 16428 (2010).
310 15 U.S.C. § 2604(a)(2) (2016).
311 *Id.* § 2604(e), (f) (2016).
312 Rob A. DeLeo, *Anticipatory Policymaking: When Government Acts to Prevent Problems and Why It Is So Difficult* 66–69 (Routledge 2016).
313 *See* Holly Doremus et al., *Environmental Policy Law: Problems, Cases, and Readings* 478 (Foundation Press 2012).
314 42 U.S.C. § 6903(5), (27) (2012); *id.* § 6921(a); 40 C.F.R. Part 261 (2014).
315 *Id.* §§ 6922–6925 (2012); DOREMUS ET AL., *supra* note 313, at 478–79.
316 42 U.S.C. § 6903(5), (27) (2012), § 6922 (2012).
317 *Id.* § 6923 (2012).
318 *Id.* §§ 6924, 6925 (2012).
319 *Id.* §§ 6928, 6972 (2012).

320 *See* RAELEN WILLARD & HOWARD E. HAINES, JR., BIOENERGY GUIDEBOOK: MONTANA 49 (1992).
321 *See* AMERICAN PHYSICAL SOCIETY, DIRECT AIR CAPTURE OF CO_2 WITH CHEMICALS 30–31 (2011).
322 42 U.S.C. § 6903(27) (2012) (defining solid waste as "any garbage, refuse, sludge from a waste treatment plant, water supply treatment plant, or air pollution control facility, and other discarded material"); 40 C.F.R. § 261.2 (2014) (regulatory definition of solid waste).
323 *See* EPA, Definition of Solid Waste, 80 Fed. Reg. 1694 (Jan. 13, 2015) (revising recycling-related provisions of the definition of solid waste under RCRA); *see also American Mining Congress v. EPA*, 824 F.2d 1177, 1185, 1193 (D.C. Cir. 1987) (holding that materials destined for immediate reuse in production process are not discarded under RCRA); *American Mining Congress v. EPA*, 907 F.2d 1179, 1186 (D.C. Cir. 1990) (deferring to the EPA determination that sludges from wastewater stored in surface impoundments were discarded even though they might later be reclaimed for treatment).
324 *See* Downie et al., *supra* note 222, at 238, 240. *Compare Safe Food & Fertilizer v. EPA*, 350 F.3d 1263 (9th Cir. 2003) (deferring to the EPA conclusion that recycled feedstocks used to make fertilizer were not a solid waste where they were substantially identical to virgin materials and were treated by market participants as valuable products), *with Howmet Corp. v. EPA*, 614 F.3d 544 (D.C. Cir. 2010) (deferring to the EPA conclusion that spent material was solid waste even though it was subsequently used to produce fertilizer).
325 Environmental Protection Agency, Hazardous Waste Management System: Identification and Listing of Hazardous Waste: Carbon Dioxide (CO_2) Streams in Geologic Sequestration Activities, 76 Fed. Reg. 48073, 48077–79 (Aug. 8, 2011).
326 Environmental Protection Agency, Hazardous Waste Management System: Conditional Exclusion for Carbon Dioxide (CO_2) Streams in Geologic Sequestration Activities, 79 Fed. Reg. 350 (Jan. 3, 2014).
327 Environmental Protection Agency, Federal Requirements Under the Underground Injection Control (UIC) Program for Carbon Dioxide (CO_2) Geologic Sequestration (GS) Wells, 75 Fed. Reg. 77230 (Dec. 10, 2010). These regulations established a new class of wells, Class VI, and requirements specific to that class. *See id.*
328 42 U.S.C. §§ 9604, 9606, 9607(a) (2012).
329 *Id.* § 9607(a)(1)–(4) (2012).
330 *Id.* § 9601(9) (2012).
331 *Id.* § 9601(14) (2012); 40 C.F.R. Table 302.4 (2014).
332 *See* Doremus et al., *supra* note 313, at 560–61; *Burlington Northern & Santa Fe Rwy Co. v. U.S.*, 556 U.S. 599, 614–15 (2009).
333 42 U.S.C. §§ 9601(16), 9607(f)(1) (2012).
334 *Cf. supra* section 4.1.4.
335 *See* Hester, *supra* note 92, at 887 n.162. At least two substances that could be used to generate stratospheric aerosols, hydrogen sulfide and sulfuric acid, are listed as CERCLA hazardous substances. 40 C.F.R. Table 302.4 (2014).
336 42 U.S.C. § 9601(29) (2012) (incorporating by reference the definition of disposal found in RCRA, 42 U.S.C. § 6903(3)).

337 Similarly, CERCLA's definition of facility includes an exception for "any consumer product in consumer use," but this exception has been held to apply only to goods normally used for personal, family, or household purposes. *See Uniroyal Chem. Co. v. Deltech Corp.*, 160 F.3d 238, 257 (5th Cir. 1998).

338 *See* Greg DeGulis & Sarah Gable, *Burlington Northern: CERCLA and Its Ever-Changing, Unpredictable Landscape*, 28 NAT. RES. & ENV'T, *Spr.* 40 (2014).

339 *Burlington Northern & Santa Fe Rwy. Co. v. United States*, 556 U.S. 599, 611–12 (2009).

340 *Id.*

341 *Dayton Indep. Sch. Dist. v U.S. Mineral Products Co.*, 906 F.2d 1059, 1065–66 (5th Cir. 1990).

342 *See* DeGulis & Gable, *supra* note 338, at 40.

343 42 U.S.C. § 9607(i) (2012). The "normal application of fertilizer" is similarly exempted from CERCLA liability. *Id.* § 9601(22)(D) (2012).

344 42 U.S.C. § 9601(10) (2012) (defining federally permitted release); 42 U.S.C. § 9607(j) (2012) (recognizing defense to CERCLA § 107 liability for federally permitted releases).

345 42 U.S.C. § 9604(a)(1) (2012) (emphasis added).

346 *See* Hester, *supra* note 92, at 887–88 & n.164. *Compare* 42 U.S.C. § 9601(33) (defining pollutant), *with id.* § 9601(14) (defining hazardous substance).

347 *See* Hester, *supra* note 92, at 888.

348 Robock et al., *supra* note 218, at 2–3 & Fig. 1; DAVID KEITH, A CASE FOR CLIMATE ENGINEERING 96–99 (2013).

349 49 U.S.C. § 40103 (2012).

350 14 C.F.R. § 1.1 (2014).

351 *Id.* § 101.1 (2014).

352 *Id.* §§ 101.5, 101.7 (2014).

353 *Id.* § 101.22 (2014). Prior to the promulgation of these regulations, the FAA relied on self-regulation by the amateur rocket community. The FAA's current regulations incorporate many of that community's practices. *See* Federal Aviation Administration, Requirements for Amateur Rocket Activities, 73 Fed. Reg. 73768, 73768–69 (Dec. 4, 2008).

354 14 C.F.R. § 101.23 (2015).

355 *Id.* § 101.25 (2015).

356 *Id.* § 101.25 (2015).

357 *Id.* § 101.27 (2015).

358 Requirements for Amateur Rocket Activities, 73 Fed. Reg. at 73774.

359 14 C.F.R. § 101.13 (2015).

360 *Id.* § 101.17 (2015).

361 *Id.* § 101.15 (2015).

362 *Id.* § 101.33 (2015).

363 *Id.* § 101.35 (2015).

364 *Id.* § 101.37 (2015).

365 *Id.* § 101.39 (2015).

366 *Id.* § 101.3 (2015).

367 *Id.* § 101.29 (2015).

368 *Id.* § 101.21(b) (2015).

369 *Id.* § 400.2 (2015).
370 51 U.S.C. § 50904(a) (2012).
371 14 C.F.R. Pt. 415 (2015); *see also* GOVERNMENT ACCOUNTABILITY OFFICE (GAO), COMMERCIAL SPACE LAUNCHES: FAA NEEDS CONTINUED PLANNING AND MONITORING TO OVERSEE THE SAFETY OF THE EMERGING SPACE TOURISM INDUSTRY 17 n.20, 52–53 (2006).
372 GAO, *supra* note 371, at 17.
373 14 C.F.R. § 437.5 (2015); *see generally* 51 U.S.C. § 50906 (2012); 14 C.F.R. Pt. 437 (2015).
374 *See* FEDERAL AVIATION ADMINISTRATION, THE ANNUAL COMPENDIUM OF COMMERCIAL SPACE TRANSPORTATION: 2014, at 90 (2015).
375 14 C.F.R. § 437.21 (2015).
376 *See* Thomas Pfitzer, *US Department of Defense Launch Safety Standards*, in *SPACE SAFETY REGULATIONS AND STANDARDS* 17–27 (Joseph N. Pelton & Ram S. Jakhu eds., Elsevier, 2010).
377 RANGE COMMANDERS COUNCIL, STANDARD 321-10: COMMON RISK CRITERIA STANDARDS FOR NATIONAL TEST RANGES 3-3 (2010), www.dtic.mil/get-tr-doc/pdf?AD=ADA639830.
378 49 U.S.C. §§ 40101, 44701 (2012) (authorizing the FAA to promulgate aviation safety standards); *see* 14 C.F.R. ch.I (2015); *see also* Michael J. Holland, *Federalism in the Twenty-First Century: Preemption in the Field of Air*, 78 DEFENSE COUNSEL *J.*, 11, 12 (2011).
379 14 C.F.R. Pts. 21–95 (2015).
380 *See* JUSTIN MCCLELLAN ET AL., AURORA FLIGHT SCIENCES CORP., GEOENGINEERING COST ANALYSIS 43–49 (2011), *available at* http://justinmcclellan.com/GeoengineeringReport.pdf (analyzing possible new aircraft design to disperse sulfur in the stratosphere); *but cf.* Robock et al., *supra* note 218, at 4–5 (analyzing options for distributing aerosols using existing military aircraft).
381 14 C.F.R. § 21.191 (2015).
382 *See* Thomas N. Tarnay, Comment, *Aircraft Designs Subjected to FAA Special Certification Review – Mitsubishi MU-2 and Beechcraft Bonanza: The Role of the SCR in Aircraft Design Certification and Implications on Federal Preemption*, 62 *J. AIR L & COM.* 591, 601–605 (1996) (summarizing certification requirements).
383 14 C.F.R. §§ 21.173, 47.3, 91.409 (2015); Tarnay, *supra* note 382, at 605.
384 49 U.S.C. § 40102(a)(6) (2012).
385 Federal Aviation Administration, Clarification of the Applicability of Aircraft Registration Requirements for Unmanned Aircraft Systems (UAS) and Request for Information Regarding Electronic Registration for UAS, 80 Fed. Reg. 63912, 63913–14 (Oct. 22, 2015).
386 Federal Aviation Administration, Registration and Marking Requirements for Small Unmanned Aircraft, 80 Fed. Reg. 78594 (Dec. 16, 2015).
387 *See id.* At present, the streamlined registration requirements apply only to drones used for hobby or recreational purposes. *See* 14 C.F.R. § 48.5(b) (providing that drone operators conducting operations other than model aircraft operations must register under ordinary procedures for aircraft registration); 14 C.F.R. § 1.1 (defining model aircraft as an unmanned aircraft that is, among other criteria, flown for hobby or recreational purposes).
388 14 C.F.R. § 91.119(a) (2015). In fixed-wing aircraft, the most common type of aircraft, the wings are attached to the fuselage and are not intended to move independently

to create lift. FEDERAL AVIATION ADMINISTRATION, AVIATION MAINTENANCE TECHNICIAN HANDBOOK – AIRFRAME, VOL. 1, at 1–5 (2012).

389 14 C.F.R. § 91.119(b),(c) (2015).

390 *See* FEDERAL AVIATION ADMINISTRATION, PILOT'S HANDBOOK OF AERONAUTICAL KNOWLEDGE, ch. 14 (2008) (describing different airspace classifications).

391 *See id.* at 14-2.

392 14 C.F.R. § 71.33 (2015); FEDERAL AVIATION ADMINISTRATION, *supra* note 390, at 14-2; J. Scott Hamilton, *Allocation of Airspace as a Scarce National Resource*, 22 *TRANSP. L.J.* 251, 272 (1994).

393 14 C.F.R. § 91.135 (2015). The instrument flight rules can be found at 14 C.F.R. §§ 91.167–91.193.

394 49 U.S.C. § 44718 (2012).

395 14 C.F.R. §§ 77.7, 77.9 (2015).

396 *Id.* § 77.9(e) (2015).

397 *Id.* § 77.7(d) (2015).

398 *Id.* §§ 77.5, 77.31 (2015).

399 *See* Wendie L. Kellington, Land Use Considerations in Siting Renewable Energy Projects (With a Focus on Wind), *SR004 ALI-ABA* 587, 619 (2009).

400 *See* Hamilton, *supra* note 392, at 267.

401 *See, e.g.*, Andy Ridgwell et al., *Tackling Regional Climate Change by Leaf Albedo Bio-Geoengineering*, 19 *CURRENT BIOLOGY* 146 (2009); Emily Boyd, *Societal Choice for Climate Change Futures: Trees, Biotechnology, and Clean Development*, 60 *BIOSCIENCE* 742 (2010).

402 The discussion here uses the common term GMO to refer to genetically modified plants, including those that might be developed to have increased albedo for geoengineering purposes.

403 Office of Science and Technology Policy, Coordinated Framework for Regulation of Biotechnology ("Framework"), 51 Fed. Reg. 23302 (June 26, 1986).

404 7 U.S.C. §§ 7711, 7712 (2012).

405 7 C.F.R. § 340.1 (2015).

406 *Id.* § 340.0(a) (2015).

407 *Id.* § 340.3 (2015).

408 *See* ALBERT C. LIN, PROMETHEUS REIMAGINED: TECHNOLOGY, ENVIRONMENT, AND LAW IN THE TWENTY-FIRST CENTURY 57–58 (2013).

409 7 C.F.R. § 340.4 (2015).

410 *Id.* § 340.4(b), (f) (2015).

411 *Id.* § 340.6 (2015).

412 *Id.* § 340.6(e)(1) (2015).

413 *See* U.S. DEPARTMENT OF AGRICULTURE (USDA), ANIMAL AND PLANT HEALTH INSPECTION SERVICE, INTRODUCTION OF GENETICALLY ENGINEERED ORGANISMS DRAFT PROGRAMMATIC ENVIRONMENTAL IMPACT STATEMENT 20 (2007), www.aphis.usda.gov/brs/pdf/complete_eis.pdf (acknowledging that APHIS does not regulate GMOs based on their potential to pose human health or environmental risks); *see* LIN, *supra* note 408, at 57.

414 Emily Montgomery, *Genetically Modified Plants and Regulatory Loopholes and Weaknesses Under the Plant Protection Act*, 37 *VT. L. REV.* 351, 351 (2012).

415 *Id.* at 351–52.

416 *See* USDA, *supra* note 413, at 20.

417 7 U.S.C. §§ 136-136y (2012). The EPA also regulates GMOs by setting limits for pesticide residues in food under Section 408 of the Federal Food, Drug, and Cosmetic Act. *See* 21 U.S.C. § 346a(a)(1) (2012).

418 40 C.F.R. Pt. 174 (2014) (setting out requirements for PIPs and reserving various subjects for future rulemaking); *see* Mary Jane Angelo, *Regulating Evolution for Sale: An Evolutionary Biology Model for Regulating the Unnatural Selection of Genetically Modified Organisms*, 42 WAKE FOREST L. REV. 93, 127–28 (2007).

419 40 C.F.R. § 172.3 (2014).

420 *Id.* § 152.1 (2014) (noting that regulations governing pesticide registration also apply to PIPs).

421 7 U.S.C. § 136a(c)(5) (2012).

422 Food and Drug Administration (FDA), Statement of Policy: Foods Derived From New Plant Varieties, 57 Fed. Reg. 22984 (May 29, 1992).

423 *Id.* at 22989–90.

424 *See* US Department of Agriculture, Economic Research Service, Adoption of Genetically Engineered Crops in the U.S., Recent Trends in GE Adoption, www.ers.usda.gov/data-products/adoption-of-genetically-engineered-crops-in-the-us/recent-trends-in-ge-adoption.aspx (last updated July 14, 2016).

425 *See* JULIAN CONRAD JUERGENSMEYER & THOMAS E. ROBERTS, LAND USE PLANNING AND DEVELOPMENT REGULATION LAW 10 (2d ed. 2007).

426 *See id.* at 69.

427 *See* Justin Berton, *Global Warming Plan That Hits the Roof*, S.F. CHRON., Feb. 20, 2009, at F1.

428 *See, e.g.*, CAL. CODE OF REGS. tit. 24, § 140.3(a) (West 2014) (governing newly constructed nonresidential buildings); CAL. CODE OF REGS. tit. 24, § 150.1(c)(11) (West 2014) (governing newly constructed low-rise residential buildings).

429 *See, e.g.*, CITY OF NEW YORK LOCAL LAW NO. 21, 2011 (setting out solar reflectance requirements for roof coverings of new or altered buildings); CHICAGO MUN. CODE, tit. 18, ch. 13, § 101.5.4 (2008); CITY OF PHILADELPHIA, PA., Bill No. 090923 (2010), http://legislation.phila.gov/attachments/10096.pdf (amending city building construction and occupancy code to require that certain new buildings have roof coverings that are white in color or Energy Star-rated as highly reflective); LOS ANGELES MUN. CODE, sec. 18, 99.04.106.5 (2013).

430 *See, e.g.*, YUCAIPA (CALIFORNIA) DEVELOPMENT CODE, div. 7, ch. 9, § 87.0920 (2014) (specifying maximum acceptable glare levels); BEVERLY HILLS MUN. CODE, art. 24, § 10-3-2424 (2014) (specifying acceptable roofing materials and prohibiting "[r]eflective, glossy, polished, and/or roll formed type metal roofing").

431 *See* Juergensmeyer & Roberts, *supra* note 425, at 574.

432 *See, e.g.*, PALM SPRINGS ZONING CODE, § 92.21.1.05(D)(16), (17) (2014) (requiring that developments in environmentally sensitive areas have exterior colors, materials, and finishes that blend with the color and texture of surrounding stone or soil and do not use reflective building materials); SANTA BARBARA MUN. CODE § 22.69.050 (2014) (requiring that proposed developments be compatible with the neighborhood and "consistent with the scenic character of the City").

433 *See* Julie A. Tappendorf, *Architectural Design Ordinances*, ALI-ABA Course of Study, SM004 ALI-ABA 497, 500–01 (2006).

434 *See* Kristina Caffrey, *The House of the Rising Sun: Homeowners' Associations, Restrictive Covenants, Solar Panels, and the Contract Clause*, 50 NAT. RESOURCES J. 721, 724 (2010).
435 *See id.* at 724–5.
436 *See id.* at 722–3.
437 *See id.* at 733.
438 *See id.* at 733.
439 *See* Ryan McCarl, *When Homeowners Associations Go Too Far: Political Responses to Unpopular Rules in Common Interest Communities*, 43 REAL ESTATE L.J. 453, 455–56, 459–60 (2015); Caffrey, *supra* note 434, at 726, 737–8 (noting that common law legal policies tend to support enforcement of restrictive covenants, as long as they are not racially motivated).
440 *See* Mark A. Pike, Note, *Green Building Red-Lighted by Homeowners' Associations*, 33 WM. & MARY ENVTL.L. & POL'Y REV. 923, 941–42 (2009).
441 *See* Caffrey, *supra* note 434, at 739–45 (discussing state statutes invalidating covenants that restrict or prohibit the use of solar panels); McCarl, *supra* note 439, at 479–81 (summarizing instances in which legislatures have enacted statutory carve-outs to restrictive covenants).
442 *See* Klaus S. Lackner, *The Use of Artificial Trees*, in GEOENGINEERING OF THE CLIMATE SYSTEM 80, 86 (Roy Harrison & Ron Hester eds., Royal Society of Chemistry, 2014).
443 For discussion of legal issues presented by the siting of wind turbines and cellular towers, *see, e.g.*, Arthur F. Coon & Sean Marciniak, *Local Matter or Federal Case? The Network of Cell Tower Regulation in California*, 21 No. 6 MILLER & STARR, REAL ESTATE NEWSALERT, July 2011; Dwight H. Merriam, *Regulating Backyard Wind Turbines*, 10 VT. J. ENVTL. L. 291 (2009); James M. McElfish Jr. & Sara Gersen, *Local Standards for Wind Power Siting: A Look at Model Ordinances*, 41 ENVTL. L. REP. NEWS & ANALYSIS 10825 (2011). Federal law explicitly preserves state and local zoning authority over cellular towers except that it prohibits unreasonable discrimination among providers, as well as zoning that would have the effect of totally prohibiting cellular services. *See* 47 U.S.C. § 332(c)(7) (2012).
444 *See* Juergensmeyer & Roberts, *supra* note 425, at 71.
445 *See id.* at 157, 170.
446 *See id.* at 172; *see also* Coon & Marciniak, *supra* note 443 (discussing concerns addressed in permitting decisions regarding cellular towers).
447 *See* Juergensmeyer & Roberts, *supra* note 425, at 157–60 (noting that some courts use a four-factor test to determine unnecessary hardship: "(1) that the land in question cannot yield a reasonable return as currently zoned, (2) that the plight of the landowner is due to unique or unusual circumstances …, (3) that the variance requested will not alter the essential character of the neighborhood, and (4) that the variance not issue if it would be contrary to the public interest").
448 Analogously, the 1996 Telecommunications Act facilitated the expansion of the wireless telephone service by restricting the authority of local bodies to regulate the zoning of cell towers and similar facilities. *See* 47 U.S.C. § 332(c)(7)(B).
449 *See* Juergensmeyer & Roberts, *supra* note 425, at 70.
450 *See* Troy A. Rule, *Renewable Energy and the Neighbors*, UTAH L. REV. 1223, 1238 (2010) (noting that "[l]ocal height restrictions are perhaps the most common obstacle to small wind turbine installations").

451 *See* Juergensmeyer & Roberts, *supra* note 425, at 158–9.

452 *E.g.*, HUMBOLDT COUNTY, CAL., HUMBOLDT COUNTY GENETIC CONTAMINATION PREVENTION ORDINANCE (Nov. 4, 2014); MARIN COUNTY, CAL., MUN. CODE ch. 6.92 (2004), www.municode.com/library/ca/marin_county/codes/code_of_ordinances?nodeId=TIT6PUPESAMO_CH6.92PRGRGEMOOR; MENDOCINO COUNTY, CAL., MENDOCINO COUNTY CODE, ch. 10A.15 (2004), www.municode.com/library/CA/mendocino_county/codes/code_of_ordinances?nodeId=MECOCO_TIT10AAG_CH10A.15PRPRCURAGRGEMOORMECO; SANTA CRUZ COUNTY, CAL., MUN. CODE, ch. 7.31 (2006); TRINITY COUNTY, CAL., MUN. CODE, ch. 8.25 (2004), www.municode.com/library/ca/trinity_county/codes/code_of_ordinances?nodeId=TIT8HESA_CH8.25GEENOR_ARTIGEENOR; ARCATA, CAL., MUN. CODE, title V, ch. 10.5; *see generally* Allison H. Scott, *Genetically Modified Crop Regulation: The Fraying of America's Patchwork Farm Lands*, 26 VILL. ENVTL. L.J. 145, 156–57 (2015).

453 *See, e.g.*, JACKSON COUNTY, OR., THE GENETICALLY ENGINEERED PLANT ORDINANCE (May 20, 2014); JOSEPHINE COUNTY, OR., THE GENETICALLY ENGINEERED PLANT ORDINANCE (May 20, 2014); SAN JUAN COUNTY, WA., CNTY. CODE, ch. 8.26 (2012), www.codepublishing.com/wa/sanjuancounty/html/SanJuanCounty08/SanJuanCounty0826.html.

454 *See* Scott, *supra* note 452, at 156 (noting that 13 states have enacted preemption laws barring local regulation); *see, e.g.*, CAL. FOOD & AGRIC. CODE § 52334 (West 2015) (barring local jurisdictions from regulating plants, crops, or seeds on or after January 1, 2015 without the consent of California's Secretary of Food and Agriculture).

455 *See, e.g.*, *Schultz Family Farms LLC v. Jackson Cnty.*, No. 1:14-CV-01975, 2015 WL 3448069 (D. Or. May 29, 2015) (rejecting a preemption challenge based on Oregon law to Jackson County's ordinance prohibiting the cultivation of genetically engineered plants); Mateusz Perkowski, *Josephine County Won't Enforce GMO Ban While Lawsuit is Pending*, CAPITAL PRESS, Sept. 11, 2015, www.capitalpress.com/Oregon/20150911/josephine-county-wont-enforce-gmo-ban-while-lawsuit-is-pending.

456 *See* Audrey McAvoy, *Federal Judge Rules Maui County Ban on GMO Crops Invalid*, MIAMI HERALD, July 1, 2015.

457 *Syngenta Seeds, Inc. v. County of Kauai*, 664 Fed. Appx. 669, 671 (9th Cir. 2016).

458 *Hawai'i Floriculture & Nursery Ass'n v. County of Hawaii*, Civ. No. 14-00267 BMK, 2014 WL 6685817 (D. Haw. 2014), appeal docketed, No. 15-15020 (9th Cir. Jan. 8, 2015).

459 *Atay v. County of Maui*, 842F.3d 688 (9th Cir. 2016).

460 *County of Kauai*, *supra* note 457, at *1-*3; *Hawaii Papaya Indus. Ass'n v. County of Hawaii*, 666 Fed. Appx. 631, 634 (9th Cir. 2016); *County of Maui*, *supra* note 459, at *13-*17.

461 *County of Hawaii*, *supra* note 460, at *1-*2; *County of Maui*, *supra* note 459, at *11. Although a federal district court held that the Kauai notification requirement was not preempted by the federal Plant Protection Act, the Ninth Circuit did not address this issue on appeal. *Syngenta Seeds, Inc. v. County of Kauai*, Civ. No. 14-00014 BMK, 2014 WL 4216022, at *12-*14 (D. Haw. 2014).

462 *See* Scott, *supra* note 452, at 155.

463 *See id.*, at 156; *see, e.g.*, MINN, STAT., ch. 18F.07 (2014); FLA. STAT. § 581.083 (2014).

5

Liability and Compensation

Tracy Hester

To actually sway the behavior of sovereign states and individuals, a climate engineering legal regime will need to set out clear and practical rules that impose liability for damages. Such rules should be adopted in advance of any large-scale deployment; it is far better for states to consider liability rules in the abstract than to try to craft them after an actual dispute has arisen. Yet climate engineering will inevitably pose difficult challenges to the typical legal approaches used to assign liability for violations or injury. For example, if the degree of liability should proportionately reflect the seriousness of an offense's threatened harm or gravity, courts may be forced to weigh charges or assessments for climate engineering projects that arguably could have unpredictable global consequences. National courts or international tribunals may also have to decide whether to impose liability for allegedly illegal climate engineering activities when the failure to take action itself may risk even graver consequences from unmitigated climate change effects.

This chapter explores possible approaches to liability that might apply to climate engineering field research or deployment. Given the early state of climate engineering research, the field of climate intervention law and liability is unsurprisingly also unshaped. As noted in earlier chapters, very few international entities or domestic governments have yet proposed or implemented new laws or regulations to explicitly regulate climate engineering projects. The lack of a binding and explicit governance framework has led governments and advocates to jury-rig regulation of climate engineering projects by adapting existing international instruments and domestic laws crafted for other purposes. As a result, the degree of liability arising from climate engineering legal violations will heavily depend on the nature of the attempted engineering project, the type and extent of damage that it allegedly can cause, and the specific law or statute triggered by the nature of the project and its attendant damages. Despite the early and unclear nature of climate engineering law, we can still forecast some possible general principles of liability that this field might adopt under current domestic, international, and transnational laws.

To explore those issues, this chapter will focus on legal liability and compensation obligations to redress alleged damages or injuries caused by climate engineering efforts. To the extent that research and early limited field tests pose differing possible risks and damages than a broader deployment of climate engineering technologies, this analysis will parse out separate liability concepts that apply solely at the research stage. In the same vein, this chapter will assess possible injunctive remedies when they seek restitution or imposition of potentially costly corrective actions, to compel future regulatory actions, or to spur government implementation of climate engineering efforts.

As a result, this chapter will not explore other possible sources of legal obligation and responsibility. In particular, it will not focus on legal efforts by citizens (1) to impose obligations on entities charged as trustees to protect natural resources (unless such actions result in monetary or financial obligations), (2) to challenge perceived violations of explicit or implied constitutional guarantees to environmental protection,[1] or (3) to pursue disclosure obligations related to financial securities or liabilities arising from false statements to governmental officials.

5.1. ASPECTS OF POSSIBLE CLIMATE ENGINEERING TECHNIQUES THAT MIGHT AFFECT FUTURE LIABILITY

As earlier chapters made clear, sizable climate engineering field experiments or deployments theoretically could result in significant and unpredictable damages. The types of damages that could result will likely turn on the particular type of climate engineering that the projects use, and the variety of damages that each type can respectively cause.

Solar Radiation Management (SRM). Solar radiation management technologies would reduce the amount of sunlight that reaches the Earth's surface as a method to reduce mean global temperatures. Some techniques, as noted in earlier chapters, would use either the dispersal of particulates at stratospheric levels to scatter incoming solar radiation, the brightening of marine clouds, or the direct deflection of sunlight by space-based satellites. Some researchers fear that SRM, if deployed on a broad or global scale, could potentially cause an array of unexpected global and regional weather consequences such as diminished monsoons or redistributed precipitation. In addition, alteration of solar radiation influx could possibly reduce agricultural productivity in some regions (and increase it in others), impair ecosystems that rely on existing levels of light and radiation, and deplete stratospheric ozone levels through the use of chemicals that could scavenge existing ambient ozone.

Beyond risks posed by SRM even when it is deployed flawlessly, this strategy would not help prevent other damaging aspects of climate change such as continued or increased ocean acidification or accelerated deposition of possible by-products of SRM chemicals that eventually migrate to the Earth's surface. More controversially, some modeling of proposed SRM approaches suggest that it could cause whitening of daytime skies and reduce the efficiency of solar power plants. SRM allegedly could also have unpredictable impacts on ecosystems and biodiversity as a whole, or at least on protected species that might prove sensitive to changes in solar radiation levels or climate engineering materials and practices. Last, even if SRM works perfectly, it poses the risk of accumulated climate momentum: If SRM deployment occurred on a broad and long-term scale, a sudden halt in its use could lead to a sudden and accelerated burst of warming and climate change effects due to a possible buildup of greenhouse gases (GHG) that would act without a counterbalance.

Direct Carbon Dioxide Removal (CDR). When compared to SRM, CDR technologies[2] pose a different and much less imminent set of risks and environmental hazards. Direct Air Capture (DAC) is the mechanical technique of scrubbing greenhouse gases from the ambient atmosphere. It is usually combined with some form of storage to permanently immobilize or reuse the captured gas in a way that prevents its re-entry into the atmosphere. The direct removal of carbon dioxide (CO_2) from the ambient atmosphere generally should neither directly damage the local environment nor disrupt regional climatic conditions.[3] Depending on the specific type of method used to remove the CO_2, however, the technologies may yield unwanted residues, side effects, and local land use disruption. For example, some promising industrial CDR approaches would rely on direct absorption of CO_2 by passing ambient air over screens impregnated with an amine solution.[4] While this approach would reuse the amine solutions through a wash cycle that removes CO_2 from a saturated amine screen, the amines ultimately would become contaminated and require replacement. Managing the large volumes of spent materials and by-products of globally deployed CDR would therefore likely pose a significant waste management challenge. In addition to waste products from the CDR process, the captured CO_2 itself would require either its reuse as a commercial product, use as a replacement for other feedstocks and commercial chemical products, or – most likely – sequestration and disposal. The volumes of CO_2 requiring management and/or disposal generated by large-scale CDR needed to affect global temperatures would be staggering.

By rough calculations, capturing enough CO_2 to reduce atmospheric levels by 1 ppm would generate 7.8 gigatons of CO_2 – over 31 times the amount of all municipal solid waste disposed in the United States in 2014.[5]

Until a dominant method of CDR emerges from research and market competition, we cannot identify likely environmental impacts and consequences with confidence. Some possibilities, however, do suggest themselves. For example, some CDR methods may require substantial amounts of energy for operation.[6] Unless that power comes from renewable sources, GHG emissions generated by CDR's power demands may therefore offset some of its benefits or pose collateral pollution risks.[7] The final choice of technology may also affect the amount of water that will be used. For example, certain versions of CDR will require water to release CO_2 from spent amine screens.[8] While preliminary descriptions of these technologies emphasize that they will reuse water as much as possible, deployment of CDR in arid regions in amounts sufficient to affect global CO_2 concentrations might impose significant demands on local water supplies.

If we consider broad-scale reforestation as a form of CDR,[9] it too might pose environmental risks and costs. Some estimates of the amount of acreage requiring reforestation to significantly offset global ambient CO_2 concentrations would exceed 500 million hectares (i.e., over half the land area of Australia).[10] Aside from the potential disruption of converting large swaths of land from its current use to this type of sequestration, the selection of the type of trees used for the reforestation could have sizable impacts on local biodiversity and the potential vulnerability of forest ecosystems to invasive species or other biological threats that prey on the vulnerabilities of monoculture ecosystems. For example, some models project that broad biological energy carbon capture and storage (BECCS) deployment would lead to species losses equivalent, at least, to a 2.8°C temperature rise.[11] While current international agreements recognize some of these risks and attempt to balance the climatic benefits of reforestation with strategies to reduce biodiversity loss and distortions caused by large-scale monoculture methods,[12] their effectiveness remains to be seen.

Ocean Iron Fertilization (OIF). This technology, as described in Chapter 2, would rely on the dispersal of significant amounts of iron oxide onto the surface of anoxic[13] ocean waters to spur the growth of algal blooms.[14] These blooms, like any other plant growth, would capture ambient CO_2 and incorporate it into the plants' mass. When the algae deplete the supplemental iron and then ultimately die, they theoretically would sink into lower ocean depths and effectively sequester the CO_2 outside of the atmospheric carbon cycle.[15]

Because this technology directly alters the local ocean ecosystem and spurs atypical growth of large algal blooms, it has already triggered the most vociferous objections and strongest legal opposition.[16] Some of the feared environmental damages from OIF deployed on a global scale include the risk of significant ocean eutrophication and creation of local hypoxic areas, the stimulation of toxic algal blooms that could injure or kill local marine life, the skewing of the distribution of species that feed on plankton and which might be vulnerable to large fluctuations in their food supplies, and the potential disruption of plankton ecosystems and the mix of plankton species that support the base levels of oceanic food chains and ecosystem webs.[17]

Other varieties of climate engineering. While SRM, CDR, and OIF have received most of the initial focus, other new climate engineering technologies are emerging and may pose yet another set of risks for different types of damages. For example, some scientists have proposed a modified form of SRM that would increase the albedo of local marine clouds (marine cloud brightening). Other proposed techniques would thin the density of cirrus clouds in a way that would allow greater amounts of infrared radiation to escape from the Earth's surface into space (cirrus cloud thinning).[18] Both marine cloud brightening and cirrus cloud thinning remain relatively new proposed approaches, and as a result most assessments have not focused on how they might pose a risk of damage that differs from other SRM approaches. The environmental assessment of other possibly promising new approaches to climate engineering – including broad deployment of biochar, enhanced weathering of olivine or calcine minerals, or aggressive modification of surface albedo through ground coverings or reflective surfaces – remains relatively unexplored. To the extent that the following discussion identifies general principles of liability that apply to SRM, CDR, and OIF, those principles will also likely apply to environmental damages or side effects of all climate engineering technologies.

Climate engineering, if it matures into a viable strategy to respond to climate change effects, would likely evolve from initial local field tests and research into increasingly broader deployments and, ultimately, global implementation.[19] As it expands the scale of its possible impact, climate engineering will likely face different sources of liability as it becomes subject to three possible fields of law: international, transnational, and domestic law. Each of these fields will now be addressed in turn.

5.2. STATE LIABILITY UNDER INTERNATIONAL LAW FOR DAMAGES FROM CLIMATE ENGINEERING

The potential application of international law to climate engineering efforts has already received an in-depth examination in Chapter 3. As a result, this chapter instead assumes (for purposes of analysis) that some of those international legal norms will apply to particular types of climate engineering, and it will instead focus on the scale and types of liability that international law may impose on those projects.

5.2.1. *Framework for State Liability Claims*

Under basic precepts of public international law, *state liability* means a state can bear liability for its actions (whether otherwise lawful or unlawful) when they cause damage to other states. This foundational principle operates in parallel with the concept of *state responsibility*, which holds sovereigns accountable when they fail to satisfy their obligations under international law. The concepts of state liability and state responsibility can operate in tandem, but they differ in important respects. For example, a government can incur state responsibility for its violations of international law, but its actions may not trigger state liability. State responsibility differs fundamentally from state liability because liability can arise even when a state takes actions that do not violate international law. In these circumstances, a state may bear liability even if it undertakes all reasonable steps available to it to prevent the damages.[20] In the absence of an affirmative international governance framework for climate engineering research or deployment, state liability may play a pivotal role to redress damages that arise from climate engineering activities that do not create state responsibility. This legal structure also suggests that international law will most likely address climate engineering liability claims against a state only when another sovereign brings the claim.[21]

5.2.2. *State Liability under Treaties and International Environmental Agreements*

Under most current environmental treaties and international agreements, states incur liability for damages caused by their violations of those agreements under a standard of strict liability. While civil liability claims against private parties still typically require proof of negligence or fault,[22] international environmental legal precepts have evolved toward strict liability because of the difficulty that claimants can face in proving culpability, establishing the applicable standard of care, or mastering complex technical issues related to environmental releases and causation.[23] For example, international agreements now impose strict state liability for environmental damages resulting from the transboundary shipments of hazardous wastes, responses to nuclear incidents, and oil spills in oceanic waters. Strict liability claims, however, still must provide proof of causation, and as a result their application may prove problematic for state liability claims arising from climate engineering activities where the causal link between the global climate engineering activity and specific local harms may be hard to establish.[24]

Typically, state liability in these circumstances is not absolute. Most international agreements provide exceptions from state liability in circumstances of acts of God, acts of war or armed insurrection, or acts of a third party unrelated to the claimant.[25] While some agreements and treaties provide a limitations time period within which a claimant must present its environmental damage claim, it is unclear as to the extent to which these time limits apply when the claimant did not know, or could not have reasonably discovered, the damages or the causes of them.[26] And finally, state liability almost always provides solely a monetary remedy. If a state bears liability for its climate engineering activity, current environmental agreements and practices would typically only determine that nations should compensate the claimants for the damages and lost economic benefits caused by that activity.[27] The doctrine of state liability does not empower claimant states to halt another state's actions or require it to take preventive steps.[28] In addition, most international environmental agreements set liability caps, compensation schemes, and/or mandatory insurance requirements in conjunction with their strict liability frameworks.[29]

5.2.3. *Application of State Liability Frameworks to Climate Engineering*

As noted previously, the international community has not yet adopted any treaty or convention that expressly regulates climate engineering research, testing, or deployment, and there is not an immediate prospect that it will undertake to negotiate such a comprehensive agreement.[30] Nonetheless, several existing international regimes may apply to climate engineering initiatives based on the geographic region affected by the project as well as the nature of the climate engineering technique. For example, states that conduct climate engineering activities by releasing sulfate particles into the lower stratosphere may need to assess their obligations under the UN Framework Convention on Climate Change (UNFCCC) and its current implementation through the Paris Agreement,[31] the Montreal Protocol to the Vienna Convention for the Protection of the Ozone Layer,[32] the UN Convention on the Law of the Sea (UNCLOS),[33] or the Long-Range Transboundary Air Pollution Treaty.[34] As discussed previously, however, these treaties may not apply squarely to many suggested climate engineering techniques.[35] Even if they do apply, these treaties and international agreements may only establish state responsibility for violations; they do not, however, provide a clear mechanism to impose state liability.[36]

For example, the Paris Agreement,[37] reached at the 21st Conference of Parties of the UNFCCC in 2016, represents the crowning achievement of decades of international diplomatic negotiation, but it offers scant grounds to divine any principles of international legal liability for climate engineering efforts. The Agreement refused to set out a liability mechanism for climate change effects (much less climate engineering), and it provides no compulsory or binding framework for enforcement efforts against ratifying parties who fail to satisfy their obligations under the Paris Agreement or the UNFCCC.[38] While the Agreement does provide an ambitious platform for future revisions to strengthen its terms and for continued negotiations over possible financial responsibilities arising from losses and damages due to climate change effects,[39] the possibility that these prospective liability mechanisms might conceivably apply to future climate engineering efforts did not arise in the Parties' discussion of loss and damage. Nor did these issues play a prominent role at the 22nd Conference of the Parties that took place in Marrakech, Morocco in November 2016. The future extension and implementation of the Paris Agreement, however, may have important implications for the growth of customary international law that might apply to climate engineering.[40]

While state liability principles under general environmental international agreements seem to have limited application to the specific state actions underlying climate engineering activities, other international agreements that protect specific regions or environmental resources may offer stronger opportunities to hold states liable for climate engineering research or deployment. For example, the Convention on Biological Diversity (CBD) arguably could create a binding legal obligation for its Parties to refrain from climate engineering activities that might undermine biodiversity or threaten protected species. As noted in earlier chapters, prior Conferences of Parties (COPs) to the Convention have passed resolutions which request CBD's Parties and other governments to refrain from undertaking climate engineering activities that "may affect biodiversity" until there is an adequate scientific basis to justify those activities and an appropriate consideration of the risks of the specific climate engineering project has taken place. The CBP COP resolutions added an exemption for "small scale scientific research studies that would be conducted in a controlled setting," but the exemption only applies when the experiments are "justified by the need to gather specific scientific data and are subject to a thorough prior assessment of the potential impacts on the environment."[41] A prior CBD COP had also promulgated a resolution that specifically urged tight restrictions on OIF experiments and deployment.[42]

The CBD resolutions consequently highlight the limitations of using such multilateral environmental agreements that protect specific regions or resources as a basis to find state liability for climate engineering. For example, none of the CBD COP resolutions are binding on either Parties to the CBD or other nations (including the United States) that have not ratified the CBD.[43] Discussions of the CBD resolutions at the 13th COP to the CBD in Cancún, Mexico on December 4–17, 2016, led to a resolution that confirmed prior COP resolutions that climate engineering research should take place only for valid scientific research with careful assessment of potential environmental impacts.[44]

Other international environmental agreements that protect specific regions – e.g., the Antarctic,[45] outer space,[46] or the surface of the ocean[47] – set out strict (or absolute) liability frameworks that could hold a state liable for its climate engineering activities in those regions or which affect specific protected resources.[48] In general, these agreements impose strict or absolute liability on noncommercial actions by state actors, and those states in turn bear a responsibility to control the actions of individuals under their jurisdiction who might violate the requirements of the agreement that pertain to the protected region. For example, the liability framework established under the Outer Space Treaty[49] and implemented under the Space Liability Convention[50] imposes international liability on a nation "that launches or procures the launching of an object into outer space … and each State Party from whose territory or facility an object is launched" if that object causes "damage to another State Party to the Treaty or to its natural or juridical persons by such object or its component parts on the Earth, in airspace or in outer space."[51] This liability exists only between nations (private parties must be represented by states), and the Convention imposes an absolute liability for orbital and airborne injuries; this liability cannot be mitigated by defenses based on negligence, fault, or due care.[52]

If the Space Liability Convention applies to certain varieties of SRM, a nation that launched satellite-based reflectors to modify the Earth's solar radiation influx would bear absolute liability to any other state injured by actions in orbit or to aircraft caused by the satellites or their launches. Should a space-based reflector system cause such harm, the injured state would initiate the claim by presenting it to the launching nation(s) and demanding compensation for the damages caused. If the launching nation refuses this demand, the Space Liability Convention requires the convening of a three-person Claims Commission to issue a binding decision on the parties' liabilities.[53] As noted in

Chapter 3, this model offers a clear potential mechanism for assessing liability for climate engineering, but it only applies to the narrow and relatively unlikely technology of space-based reflectors.[54] The Space Liability Convention will likely have little or no applicability to other climate engineering technologies such as stratospheric aerosol deployment, cirrus cloud stripping, or marine cloud brightening (even if the dispersals or cloud management occur at high altitudes) because of their lower altitude for deployment.[55] And even in the sole historical instance where a nation presented a claim under the Space Liability Convention's liability mechanisms, the results were decidedly mixed.[56]

If these multilateral agreements do impose binding obligations on their Parties that would affect possible climate engineering actions,[57] they often expressly assure enforcement of judgments or rulings against a participating state that violates their terms. For example, most international agreements that establish state liability pursuant to dispute resolution provisions also include waivers of sovereign immunity defenses. They also often require member states to consent to the execution of judgments rendered by a designated court or arbitral panel without objecting to jurisdiction or invoking domestic policy commitments.[58] Despite these assurances of enforcement of state liability under international conventions, however, some agreements still preserve the ability of a sovereign state to refuse enforcement on public policy grounds or require the consent of the member state to execute the decision.[59]

5.2.4. *State Liability under Customary International Law for Climate Engineering Activities*

In addition to international environmental agreements, states can incur liability under dictates of customary international law. These legal obligations, as noted previously, arise from international legal norms created by the consistent pattern of conduct by nations undertaken through a sense of legal obligation.[60] Some customary international norms, for example, admonish nations from allowing actions within their jurisdiction that injure the sovereign interests or resources of another nation.[61] Other customary international principles have incorporated various versions of the precautionary principle to inform the obligations of nations to act even in the face of residual uncertainty about the degree of risk posed by the situation.[62]

While many commentators have argued that customary international law may impose responsibility on states that wish to conduct climate engineering research or deployment, relatively few scholars have discussed how a state might incur liability under customary international law for its actions. To date, no decision by an international tribunal, arbitral panel, or other international body has concluded that customary international legal norms have developed specifically for climate engineering. In addition, if a state wished to undertake climate engineering actions that arguably conflicted with a developing customary international legal norm, that nation could probably preserve its legal ability to undertake those actions without state liability or responsibility under the norm by consistently voicing its unwillingness to accede to the norm or to acknowledge that it acted under any sense of *opinio juris*.[63]

Even if customary international law did ultimately apply to particular climate engineering activities – for example, by conflicting with an international norm not to take actions that threaten another nation's fundamental right to existence, or a developing norm of *jus cogens* or *erga omnes* for actions that potentially affect the entire community of nations – it typically does not create a compulsory obligation to satisfy that liability. As with treaties that set out general obligations, customary international law would only establish the state's liability without any corresponding mandatory legal obligation to compensate the claimant for their damages.[64] If a state has acceded to the mandatory jurisdiction of the International Court of Justice (ICJ)[65] or another international decisional body, it must respect the court's determination of state liability. But that obligation does not extend to a mandatory obligation to satisfy any legal damages through payment or compensation absent that nation's consent.

5.2.5. *Liability for Violations of Internationally Protected Human Rights*

As noted previously, climate engineering projects may arguably impinge on human rights protected under international conventions or customary international law.[66] In part, this risk rests on the dramatic growth in the role of human rights law in climate change discussions and activities in recent years. For example, the United Nations has appointed a Special Rapporteur on Human Rights and the Environment, and the Special Rapporteur has pointed out the strong connection between climate change legal obligations and human rights.[67] To that end, the Paris Agreement's Preamble provides the first explicit acknowledgment in the UNFCCC agreements of the relevance of human rights to international climate efforts. The Accords specifically state that:

> [p]arties should, when taking action to address climate change, respect, promote and consider their respective obligations on human rights, the right to health, the rights of indigenous people, local communities, migrants,

> children, persons with disabilities and people in vulnerable situations and the right to development, as well as gender equality, empowerment of women and intergenerational equity.[68]

This language, however, appears solely in the Preamble, and the Paris Accords do not contain any binding obligation on the Parties to act in regard to human rights in a way that might create liability for them if they fail to do so.

The prospect of liability created by violations of human rights obligations immediately raises fundamental difficulties. First, it remains unclear whether international human rights norms have extraterritorial effect. In other words, current human rights law focuses on how a nation deals with its own citizens, but it is unclear whether those same norms would apply to a state's actions which affect individuals outside its territory or effective control. To the extent that climate engineering projects would almost by definition affect persons outside the nation that engages in them, the unsettled extraterritorial reach of human rights norms would pose an immediate barrier. At first blush, it appears that some international conventions contemplate that certain human rights would have extraterritorial effect, but others don't address the issue or seem to focus solely on internal application.

Second, the enforceability of human rights obligations generally has proven problematic. In general, the failure of a nation to meet its human rights obligations toward its citizens has proven notoriously difficult to pursue in international fora.[69] The creation of special-purpose human rights tribunals, arguably, may offer a more hospitable setting to redress claims that one nation's climate engineering efforts have injured the human rights of its own citizens or, more tellingly, individuals outside that state's territory or jurisdiction.[70] This approach, however, would require the consent of the defendant nation, which would almost certainly be withheld.

Last, and least likely, individuals might face personal liability for egregious violations of human rights caused by climate engineering actions. Suppose, for example, the leader of a state chose to pursue a climate engineering project that would benefit her nation, even though that action foreseeably could also disrupt water supplies and jeopardize the food supply and cultural practices of a large indigenous population in an adjoining nation. If that disruption occurs and causes widespread death, disruption, and suffering, it might rise to the level of an international crime. To do so, that leader's actions would have to also constitute a crime subject to universal jurisdiction to provide for individual liability. This type of liability has traditionally been reserved for war crimes, piracy, genocide, and other notorious actions set out in customary international law. This scenario, of course, appears exceedingly unlikely, and it is entirely unclear whether grossly negligent or willful climate engineering projects would qualify as crimes subject to personal and universal jurisdiction under existing international laws.[71]

5.3. CLIMATE ENGINEERING LIABILITY UNDER DOMESTIC LAW

While international and domestic environmental laws usually impose their own distinct forms of liability, the difference between the two is not always sharp. For example, the United States incorporates some international environmental obligations directly into its domestic federal laws, and it satisfies others through passage of domestic laws.[72] Self-executing treaties receive the same legal status and preemptive effect as federal statutes, and federal common law generally incorporates developing principles of customary international law.[73] A few other nations' judiciaries have recently shown a greater willingness to turn to international law principles and obligations to derive domestic legal obligations to compel their national governments to take more aggressive action to address climate change.[74] In addition, international law may constrain the enforceability of liability determinations reached by other nations' judicial systems.[75] While a full assessment of the potential application of individual domestic laws of numerous sovereign nations lies beyond the scope of our analysis, this section will highlight important common features and doctrines that might drive the application of existing national environmental or procedural laws to potential climate engineering experiments or deployment. This analysis will focus predominantly on the United States, but it will also include reviews of other nations' domestic legal systems where relevant.

To the extent that the treaty and customary international law obligations described above can apply to climate engineering efforts, domestic laws which implement those international legal obligations could also apply to climate engineering efforts. For example, the United States has promulgated several statutes to implement its international environmental treaty obligations, and those statutes might apply to climate engineering activities under US jurisdiction if the underlying treaties would apply to climate engineering on a larger scale. For example, the federal Migratory Bird Treaty Act implements the United States' obligations under the Migratory Bird Treaty, and the federal Ocean Dumping Act carries out the United States' duties under the London Convention. To the extent that each statute sets out civil and criminal liability for violations by private actors subject to national domestic jurisdiction, they can translate international legal obligations incurred by states into binding legal duties for their private citizens, who would face direct legal liability for any violations that they commit.

5.3.1. *Liability for Violations of Domestic Environmental Statutes*

5.3.1.1. Liability for Undertaking Actions that Violate Domestic Environmental Laws

While local and domestic liability for actions designed to alter the atmosphere on a planetary scale may seem mismatched, they arise from the likelihood that many actions to implement climate engineering research or deployment will necessarily take place within the sovereign jurisdiction of a nation[76] which may have applicable statutory programs. Alternatively, a nation may have domestic statutory obligations, rights, or remedies that could respond to local effects or damages even if they result from climate engineering actions which take place outside its borders or jurisdiction. This assessment is necessarily heavily dependent on the specific facts and circumstances of the particular climate engineering method deployed; the location, nature and scope of the alleged climate effects and their damages; and the source and extent of the statutory obligation triggered by the actions or their local effects.

As an illustrative example, the United States could apply several domestic statutes to some climate engineering research or deployment activities that would create liability for persons who violate them.[77] Importantly, most of these statutes – including the federal Clean Air Act, the Ocean Dumping Act, the Endangered Species Act, and the Migratory Bird Treaty Act – carry significant civil and criminal liability for violations. This statutory liability can cover a wide variety of climate engineering activities if US courts find that these federal statutes apply to them. For example, all of these federal statutes would impose liability for violations that occur within the territorial jurisdiction of the United States.[78] But they also can hold persons liable for actions they take outside the physical boundaries of the United States, but which cause environmental consequences within the United States.[79] In doing so, a US federal or state court could rely on the full extent of the long-arm jurisdictional powers granted to them under either US federal statutes or their state analogs. For example, the federal statutory long-arm jurisdiction code provision authorizes federal courts to exercise jurisdiction over any person to the full extent allowed under the US Constitution.[80] This statutory language has already allowed federal courts to hold parties accountable in judicial actions to assess liability for conduct outside the United States which caused environmental damages within US borders.[81] Finally, the United States has also asserted jurisdiction over US citizens who threaten to violate statutory obligations under these statutes solely because of their citizenship, even if they act outside national boundaries.[82]

To the extent that federal or state environmental laws impose liability on climate engineering activities, the degree of liability will depend on the statute's specifications. For example, most federal environmental statutes provide strict civil liability for failures to comply with permit terms or regulatory requirements. Several of those statutes also hold persons who spill or emit hazardous substances, dangerous materials, pollutants, or petroleum strictly liable for the costs of clean-up and restoration. To the extent that federal environmental statutes allow consideration of traditional negligence or fault in assessing civil liability, these concepts tend to arise when a federal statute provides for heightened levels of financial liability for degrees of extraordinary fault or negligence, the allocation of liability among responsible parties, or the factors to govern exercise of enforcement discretion.

Federal environmental laws also establish standards of criminal responsibility that depart from statutory liability in other subject areas. For example, many federal environmental laws allow criminal punishment for relatively low degrees of culpability or fault. In some instances, the federal Clean Water Act provides for criminal liability arising from simple (rather than gross or reckless) negligence,[83] and federal penalties for environmental crimes can reach up to double the value of the damages caused by the violation.[84] Other federal environmental statutes provide for strict criminal liability for the taking or killing of certain species protected under the Migratory Bird Treaty Act, the Bald and Golden Eagle Protection Act, and the Endangered Species Act.[85]

Most importantly, no nation or subnational jurisdiction has yet implemented domestic legislation that expressly governs or regulates climate engineering research or implementation.[86] This situation may change in the future, however, as local and state governments express more interest in climate engineering activities within their jurisdiction. For example, Rhode Island's legislature has considered bills during its past sessions that would restrict or ban climate engineering activities occurring within its borders.[87] Even if the US Congress and state legislatures decline to enact new legislation that might impose liability on unpermitted climate engineering activities, parties opposed to ungoverned climate engineering research might nonetheless seek to apply existing federal and state environmental laws in a novel and expansive fashion to climate engineering programs.[88]

5.3.1.2. Liability for Failure to Take Actions Required under Domestic Environmental Laws

While initial legal challenges will likely seek to halt or constrain proposed climate engineering research efforts, other legal actions may seek instead to compel a governmental party to take affirmative actions to undertake, regulate, or forbid climate engineering. For example, an environmental nongovernmental organization might bring a lawsuit to force the US Environmental Protection Agency (EPA) to issue regulations to tightly control any prospective climate engineering. Alternatively, a state or industry party might take actions to force the EPA to allow the use of climate engineering technologies as an alternative means to comply with other statutory or regulatory mandates under the federal Clean Air Act or other statutes.[89]

In general, this issue has not received as much attention as other possible legal challenges because federal and state governments typically enjoy great discretion over their choice of regulatory initiatives and regulations to pursue. The US Supreme Court, for example, has given federal agencies virtually untrammeled discretion over the enforcement actions that they elect to bring.[90] This discretion, however, has some limits that might prove relevant for future litigation regarding governmental inaction over climate engineering. First, Congress or a state legislature can statutorily constrain the agency's discretion by mandating enforcement or regulatory action within a specific timeframe. And second, if a person submits an administrative petition that requests the agency to promulgate regulations, the agency's failure to respond or to agree can supply an independent final agency action that a court can review as a potentially arbitrary or capricious action.[91]

It should be noted, however, that one court has taken a different approach. In the landmark *Urgenda* decision, a Dutch trial court ruled that the Netherlands' failure to effectively reduce GHG emissions from sources within its control constituted a failure to meet the nation's obligations under international law.[92] (This decision is now under appeal.) If US courts look to this case, it may presage a more flexible approach by US trial courts that receive petitions to compel the federal or state governments to take action to forbid, regulate, or undertake climate engineering efforts or laws.[93]

5.3.1.3. Liability for Tortious Conduct under Domestic Law

In addition to any statutory authorities that potentially apply to climate engineering projects, tort law likely offers another vehicle for attempts to impose liability on climate engineering efforts that result in damages to other persons or nations. In fact, in 2016 an advocacy group in Canada filed the first class action lawsuit to challenge alleged clandestine climate engineering activities through chemical contrails from jets (known as "chemtrails") and they have announced plans to file a parallel lawsuit in the United States.[94] While many national legal systems rely on tort actions and concepts to provide redress to injured parties, this section will focus on US federal and state tort laws and how they may apply to proposed climate engineering approaches.

While the legal challenge to alleged chemtrails may constitute the first tort challenge to alleged field deployment of a putative climate engineering technology, this lawsuit should be viewed in context of the larger chemtrail movement. Chemtrail adherents energetically contend that the United States and other nations have engaged in a vast clandestine program to test and deploy chemicals to control the climate through surreptitiously impregnating them into commercial and military jet fuels. To date, no independent credible investigation has uncovered any evidence of such a global or national conspiracy, or indeed of any such practice at all. Absent evidence of such secret climate engineering efforts, the pending lawsuit may face dismissal in the Canadian courts.

While US federal and state common law both provide for tort liability, the vast majority of tort actions arise under state law and proceed within state court systems.[95] If climate engineering research activities or projects result in harm to the other persons or their property, those damages could spark the filing of common law tort actions under the theories of private or public nuisance, trespass, negligence, failure to warn, and strict liability for ultrahazardous activities. Claimants alleging damages from general climate change effects have brought actions under each of these theories against public and private institutions seeking both injunctive relief and legal damages. Notably, to date each effort to recover climate change damages under a tort claim has failed,[96] and those actions will continue to face daunting hurdles to satisfy standing requirements, prove causation and redressability of injuries, and overcome objections based on political question doctrines. They would also have to navigate likely equitable defenses such as laches, contributory negligence, and allocation of responsibility among a vast array of potential co-defendants.

5.3.1.4. Public and Private Nuisance Actions

Under state common law private nuisance actions, property owners can seek compensation if another person's actions have unreasonably interfered with their use and enjoyment of their real property. These lawsuits tend to focus on whether the alleged substantial and unreasonable interference rises to a level that outweighs the social utility of the defendant's activities. In addition to private nuisance actions, a governmental plaintiff (or a private party suffering harms beyond those shared by the general public) can bring a public nuisance action against persons who interfere with the enjoyment of a right held in common by the general public. For example, states have used public nuisance actions to halt, or obtain damages for, a wide variety of activities that injure public interests such as blocking public roads or waterways, maintaining a public bawdy house, or – most relevant – polluting or damaging an environmental resource (such as a waterway) held in common by the public. Other public nuisance claims have challenged activities that caused odors or noise which impaired the public's ability to quiet enjoyment of public resources.[97]

Given the many uncertainties surrounding their effects and operational consequences, climate engineering research and projects may create unanticipated damages that could lead to public and private nuisance claims. For example, SRM techniques will likely rely on the dispersal of aerosolized sulfates that could affect rainfall distribution and amounts, precipitate back to the ground after spending several months in the stratosphere, or materially reduce solar insolation to ground-level facilities that rely on constant solar influx (e.g., solar power facilities and large-scale open-air agriculture). Alternatively, OIF projects that disperse large amounts of iron onto marine bodies could allegedly cause water pollution, ecosystem damage, algae blooms that lead to hypoxic conditions, and other chemical imbalances in ambient water conditions. Private claimants whose use of their real property suffers because of these side effects could bring a private nuisance action that seeks damages or injunctive relief because the climate engineering has substantially and unreasonably interfered with their property enjoyment. Alternatively, other claimants could bring a public nuisance tort action to allege that these unintended consequences interfered with the use and enjoyment of a right held in common by the public to a healthy and stable climate and environment. To do so, private plaintiffs contesting a public nuisance would also have to demonstrate that they had suffered a special injury beyond the general damage suffered by the public at large[98] – a requirement which may pose considerable difficulties for many members of the public who want to challenge climate engineering projects.

5.3.1.5. Trespass

Trespass occurs when a person intentionally and tangibly interferes with a property owner's exclusive possession and use of their land. As opposed to nuisance and negligence, which require some demonstration of harm, a trespass occurs with the invasion or interference itself without proof of any additional injury. Common law trespass actions would typically arise when one person made a tangible entry upon the other party's land. In modern cases, some state and federal courts have found viable trespass claims based on the physical intrusion of smoke, particulate matter, or vibration and seismic waves, especially if that trespass resulted in substantial physical damage to the property.

Some climate engineering technologies might create an opportunity for actions that could constitute a trespass. For example, some SRM methods might disseminate sulfate particulate matter that eventually precipitates onto the claimant's land. Some CDR approaches might also require machinery and supplies that create smoke, dust, vibrations, and particulate matter that crosses onto adjoining properties. Trespass might also occur if a climate engineering project resulted in the creation of clouds or enhanced precipitation that then physically passed onto the plaintiff's property.[99] But the ultimate application of the trespass doctrine to climate engineering seems limited, however, because the leading proposed technologies for SRM and CDR require either overflights at an altitude that typically would not create a trespass, intrusions at depths below thresholds that have historically constituted trespass, or do not require any impact or access to adjoining properties at all (except for potentially altering the composition of the ambient atmosphere over that land area).

5.3.1.6. Negligence

Persons who perform climate engineering may face liability if they act in a way that creates an unreasonable risk of harm to others.[100] In general, a claimant alleging negligence must demonstrate that the defendant's conduct breached a duty of care that the defendant owed to the plaintiff, and that the breach proximately caused harm to the plaintiff. Because negligence focuses on a broad duty of responsible care that can apply flexibly to numerous situations, claimants often allege negligence claims first when faced with novel or unfamiliar technologies or risks. To the extent that persons who conduct climate engineering projects owe a duty of care to others who might suffer injury if the project goes awry, negligence claims would provide an attractive legal vehicle for persons who wish to challenge climate engineering activities. This argument would require the plaintiffs to demonstrate that persons who conduct climate engineering owe a duty to them to conduct it in a reasonable and safe manner that meets a level of care reasonably expected from anyone who undertakes this action. If the climate engineering project arguably fails to comply with any statutory or regulatory standard, plaintiffs could bolster their claims by arguing that those violations render the climate engineering project negligent per se.

5.3.1.7. Strict Liability

A person can incur strict liability for damages caused by their conduct if they engage in an abnormally dangerous action. If so, plaintiffs could recover damages without needing to identify any fault by the defendant. While the common law does not set out a bright-line standard to identify abnormally dangerous conduct, courts have imposed strict liability if the activity poses a high degree of risk, that risk cannot be eliminated through reasonable care, the activity is relatively uncommon, the activity takes place in an inappropriate location, and the risks posed by the activity outweigh its value to the community. While some states have imposed liability on persons who conduct weather modification efforts that cause damage, the case law remains divided and relatively old.[101] Climate engineering activities, by contrast, pose a relatively novel technology that raises a risk of unpredictable harm to other persons. In this context, a court might be more likely to subject climate engineering to strict liability to assure that injured parties can successfully obtain compensation.

5.3.1.8. Possible Obstacles to Climate Engineering Liability Actions

While several tort theories could arguably apply to climate engineering activities that allegedly injure other persons, those tort claims will face a daunting list of jurisdictional and procedural obstacles in federal and state courts in the United States. For example, federal common law tort lawsuits seeking damages or injunctive relief against persons who have allegedly contributed to climate change have faced dismissal on multiple fronts, including political questions, sovereign immunity, laches, statutes of limitation, standing, and justiciability.[102] For tort actions filed under federal common law, the US Supreme Court has ruled that all such climate change claims were displaced by Congress' grant of power to the EPA to regulate greenhouse gas emissions.[103] Beyond these jurisdictional threshold challenges, common law tort actions seeking to impose liability on climate engineering project operators would also have to prove causation (i.e., that the climate engineering project directly and proximately caused their injury). Attempts to prove similar causation claims for climate nuisance tort actions have faced great judicial skepticism, but climate engineering projects notably would overtly seek to produce detectable changes to the climate that could be attributed to the project. In addition, at least one state has created an affirmative defense against nuisance and trespass actions premised on greenhouse gas emissions from permitted sources, and other states have considered similar legislation.[104] The language of those laws may also limit the availability of tort actions against climate engineering projects.

Beyond the displacement defense to federal common law tort actions, claimants bringing tort or damage actions under the domestic laws of their respective nations will likely face numerous jurisdictional and procedural hurdles. For example, the plaintiffs will have to navigate varying legal standards of responsibility and liability among different nations, inconsistent transnational standards for enforcement of foreign judgments, and harmonizing potentially inconsistent liability verdicts by multiple national courts for the same action. This patchwork puzzle also may raise the risk of paralyzing any significant climate engineering research or deployment simply because of the open-ended risk of possible liability from overlapping and inconsistent jurisdictions of differing national courts and laws.

The key jurisdictional challenges will likely begin with the ability of claimants to find a court system that can exercise personal jurisdiction over the defendants.[105] As discussed previously, lawsuits against sovereigns will need to find a forum where the responsible state parties have given consent to the International Court of Justice, an international arbitral tribunal, or other international decisional body. Litigation against private defendants similarly will need to identify judicial fora where the allegedly responsible parties have sufficient contacts (or have consented) to allow that forum's judiciary to exercise jurisdiction over the defendants. US climate change tort actions under federal common law squarely faced this constraint because they typically targeted a large number of defendants,[106] but domestic litigation challenging climate engineering activities will likely need to target only a small number of defendants who participated in the climate engineering project themselves.

In addition to demonstrating personal jurisdiction over the defendants, climate engineering litigants will also face steep challenges to the courts' subject matter jurisdiction over projects where no domestic laws (as yet) explicitly impose duties on climate engineering activities. They will also have to demonstrate that their damages were caused by the climate engineering research or deployment under either a strict liability regime under public international law or, more problematically, a proximate causation standard under most domestic tort or statutory environmental laws. These difficulties, however, may be offset by the facts that climate engineering projects will include a dramatically smaller pool of defendants, will focus on activities that expressly intend to produce a discernible effect on climate systems which can be attributed to the engineering project, and can point to data and statements by the defendants about the results of their work as an admission against their own interests.[107]

Similar challenges await climate engineering litigants against private defendants in the United States federal court system. As an initial step, claimants would have to prove that they can sue the climate engineering operator in the first place. US federal courts require, for example, that plaintiffs demonstrate that they have standing to pursue their specific claims in the federal court. Absent a showing that they have suffered a concrete, particularized injury-in-fact that can be fairly traced to conduct of the defendants which the courts can redress, climate engineering plaintiffs would not satisfy the irreducible requirements mandated by the US Constitution to bring a claim before the federal courts (although redressability obviously is less problematic if the plaintiffs simply seek monetary damages or specific injunctive relief to halt ongoing activity).[108] Even if they prove that they have standing to bring their claim, climate change litigants in US federal courts would still have to navigate objections to their claims based on prudential justiciability doctrines such as the political question doctrine,[109] ripeness (if the alleged damages from the climate engineering action have not yet fully manifested themselves), and deference to decisions that affect the conduct of foreign affairs by the US president or other executive officials.[110]

For a nuisance or negligence claim, the plaintiffs would also have to show that the climate engineering project was conducted in an unreasonable way – a daunting proposition if domestic environmental laws do not require an operator to obtain a permit or authorization for certain types of climate engineering.[111] To the extent that a climate engineering project requires an environmental permit, that permit (once issued) might also act as a shield to potential tort or statutory liability if the operator complies with its terms. Finally, after they have surmounted all of these barriers, climate engineering litigants in the US federal court would have to answer challenges that Congress has already displaced any possible federal common law tort claims arising from climate change effects.[112]

In addition to identifying which individual parties can bring actions to impose liability for damages from climate engineering projects, courts will also have to designate which defendants can be sued. For example, plaintiffs may seek damages for a governmental agency's failure to anticipate foreseeable damages caused by a climate engineering project that the agency undertakes or approves. This type of action would parallel arguments that federal governmental agencies potentially bear responsibility if they build public facilities that fail to account for likely climate change effects such as either increased rainfall and flooding or higher peak surface temperatures. To do so, the plaintiffs would also have to prove that the governmental defendants had not waived their sovereign immunity against such a lawsuit. Despite these challenges, at least one case in the United States has already successfully held at least one federal agency liable for failing to plan for increased flooding, structural damage, and water disruption from climate change.[113]

Last, even if a plaintiff successfully navigates a claim past all of these shoals, the final burden of causation will likely loom. Admittedly, a lawsuit seeking to impose liability on a climate engineering project likely will not face the notorious difficulties of proof of causation that challenge tort lawsuits against historical and current emitters of greenhouse gases. Climate engineering plaintiffs, however, will still need to demonstrate that the events that damaged them arose from a climate engineering action rather than the typical variability of weather (even severe weather) and natural climate systems. While climate engineering activities will often inherently seek to produce exactly that type of discernible effect or "fingerprint" which can support causation, a plaintiff may face difficult evidentiary challenges in linking an unexpected and undesired side effect to the underlying climate engineering action.

Presumably a successful climate engineering lawsuit would request the typical methods to satisfy a liability judgment. For example, plaintiffs could likely seek monetary damages to compensate them for their injuries, injunctions against continuation of those activities,[114] revocation or remand of any governmental permits or approvals for the climate engineering action, and restitution for their diminished assets or nonmonetary losses. They could also seek a declaratory judgment that would lay out clear principles of liability to guide future climate engineering activities that involve the same parties.[115]

While a court system can draw on the panoply of traditional remedies to actions that violate rights or obligations to third parties, courts would still face novel and difficult issues when they apply those familiar tools to climate engineering. They would first have to parse damages from climate engineering from damages caused by underlying climate change itself. To the extent that overall climate change continues as a backdrop to climate engineering efforts, it may drown out the signal of causation normally attributable to climate engineering actions. In addition, the climate engineering action itself may also generate benefits that arguably should offset the damages due to any litigant that the engineering may have aided.[116]

5.4. LIABILITY UNDER TRANSNATIONAL LAW

Liability for climate engineering efforts with cross-border effects will likely raise issues of enforcement of domestic laws within transnational legal regimes. One emerging area of potential liability for climate engineering activities may result from actions brought within one nation's domestic courts under its own laws to obtain a liability judgment against a person or state conducting climate engineering activities that allegedly caused them. Once that domestic court has rendered its judgment, the claimants could then seek to enforce that judgment in the nation where the defendants reside or operate. As a result, climate engineering projects that allegedly cause damage in another nation might become the target of domestic litigation and transnational collection efforts. The broad geographic scope of certain climate engineering efforts may naturally allow claimants to initiate litigation in nations that offer particularly favorable domestic laws and hospitable judicial systems.

This dynamic could lead climate engineering into the relatively obscure body of laws that govern enforcement of foreign judgments and arbitral decrees. Because courts faced with a request to enforce another sovereign's judicial decrees will usually only assess whether those judgments meet basic requirements of due process, fairness, and public policy,[117] climate engineering challenges may include a vigorous set of transnational challenges and collection efforts. While no party has brought a judicial challenge to climate engineering work in a domestic court that has matured into a potential cross-border attempt to enforce that judgment, other environmental tort and damage actions have resulted in titanic battles before multiple national court systems and arbitral panels to enforce and collect multi-billion-dollar judgments rendered by an aggrieved nation's domestic courts under its own laws.

The paradigmatic example of this dynamic is the ongoing *Chevron–Ecuador* dispute over damages allegedly caused by Chevron's corporate predecessor while extracting petroleum from Ecuador's Amazonian regions. The initial lawsuit was brought by indigenous tribes claiming that historical petroleum and chemical contamination had damaged the environment surrounding their homes and had injured their health. The case was filed in the US federal court system and rejected on grounds of *forum non conveniens* and subsequently dismissed without prejudice. The dismissal proved to be a pyrrhic victory for Chevron. Importantly, the federal district court that dismissed the action required Chevron to forego any challenges to jurisdiction that it might otherwise have raised before the Ecuadorian courts.[118] Upon its return to the Ecuadorian court system, the lawsuit proceeded under Ecuadorian domestic laws that favored the claimants and yielded an initial verdict exceeding $19 billion. The Ecuadorian Supreme Court ultimately upheld the verdict against all appeals within the domestic court system (although the Court also reduced the damage award to $9.5 billion).

At that point, the battle shifted to the collection of the Ecuadorian judgment, and the barriers to enforcement of foreign judgments began to loom large. The US federal district court for the Southern District of New York found that the plaintiffs had obtained the judgment through fraudulent actions that violated the Racketeer Influenced and Corrupt Organizations (RICO) Act, and the court enjoined any actions to collect the verdict within the United States.[119] By contrast, the Canadian courts have shown a greater willingness to entertain actions to enforce the verdict, and collection efforts remain ongoing in Australia as well.[120] As long as the original Ecuadorian judgment remains viable, Chevron faces a potential hydra of collection efforts in any nation where it maintains a presence and significant assets, and those collection efforts will each proceed on that nation's domestic law governing the collection of foreign judgments.

The Chevron imbroglio offers a preview of potential environmental litigation risks for climate engineering projects under transnational law. A key aspect of transnational liability actions lies in the ratcheting effect of favorable domestic laws. If a particular nation's domestic laws favor (or can be made to favor) the award of a liability judgment based on a climate engineering project's effects within that nation, plaintiffs can seek the most favorable forum. Once they obtain the judgment, they can then seek to enforce it in the nation with the most favorable laws allowing enforcement of foreign judgments. This strategy assumes, of course, that the nation hosting the collection action would also contain significant assets owned by the targeted defendant. The combined effects of forum selection for both the liability judgment and the foreign enforcement action could place climate engineering defendants at a significant disadvantage. The countervailing doctrines of *forum non conveniens* and limits on personal jurisdiction may have to play a larger role in future litigation.

5.5. POSSIBLE MODELS FOR COMPENSATION AND LIABILITY STRATEGIES

The lack of a clear mechanism to determine and assess liability may deter climate engineering experimentation or deployment by private parties or smaller governmental entities. Of course, that outcome may be desirable in light of the unknown risks and side effects of large-scale climate engineering actions (especially SRM), but it reflects a default acceptance of an unplanned outcome from our existing liability frameworks. To the extent that climate engineering research or deployment becomes desirable in the future, governments at that time may wish to explore and specify possible avenues to impose liability for climate engineering activities that cause harm.

If a larger community of nations wishes to promote climate engineering, it can take several steps that would allow a coherent, transparent, and predictable set of liability rules that would govern these projects. In particular, nations could extend and modify existing international legal approaches from other types of multinational environmental liability that could arguably best fit the potential risks and benefits of climate engineering projects. This section discusses several existing liability schemes that, while not directly applicable to climate engineering, have bearing on the sort of system that might be developed.

5.5.1. *Absolute State Liability: The Space Liability Convention*

In the arena of public international environmental law, the default mechanism of liability is strict liability. In other words, neither negligence nor intent to do harm must be shown. This regime has allowed the development of protective environmental standards without the hobble of difficult burdens of proof or demonstrations of standards for culpable conduct. To the extent that nations wish to engage in climate engineering activities, strict liability would force them to internalize the costs that such work would impose on other nations or parties and act with according care.

The Space Liability Convention offers one of the clearest and most comprehensive examples of a strict liability approach to the imposition of state liability and responsibility for activities within a specific sphere. As discussed previously, the Space Liability Convention imposes a regime of absolute and strict liability on member states whose activities in space cause damage to another member state's territory or interests. This treaty, however, will likely not apply to climate engineering (in particular, solar radiation management) unless the activities take place above the Earth's atmosphere. Other than space-based reflector arrays, all other climate engineering techniques would take place at locations outside the scope of the Space Liability Convention.[121]

The Space Liability Convention, unlike other international liability agreements in the environmental arena, focuses on state-only activities rather than commercial ventures by private parties. As a result, this liability strategy seems best suited for climate engineering approaches that will likely require action by governmental parties. Global-scale SRM, for example, or broadly implemented OIF will likely require some action by governmental actors on a sizable scale rather than targeted actions by individual parties.

5.5.2. *Civil Liability Systems for Private Party Pollution Under Polluter Pays Principles*

When faced with environmental damages caused by private parties' activities across borders rather than state actions, the vast majority of international conventions have typically not adopted an absolute state liability standard. Multilateral agreements targeting private commercial activities have instead chosen a strict civil liability system that incorporates "polluter pays" precepts. Under this approach, member states agree to establish a civil liability framework that holds private parties accountable for environmental damages that they have caused beyond national borders.[122] Under such a framework, each of the member states agrees to impose liability on private parties through legal actions in their domestic courts, and they then agree to mutually recognize and enforce each other's judicial decisions.[123] These agreements usually provide for joint and several liability if multiple defendants cause a single indivisible harm.[124]

Most international agreements providing liability for environmental injuries have focused on private actions within civil liability frameworks – in fact, only the Space Liability Convention has adopted a solely state liability approach.[125] For example, the multilateral agreement addressing pollution from shipborne bunker fuels has specified that the shipowners should bear sole liability for those damages,[126] and the international agreements dealing with civilian nuclear accidents identify owners of nuclear facilities as responsible for transboundary damages caused by nuclear incidents.[127]

Last, civil liability frameworks that impose private liability under polluter pays principles almost always provide for exemptions from responsibility in certain circumstances. These exceptions can include events occurring during (1) warfare, civil insurrection, or rebellion, (2) natural occurrences of "exceptional, inevitable and irresistible character" that rise to the level of Act of God or *force majeure*, (3) acts by independent third parties, (4) damages caused in part by the party claiming injury, and (5) damages that fall outside a period of limitations established by the agreement.[128] Notably, these exclusions do not exempt a party from liability if it takes actions to prevent harm from a pre-existing environmental condition or to limit significant damages to a third party without the damaged party's consent.[129]

Most of the environmental multilateral regimes that use civil liability precepts also provide liability limits in certain circumstances. In exchange for establishing strict liability for environmental damages without the typical burden of proving specific legal causation or overcoming equitable defenses, these agreements often will provide a ceiling for the total exposure to liability that private parties or state actors might face.[130] For example, the Convention on Civil Liability for Oil Pollution Damage resulting from Exploration for and Exploitation of Seabed Mineral Resources limits the liability of operators unless the operator caused the pollution damage through an act or omission "done deliberately with actual knowledge that pollution damage will result."[131] While liability limitations provide certainty on the maximum exposure that parties might face and thereby allow valuable commercial activities to take place, agreements establishing liability limits are often criticized for setting liability ceilings that are too low.[132]

5.5.3. *Mandatory Compensation Funds and Financial Assurance Mechanisms*

To assure adequate resources to compensate victims, some multilateral agreements require that operators either demonstrate that they have sufficient financial resources to satisfy demands, obtain insurance to fulfill claims, or pay into a common shared fund that could cover damages claims within an industrial sector or type of activity. Rather than rely solely on liability regimes that require proof of claims and compensation after damage has occurred, these conventions proactively require the creation of pools of resources to cover damages before any events occur. By providing a ready source of funds and a clear procedure to seek compensation, these agreements attempt to provide certainty on liability issues that might otherwise discourage valuable commercial or social activities. This approach is also often paired with liability limits that match the pooled resource or financial assurance requirement unless certain conditions are met (such as proof of intentional infliction of damage).

This approach, for example, has played a prominent role in liability regimes for the nuclear power industry. Under the 1963 Supplementary Convention to the Convention on Third Party Liability in the Field of Nuclear Energy (also known as the Paris Convention of 1960), operators face three layers of potential compensation for nuclear accidents in Europe. While the first tier establishes liability for operators within a specified limited liability cap, the second tier provides liability coverage by states where the accident occurred. The Paris Convention's third tier establishes a compensation fund supported by contributions by all of the parties to the Convention.[133] Other treaties on damages from nuclear incidents provide a similar integrated framework of absolute civil liability for the owner or operator, limitations on total liability, supplemental state resources to cover excess damages, and a compensation fund to assure adequate funds to cover liability obligations.[134]

In a similar vein, some international conventions require compulsory insurance for activities that might cause environmental damages. Typically these schemes require anyone who engages in covered activities to obtain authorization from a signatory state, and as a condition of receiving that authorization the applicant must demonstrate that it has adequate resources to satisfy any liability claims. The Convention on Civil Liability of Damage caused during Carriage of Dangerous Goods by Road, Rail and Inland Navigation Vessels (CRTD), for example, provides that every carrier must obtain insurance sufficient to cover the maximum extent of liability, and it also requires each State Party to designate a competent authority to issue certificates to confirm the carrier's insurance.[135] Conventions governing liability for civil aviation accidents provide similar combinations of compulsory insurance and financial security requirements on aviation carriers.[136]

One of the most prominent uses of compulsory insurance and liability compensation funds is the 1971 Fund Convention's establishment of International Oil Pollution Compensation Funds (IOPC Funds). Under this system, private and public entities that import petroleum pay compulsory levies for IOPC Funds that provide additional compensation of last resort for damages caused by parties who have exhausted their liability limits, have become insolvent, or have failed to maintain adequate insurance, or when no one can identify a liable party.[137] These agreements have extended and elaborated on IOPC Funds to make them a complex and sophisticated liability regime with significant industry integration and a flexible institutional structure.[138]

The use of mandatory compensation funds and insurance obligations extends beyond the environmental arena, and it has proven effective in other fields. For example, the United States has established a compensation fund that pays damages to persons who suffer injury from vaccines. The fund is supported by taxes on vaccine sales, and claimants seeking compensation from the fund do not need to prove fault on the part of vaccine researchers, manufacturers, or distributors.[139]

5.6. FUTURE DIRECTIONS

Given its potentially sweeping role in responding to climate change disruption, the prospect of climate engineering has understandably generated a buzz of commentary on and analysis of its governance. Many initial comments have centered on the challenge of regulating global SRM, which raises the largest concerns about potentially harmful side effects, equity concerns, and allocation of risks onto objecting states and parties. As a result, much of this initial analysis has sought to define a clear and forceful set of rules that would govern most types of climate engineering in a flexible, fair, and effective fashion. Given the complex interactions that SRM liability would raise between domestic and international law as well as the challenge of prospectively regulating a complicated emerging technology, the range of suggested approaches has spanned a complete ban on all climate engineering[140] to establishment of a *de minimis* liability threshold that would protect researchers and small-scale field deployments of SRM from liability.[141]

Despite their nascent state, the liability principles described in this chapter offer some initial directions for further consideration. First, liability for climate engineering will depend heavily on the type of technology at issue. The risks and benefits posed by SRM are so vastly different from the concerns evoked by direct air capture CDR or marine cloud brightening that it is probably misleading and unproductive to define a single liability framework that treats all climate engineering technologies in the same way.[142]

That said, some initial possible distinctions may point to differing liability approaches for each technology. For example, the deployment of SRM on a global scale to regulate planetary surface temperatures would likely mandate the participation and resources of national governments or international organizations. As a result, a liability framework crafted for state actors conducting global SRM would make the most sense. The absolute liability standard and dispute resolution approaches set out by the Space Liability Convention could provide a useful starting point for analysis. While initial attempts to spur discussions about regulation of ocean iron fertilization under the London Convention and Protocol or the Convention of Biological Diversity have had limited success, they highlight a possible approach that could seek the development of either a new Protocol under the UNFCCC or a new multilateral convention. This approach, however, will likely require an enormous amount of diplomatic effort, time, and financial resources to reach a resolution acceptable to the world community.

If the world community seeks to craft a liability standard to govern SRM, it could draw on other existing principles from current international agreements. First, international law now relies almost exclusively on strict liability as a tool to impose liability under either state actor or civil liability approaches. This burden of proof would allow claimants to focus on factual causation and damage quantification issues without wrestling with nettlesome questions of fault and equity. Second, states should be required to assure that persons or organizations within its jurisdiction comply with fundamental norms that would prevent untrammeled deployment without careful consideration and research into potential side effects. This approach would likely rely on existing international norms incorporating the precautionary principle as well as an escalating scale of deployment that allows investigation into SRM's effects and consequences on an interim basis before full deployment.[143] And last, given the potentially vast scale of unintentional damages caused by side effects from SRM,[144] an SRM governance framework should also incorporate steps to assure that adequate funds will be in place to compensate aggrieved parties for demonstrable damages. This need could potentially be met with mandatory taxation on activities that generate GHG emissions or a broad carbon tax.

By contrast, the hazards posed by direct air capture or CDR are vastly different from SRM's risks. CDR will likely require local land use decisions, minimal impact on global surface temperatures from any single CDR or direct air capture project, and great challenges in discerning any causation between particular CDR efforts and unexpected climate side effects or disruptions. CDR could have local land use or environmental impacts, but those are well suited for domestic resolution. As a result, reliance on domestic laws and liability principles could effectively regulate the environmental consequences of CDR until it reaches a stage where collective CDR actions potentially disrupt climate systems or large-scale land use.[145]

In truth, liability for climate engineering will likely need to reflect the broad array of potential climate engineering technologies, their disparate scales of deployment, and their varying degrees of development and readiness. The field remains new, and it is highly likely that variations of SRM, CDR, accelerated weathering, biochar deployment, BECCS, cirrus cloud stripping, and other new approaches will emerge from ongoing research initiatives. For example, some investigators have proposed that lower-altitude tropospheric dispersion of sulfate particles can moderate solar radiation on a regional scale instead of having a global effect. As a result, this targeted variety of SRM could address immediate climate disruption concerns on a local scale such as severe heat waves or disrupted precipitation cycles. If this approach proves viable, it would allow smaller nations, state governments, local governmental authorities, or even private parties to undertake short-term regional climate engineering. Any governance system that dictates liability solely on the assumption that SRM must take place on a global scale and be performed by sovereign states would likely fail to reach or properly regulate the benefits and risks of this emerging technology.

As a result, the most fruitful approach to crafting liability standards for climate engineering will likely be a careful delineation of multiple standards that would apply to the different risks and challenges of the growing suite of climate engineering technologies. By tailoring liability rules to better fit the risk and promise of different technologies, emerging approaches that pose lesser risk or greater benefit could proceed without cumbersome governance requirements or the chilling liability requirements that larger and more risky approaches would demand. Although automobiles, motorcycles, trucks, ships, and aircraft all provide the basic service of transportation, they each operate under very different liability regimes that reflect their global and local impacts, risks and benefits, and societal demands. As climate engineering evolves, we may see a similar proliferation of liability and governance approaches to fit the growing variety of climate technologies.

NOTES

1 Many nations now recognize a constitutional right to a clean environment either through express constitutional terms or through implied rights provided by a broader right to life or liberty. *See* E. Daly and J. May, GLOBAL ENVIRONMENTAL CONSTITUTIONALISM at 13, 55–84 (Cambridge University Press 2016). Citizens of other nations may attempt to use those constitutional provisions as a basis to force

or stop climate engineering projects. That analysis, however, lies outside the scope of this chapter.

2 E. Kintisch, Chapter 2 this volume, 2.2: Carbon Dioxide Removal.

3 The current state of CDR technologies would remove CO_2 at a relatively steady and comparatively low rate that would not exceed local atmospheric mixing conditions for CO_2 in ambient conditions. As a result, the prospect of accelerated CO_2 removal resulting in a local "pocket" of ambient air with disruptively low greenhouse gas levels appears exceedingly unlikely in the near future. *See* Intergovernmental Panel on Climate Change, *Changes in Atmospheric Constituents and in Radiative Forcing*, in *CLIMATE CHANGE 2007: THE PHYSICAL SCIENCE BASIS* at 138–39 (Susan Solomon et al. eds., Cambridge University Press, 2007) (discussing CO_2 mixing and radiative forcing rates); R. Lazarus, *Super Wicked Problems and Climate Change: Restraining the Present to Liberate the Future*, 94 CORNELL L.R. 1153, 1163–64 (2009) (global warming effect of CO_2 emissions not dependent on location of emission because of fast mixing rates).

4 *See, e.g.*, T. Hester, *Direct Air Capture of Greenhouse Gases*, in *LEGAL PATHWAYS FOR DEEP DECARBONIZATION IN THE UNITED STATES* (Michael B. Gerrard & John Dernbach, eds. (Environmental Law Institute, forthcoming 2018) (describing various direct air capture technologies and strategies).

5 According to the Carbon Dioxide Information Analysis Center at the Oak Ridge National Laboratory, the removal of 1 ppm by volume CO_2 of from the ambient atmosphere would generate 2.13 gigatons (Gt) of carbon, or 7.8 Gt of CO_2. *See* F. O'Hara, Jr. (ed.), Carbon Dioxide Information Analysis Center, Oak Ridge National Laboratory, *Carbon Dioxide and Climate*, ORNL/CDIAC-39 (3d ed. 1990). Removing sufficient CO_2 to reduce ambient levels from 400 ppm to 350 ppm would therefore generate approximately 390 Gt of CO_2 that would require either sequestration or reuse. By comparison, all anthropogenic GHG emissions in 2010 totaled 49 Gt of CO_2e (±4.5 Gt). *See* Intergovernmental Panel on Climate Change, *Summary for Policymakers*, in *CLIMATE CHANGE 2014: MITIGATION OF CLIMATE CHANGE, CONTRIBUTION OF WORKING GROUP III TO THE FIFTH ASSESSMENT REPORT OF THE INTERGOVERNMENTAL PANEL ON CLIMATE CHANGE* at p. 6 (Cambridge University Press 2014). For further comparison, the total amount of all solid waste generated in the United States in 2014 was only 249 million tons – a vanishingly small fraction of the weight of such potentially captured CO_2. *See* US Environmental Protection Agency, *Advancing Sustainable Materials Management: Facts and Figures*, www.epa.gov/smm/advancing-sustainable-materials-management-facts-and-figures.

6 *See, e.g.*, P. Smith, R. Hazeldine, and S. Smith, *Preliminary Assessment of the Potential for, and Limitations to, Terrestrial Negative Emission Technologies in the UK*, 18 ENVIRON. SCI.: PROCESSES IMPACTS 1400, 1403 (2016) (alongside main technical limitations of negative emissions technologies are "high cost and energy requirements for DAC"). The energy demands will vary substantially according to the technology involved. *See* K. House, *Economic and Energetic Analysis of Capturing CO2 from Ambient Air*, 108 PNAS 20428, 20428–29 (Dec. 11, 2011), at www.pnas.org/cgi/doi/10.1073/pnas.1012253108; L. Joos, *Cutting the Cost of Carbon Capture: A Case for Carbon Capture and Utilization*, 192 FARADAY DISCUSS. 391, 391–93 (Dec. 20, 2011). The high cost of post-combustion carbon capture has long

been noted. *See, e.g.*, L. Joos, J. Huck, V. Speybroeck, and B. Smit, *Cutting the Cost of Carbon Capture: A Case for Carbon Capture and Utilization*, 192 FARADY DICUSS., 391 (2016).

7 American Physical Society, Panel on Public Affairs, *Direct Air Capture of CO_2 with Chemicals: A Technology Assessment for the APS Panel on Public Affairs* at iv, 72–73 (June 1, 2011) (concluding that high-carbon energy sources are not viable options to power DAC systems because their emissions may exceed the amount of CO_2 captured by the DAC; further, using low-carbon energy sources to power DAC would divert the more logical use of those energy sources to directly displace high-emitting centralized carbon sources).

8 *See* Kintisch, Chapter 2 this volume.

9 While the editors of this text have excluded afforestation and reforestation from its definition of climate engineering, the development of biological energy carbon capture and storage (BECCS) as a negative emissions technology would likely spur a large increase in the amount of land devoted to forests and crops as a fuel source to the biogenic energy facility. *See* K. Tokimatsu, R. Yasuoka, and M. Nishio, *Global Zero Emissions Scenarios: The Role of Biomass Energy with Carbon Capture and Storage by Forested Land Use,* 185 APPLIED ENERGY, 1899 (2017) ("[o]ne of the key requirements of BECCS is land for biomass production, with 200 Exa Joule (EJ) per year (EJ/yr) of bioenergy estimated to require approximately 500 Mha of land, corresponding to one-third of global crop land"); M. Ranjan and H. Herzog, *Feasibility of Air Capture*, 4 *Energy Procedia* 2869, 2873 (2011) ("[t]he land area required to capture and store 1 [gigaton] of CO_2 through this route is 203,125 square miles, which is a third more than the land area of California"), at doi:10.1016/j.egypro.2011.02.193; L. R. Boysen, W. Lucht, D. Gerten, and V. Heck, *Impacts Devalue the Potential of Large-scale Terrestrial CO_2 Removal through Biomass Plantations*, 11 ENVIRON. RES. LTRS 09510 (2016) (changes in land carbon uptake, albedo modification, and nitrogen fertilizer use would reduce the efficacy of BECCS and require disruptively large amounts of surface territory to achieve targets of 1.5, or even 2.0, degrees in the absence of aggressive emissions mitigation).

10 Kintisch, Chapter 2 this volume, 2.2.2 (citing 2015 National Research Council report).

11 P. Williamson, *Scrutinize CO_2 Removal Methods*, 530 NATURE 153 (2016); K. Anderson, *The Trouble with Negative Emissions*, 354 SCIENCE 182, 183 (Oct. 14, 2016). *See also* T. Powell and T. Lenton, *Scenarios for Future Biodiversity Loss Due to Multiple Drivers Reveal Conflict between Mitigation, Climate Change and Preserving Biodiversity*, 8 ENVIRON. RES. LETT. 8 (2013), doi:10.1088/1748-9326/8/2/025024 (withdrawal of energy via BECCS has a "large negative effect on biodiversity" on agricultural systems under four scenario models to 2050).

12 *See, e.g.*, T. Gardner et al., A *Framework for Integrating Biodiversity Concerns into National REDD+ Programs*, 154 BIOLOGICAL CONSERVATION 61, 62 (2012).

13 *Anoxic* waters lack dissolved oxygen and, as a result, cannot support species that respire oxygen from the water through surface filtration, diffusion, or extraction (e.g., fish with gills). Waters can become anoxic through the process of *eutrophication,* where the decomposition of eukaryotic or prokaryotic organisms in the water consumes available oxygen. By contrast, waters that still contain some dissolved oxygen (even if at a low level that cannot support life) are *hypoxic*. The

infamous zone of oxygen-depleted waters that recurs annually at the mouth of the Mississippi River, for example, is hypoxic.

14 *See* Kintisch Chapter 2 this volume, 2.2.3; Burger and Gundlach, Chapter 6 this volume, 6.1.1.1 and 6.1.1.5.

15 In other words, the carbon contained in the algal bloom sinks below the thermocline and into deep ocean depths where circulation and refresh rates are measured in centuries. Burger and Gundlach, Chapter 6 this volume, at 6.1.1.1. *But see* Kintisch Chapter 2 this volume, 2.2.4 (noting controversy over the actual duration of subsea sequestration of carbon).

16 *See* discussion by Reynolds, Chapter 3 this volume, 3.3.2 (review of legal responses to ocean iron fertilization proposals under the London Convention, London Protocol, and the Convention on Biological Diversity) and by Burger and Gundlach, Chapter 6 this volume, 6.3.2 (legal issues raised by research projects on the open seas to examine iron fertilization effects).

17 R. Abate and A. Greenlee, *Sowing Seeds Uncertain: Ocean Iron Fertilization, Climate Change, and the International Environmental Law Framework*, 27 PACE ENVT'L L. REV. 555, 566–572 (2010); C. Trick, B. Bill, W. Cochlan, M. Wells, V. Trainer, and L. Pickell, *Iron Enrichment Stimulates Toxic Diatom Production in High-Nitrate, Low-Chlorophyll Areas*, 107 PNAS 5887, 5891 (2010).

18 *See* J. Latham et al., *Marine Cloud Brightening: Regional Applications*, 372 PHIL. TRANS. R. SOC. A 20140053 (2014); T. Storelvmo, W. R. Boos, and N. Herger, *Cirrus Cloud Seeding: a Climate Engineering Mechanism with Reduced Side Effects?*, 372 *Phil. Trans. R. Soc.* A 20140116 (2014); A. Cirisan et al., *Microphysical and Radiative Changes in Cirrus Clouds by Geoengineering the Stratosphere*, 118 AMERICAN GEOPHYSICAL UNION 4533 (2013), doi:10.100/jgrd.50388.

19 There is some possibility of regional deployment, which could pose an entirely new set of legal questions and dictates. *See* T. Hester, *A Matter of Scale: Regional Climate Engineering and the Shortfalls of Multinational Governance*, 7 CLIMATE CHANGE L.R. 168 (2013). Because the prospect of meso-scale climate engineering remains relatively undeveloped, this chapter will not directly assess liabilities related to these approaches, but many of the same general principles of legal liability identified for global climate engineering may also apply to regional initiatives.

20 *See* J. Horton, A. Parker, and D. Keith, *Liability for Solar Geoengineering: Historical Precedents, Contemporary Innovations, and Governance Possibilities*, 22 N.Y.U. ENVT'L L.J. 225, 234 (2015). *See also* J. Brunée, *Of Sense and Sensibility: Reflections on International Liability Regimes as Tools for Environmental Protection*, 53 INT'L AND COMP. L.Q. 351, 351–52 (2004).

21 Any liability for individuals, corporations, or other international actors that engage in climate engineering research or initiatives which violate international law would most likely ultimately turn on the domestic laws that implement those international obligations or, in limited cases, possibly directly under human rights conventions and laws. While some nations allow domestic enforcement of self-implementing international agreements without the need for implementing domestic legislation, that enforceability ultimately still rests on precepts of that nation's domestic constitutional or statutory laws.

22 *See* discussion by Reynolds, Chapter 3 this volume, 3.6.3.

23 Horton et al., *supra* note 20, at 229; *Trail Smelter Case (U.S. v. Can.)*, 3 R.I.A.A. 1905 (1941); *Corfu Channel Case (U.K. v. Alb.)*, 1949 I.C.J. 4 (Apr. 9).

24 *See* discussion by Reynolds, Chapter 3 this volume, 3.6.3.

25 Horton et al., *supra* note 20, at 234 n.32. *See also* F. Paddeu, A *Genealogy of Force Majeure in International Law*, 82 THE BRITISH YEARBOOK OF INT'L L. 381, 385–98 (2012).

26 Several treaties provide periods of limitation that begin when claimants know "or reasonably should have known" of the facts that underlay their claims of injury. *See* UN Secretariat, *Survey on Liability Regimes Relevant to the Topic International Liability for Injurious Consequences Arising out of Acts Not Prohibited by International Law*, 60 Y.B. INT'L L. COMM'N 111–14 [1995], UN Doc. A/CN.4/471 [hereinafter *Survey on International Liability Regimes*] (describing tolling exemptions to limitation periods in multiple international treaties and agreements based on knowledge qualifiers, but noting that many of them nonetheless also impose additional absolute periods of prescription that the claimant's lack of knowledge cannot extend).

27 Notably, state liability findings do not carry any compulsory enforcement authority. A state found liable for its actions will only pay those damages if it chooses to do so. As a result, many state liability payments reflect a political judgment resting on dictates of international relations or comity. *See* Horton et al., *supra* note 20, at 234.

28 *See Survey on International Liability Regimes*, *supra* note 26, at 104 (satisfaction of international liability claims are typically accomplished through monetary compensation, although compensation may occasionally take the form of removing the danger or effecting *restitutio in integrum*). As noted above, the International Court of Justice and other international tribunals can declare that a nation should take appropriate action to come into compliance with international law. *See, e.g.*, International Court of Justice, *Case Concerning Pulp Mills on the River Uruguay (Argentina v. Uruguay)* (decision April 29, 2010) (directing Uruguay to undertake supplemental environmental impact assessment); Award, *In the Matter of the South China Sea Arbitration before an Arbitral Panel Constituted Under Annex XII under the 1992 United Nation Convention of the Law of the Sea between the Republic of the Philippines and the People's Republic of China* (July 12, 2016). In such circumstances the nation's refusal to comply with the directive would only create state responsibility, however, because the ICJ and other international tribunals lack the power to compel compliance akin to injunctive relief or a compulsory order.

29 *See Survey on International Liability Regimes*, *supra* note 26, at 114–17 ("Insurance and other anticipatory financial schemes to guarantee compensation"). Notably, the Vienna Convention on Treaties addresses the responsibility of states that breach treaties which do not contain liability provisions, but that Convention does not provide for state liability unless the treaty between the parties imposes it. The Draft Determinations on State Liability by the International Law Commission likewise do not impute liability to a state for simply violating a treaty unless that treaty provides for liability of a party upon breach.

30 Reynolds, Chapter 3 this volume, 3.1.

31 United Nations Framework Convention on Climate Change (2015), Adoption of the Paris Agreement, 21st Conference of the Parties, Paris: United Nations, entry into force Nov. 4, 2016 pursuant to art. 21(1).

32 Montreal Protocol on Substances that Deplete the Ozone Layer (1987).
33 United Nations Convention on the Law of the Sea (UNCLOS) (1982).
34 Convention on Long-Range Transboundary Air Pollution (1979).
35 Reynolds, Chapter 3 this volume, 3.3.1, 3.3.2 (review of UNFCCC and UNCLOS applicability to potential climate engineering techniques).
36 For example, it is unclear whether the Long Range Transboundary Air Pollution Treaty would apply to types of climate engineering such as solar radiation management, and only a limited portion of the community of nations has entered into the Treaty (*e.g.*, the United States and European nations). The Montreal Protocol to the Vienna Convention on Ozone-Depleting Substances may apply to SRM methods that potentially damage stratospheric ozone, but that Protocol has limited applicability outside this particular climate engineering method and has relatively restricted enforcement and damage mechanisms compared to the scale and scope of possible damages from global climate engineering activities.
37 The Paris Agreement took effect on November 4, 2016 (30 days after which at least 55 parties to the Convention, who accounted in total for at least an estimated 55 percent of total global greenhouse gas emissions, had deposited their instruments of ratification, acceptance, approval, or accession to the Paris Agreement).
38 *See* D. Bodansky, *The Legal Character of the Paris Agreement*, 25 *Rev. of European, Int'l and Comp. L.* 142, 143–47 (July 2016); A. N. Craik and W. C. G. Burns, *Climate Engineering under the Paris Agreement: A Legal and Policy Primer* at p. 13 (Centre for International Governance Innovation, 2016).
39 *See* Paris Agreement at Art. 8 (Loss and Damages provisions); Art. 4(9) (submittals of independent nationally determined contributions by signatories every five years); Art. 14 (periodic global stocktaking meetings every five years).
40 Reynolds, Chapter 3 this volume, 3.3.1.1, 3.7. The Trump Administration's announced intention to withdraw the United States from the Paris Agreement of course makes the future growth and success of the Paris Agreement much more contentious and unpredictable.
41 Report of the Tenth Meeting of the Conference of Parties to the Convention on Biological Diversity, Decision X/33.8(w) (2010).
42 *See* Decision IX/16, Ninth Meeting of the Conference of Parties to the Convention on Biological Diversity (May 2008).
43 *See* discussion by Reynolds, Chapter 3 this volume, 3.3.4.
44 *See* Conference of the Parties to the Convention on Biological Diversity, *Decision Adopted by the Conference of the Parties to the Convention on Biological Diversity, XIII/14, Climate-related Geoengineering,* CBD/COP/DEC/XIII/14 (Dec. 8, 2016) (advance unedited version). The decision simply reaffirms that decisions X/33 and XI/20 called for a precautionary approach to climate-related geoengineering, and it urges members to note the socio-economic, cultural, and ethical issues, and regulatory options created by geoengineering research (including the importance of taking into account the knowledge, experience and perspectives of indigenous peoples and local communities). *See id.* at ¶¶ 5, 6.
45 The Antarctic Treaty (Dec. 1, 1959). This treaty may prove particularly relevant for climate engineering proposals that could target particular geographic regions, including approaches which focus their effects on polar regions (for example, local marine cloud brightening, dispersal of engineered particles into the stratosphere that preferentially migrate to polar regions, or local ice albedo modification).

46 Treaty on Principles Governing the Activities of States in the Exploration and Use of Outer Space Including the Moon and Other Celestial Bodies, Jan. 27, 1967, 18 U.S.T. 2410, 610 U.N.T.S. 205 (hereinafter "Outer Space Treaty").

47 In addition to UNCLOS, the London Convention (and its subsequent London Protocol) address activities that may cause environmental damage through releases of pollutants or wastes into the high seas. While these agreements have little relevance for SRM, they have already played a key role in attempts to regulate or ban ocean iron fertilization as a climate engineering tool. *See* discussion by Reynolds, Chapter 3 this volume, 3.3.2.2.

48 Reynolds, Chapter 3 this volume, 3.3.4.2, 3.4.2, 3.6.3.

49 Outer Space Treaty, *supra* note 46.

50 Convention on International Liability for Damage Caused by Space Objects, Mar. 29, 1972, 24 U.S.T. 2389, 961 U.N.T.S. 187 ("Space Liability Convention").

51 Outer Space Treaty, *supra* note 46.

52 Notably, the Space Liability Convention imposes absolute liability only for damages caused by an object's impact to the Earth or aircraft. For accidents above the atmosphere or in orbit, the Space Liability Convention uses a fault standard to allocate liability between the launching states. *See id.* at art. V.

53 *Id.* at arts. XIV–XV.

54 *See* discussion by Reynolds, Chapter 3 this volume, 3.4.2.

55 While the definitions under international treaties and domestic laws remain murky, the most generally accepted demarcation of when outer space begins is 100 kilometers, or 62 miles. This altitude, referred to as the Kármán Line, is roughly where an aeronautic vehicle must travel at orbital velocity to generate enough aeronautic lift to support itself in the thin upper atmosphere. This altitude lies far above any proposed solar radiation management approaches that rely on dispersal of reflective materials in the stratosphere. *See* G. Oduntan, *The Never-Ending Dispute: Legal Theories on the Spatial Demarcation Boundary Plane Between Airspace and Outer Space*, 1(2) HERTFORDSHIRE *L.J.* 64, 72–74 (2003) (summarizing this functional definition of the boundary of space, while noting several shortcomings as well).

56 When Canada invoked the Space Liability Convention to present a claim to the Union of Soviet Socialist Republics (USSR) in 1979 to recover clean-up costs from the landfall of the Cosmos 954 satellite, the USSR initially denied responsibility for the vast bulk of Canada's costs. After bilateral negotiations, the USSR ultimately agreed to pay only $3 million towards Canada's initial $14 million claim. *See* Horton et al., *supra* note 20, at 249–50.

57 Notably, the use of climate engineering as an offensive weapon in a military setting poses a special case. The Convention on the Prohibition of Military or Any Other Hostile Use of Environmental Modification Techniques (ENMOD) would apply to hostile deployments of climate engineering, and it sets out a clear framework to hold liable states who engage in such conduct. *See* discussion by Reynolds, Chapter 3 this volume, 3.3.4.3.

58 *See Survey on International Liability Regimes*, supra note 26, at 117–19 ("[m]ultilateral agreements generally contain provisions ... [which] provide that, once a final judgement [*sic*] on compensation has been rendered, it shall be enforced in the territories of the contracting parties and that parties may not invoke jurisdictional immunity").

59 *Id.* at 117–19 (The Space Liability Convention requires the consent of member states to any liability judgment; article 23 of the Convention on Civil Liability for Damage resulting from Activities Dangerous to the Environment provides that signatories will accept judgments of domestic courts of other signatory parties unless "such recognition is contrary to the public policy in the Party in which recognition is sought").

60 *See* discussion by Reynolds, Chapter 3 this volume, 3.6.

61 *See id.* at 3.6.2.

62 *See id* at 3.7.3.

63 *See* C. Bradley and M. Gulati, *Withdrawing from International Custom*, 120 YALE L.J. 202 (2010); C. L. Lim and O. Elias, *Withdrawing from Custom and the Paradox of Consensualism in International Law*, 21 DUKE J. COMP. & INT'L L. 143, 143–44 (Fall 2010).

64 *See* discussion by Reynolds, Chapter 3 this volume, 3.6.3.

65 *See id.* at 3.4.5 (reviewing bases for jurisdiction of the ICJ, submittal by states which have acceded to the ICJ's mandatory jurisdiction, and ability of the UN General Assembly to request advisory opinions).

66 *See id.* at 3.4.3.

67 *See, e.g.*, UN General Assembly, Human Rights Council, REPORT OF THE SPECIAL RAPPORTEUR ON THE ISSUE OF HUMAN RIGHTS OBLIGATIONS RELATING TO THE ENJOYMENT OF A SAFE, CLEAN, HEALTHY AND SUSTAINABLE ENVIRONMENT, A/HRC/31/52 at pp. 20–21 (Feb. 1, 2016).

68 Preamble to United Nations Framework Convention on Climate Change, Conference of the Parties, Twenty-First Session, Paris, 30 November–11 December, 2015, Adoption of the Paris Agreement, Draft Decision, – CP.21, 12 December 2015, FCCC/CP/2015/L.9/Rev.1.

69 *See* J. Crook, *The International Court of Justice and Human Rights*, 1 N.W.U. J. INT'L HMN. RTS. 2, 4–5 (2004).

70 For example, the InterAmerican Commission on Human Rights received a petition from an Inuit tribe which alleged that greenhouse gas emissions by private parties and the United States had caused climate change that injured their human rights. *See* S. Watt-Cloutier, *Petition to the Inter-American Commission on Human Rights Seeking Relief from Violations Resulting from Global Warming Caused by Acts and Omissions of the United States* (Dec. 7, 2005). While the Commission ultimately dismissed the petition without prejudice, it proceeded with hearings to investigate allegations raised in the petition. *See* M. Chapman, *Climate Change and the Regional Human Rights Systems*, 10 SUSTAINABLE DEVELOPMENT L. & POL. 32, 33 (Winter 2010). It should be noted, however, that this Commission existed prior to the claim pursuant to a broad international treaty, and consequently the Inter-American Commission on Human Rights did not operate as a special purpose human rights tribunal convened to resolve a specific claim.

71 In addition to the unlikelihood of this scenario, it also raises fundamental legal questions. For example, most universal crimes under international law require a clear *mens rea* to intentionally engage in the conduct and cause the prohibited harm. The crime of piracy, for example, typically requires that the actor intend to engage in unauthorized predation of ships on the high seas. It is unclear whether collateral harm caused by climate engineering, even if caused through

gross negligence or willful neglect, would rise to the level of culpability that would trigger criminal liability unless the actor actively intended to harm the victim.

72 For example, the Lacey Act and the Endangered Species Act each directly incorporate by reference obligations that the United States incurred by entering into the Convention Prohibiting the International Trade of Endangered Species (CITES). *See* 16 U.S.C. § 3371(f)(A) (the Lacey Act defines "plant" to include flora "listed on an appendix to the Convention on international Trade in Endangered Species of Wild Fauna and Flora"); 16 U.S.C. § 1538(c)(1) (the Endangered Species Act prohibits trade in any species "contrary to the provisions of the Convention [on International Trade in Endangered Species]"). By contrast, other federal statutes such as Title VI of the federal Clean Air Act and the federal Ocean Dumping Act impose independent federal statutory obligations designed to satisfy international obligations without expressly incorporating their terms. *See, e.g.*, 42 U.S.C. § 7671(a)(1),(2) (Title VI of the federal Clean Air Act requires the EPA Administrator to list ozone-depleting substances subject to restriction under federal listing of Class 1 and Class 2 substances, which must be consistent with – but does not incorporate by reference – the terms of the Montreal Protocol).

73 *See* L. Henkin, *Foreign Affairs and the US Constitution* at pp. 232–39 (2nd edn Oxford University Press 1997).

74 *See, e.g.*, *Urgenda Foundation v. State of the Netherlands* (2015) (finding that international law required the Netherlands government to undertake actions to reduce greenhouse gas emissions by 25 percent by 2020), https://elaw.org/nl.urgenda.15; *Leghari v. Federation of Pakistan*, Case No. 25501/2015 (2015) (ordering the Pakistani federal government to undertake immediate comprehensive planning efforts to reduce greenhouse gas emissions); *Our Children's Trust v. United States*, – F. Supp.3d – (D. Ore. 2016) (Case No. 6:15-cv-01517-TC) (declining to dismiss a petition to compel the United States to take action to protect the interests of minors held within the federal public trust doctrine); *Bin re Vienna-Schwechat Airport Expansion*, W109 2000179-1/291E (Feb. 2, 2017) (reversing administrative approval of the proposed construction of an additional runway based on human rights concerns invoked by unchecked climate change).

75 *See* discussion *infra* at 5.4 regarding transnational liability and enforcement of foreign judgments seeking to impose climate liability under domestic laws.

76 The "sovereign jurisdiction" of a nation extends beyond its simple geographic borders. For example, a state may exert its jurisdiction over the full extent of its territories (including its territorial waters and its exclusive economic zone under UNCLOS or customary international law), its jurisdictional airspace, its citizens and residents for actions conducted abroad, and its agents and representatives who follow the nation's directions or act under its control. *See* RESTATEMENT (THIRD) OF THE FOREIGN RELATIONS LAW OF THE UNITED STATES §§ 401–403. It should be noted, however, that some climate engineering research technologies may be tested outside domestic borders. For example, ocean-based iron fertilization test projects will likely occur on the high seas outside any nation's territorial waters or exclusive economic zone, and marine cloud brightening projects may also be conducted outside the territory of any specific nation.

77 *See* discussion by Lin, Chapter 4 this volume, 4.1, 4.2 (domestic US statutes that might apply to domestic climate engineering actions).

78 *See* Tracy Hester, *Remaking the World to Save It: Applying U.S. Environmental Law to Climate Engineering Projects*, 38 ECOLOGY L. Q. 851, 861–89 (2011).
79 *See id.* at 875.
80 *See, e.g.*, N.Y. Civ. Prac. L. & Rules § 302 (Personal jurisdiction by acts of non-domiciliaries). Federal courts will limit the scope of this exercise of extraterritorial jurisdiction over a foreign defendant to assure that the defendant had the minimum contacts required to satisfy requirements of due process under the US constitution. *See World-Wide Volkswagen Corp. v. Woodson*, 444 U.S. 286 (1980). The US Supreme Court has noted special concerns over the effect that broad extraterritorial jurisdiction claims could have on international comity and diplomatic relations. *See Kiobel v. Royal Dutch Shell Petroleum Co.*, 569 U.S. 12, slip. op. 5–6 (2013). *Cf. Pakoostas v. Teck Cominco Metals, Ltd.*, 452 F.3d 1066 (9th Cir. 2006) (holding that CERCLA liability does not reach beyond the US border into Canada, but finding that a Canadian factory created a CERCLA "facility" within the United States because its discharges into a river flowed directly into a US water body and created a "release" within the United States). *See* W. Smith, *Endangered Species Act or Extradition? Protecting Foreign Species in the Aftermath of the Cecil the Lion Controversy*, 18 VERMONT J. ENVT'L L. 55 (2016) (discussing the limited extraterritorial reach of the Endangered Species Act).
81 Fed.R.Civ.P. 4(k)(2). *See* J. West, *The Ixtoc I Spill Litigation: Jurisdictional Disputes at the Threshold of Transnational Pollution Responsibility*, 16 TEX. INT'L L. J. 475 (1981).
82 Hester, *supra* note 78, at 862.
83 *See, e.g.*, *U.S. v. Pruett*, 681 F.3d 232 (2012); *U.S. v. Hanousek*, 176 F.3d 1116 (9th Cir. 1999).
84 *See* 18 U.S.C. § 3571(b)(2)(d).
85 *See* D. Uhlmann, *Environmental Crime Comes of Age: The Evolution of Criminal Enforcement in the Environmental Regulatory State*, 4 UTAH L. REV. 1223, 1238 (2009).
86 Our research has identified only one state legislative effort to regulate climate engineering research of deployment. The Rhode Island legislature has considered, and rejected, a bill to control climate engineering projects that take place within the state. *An Act Relating to Health and Safety – Geoengineering*, H.R. 7578, Rhode Island Legislature (introduced 2016), http://webserver.rilin.state.ri.us/BillText16/HouseText16/H7578.pdf.
87 *See id.*
88 For example, the current National Weather Modification Reporting Act may apply to climate engineering activities even though Congress originally intended it to regulate rain-making and other local weather modification activities. *See* discussion by Lin, Chapter 4 this volume, 4.2; Hester, *supra* note 78, at 873.
89 An applicant seeking a Non-Attainment New Source Review permit might need to control its greenhouse gas emissions as part of the control technology selected by a state as the Best Achievable Control Technology (BACT). *See Utility Air Regulatory Group v. EPA*, 134 S. Ct. 2427 (2014) (Part II). If the selected BACT technology imposed tight restrictions on the GHG emissions allowed at the facility, the permit applicant might challenge the state's BACT choice if it ruled out the option to offset GHG reductions through carbon removal technologies (e.g., direct air capture devices or ocean iron fertilization) or solar radiation

management that achieved comparable reductions in surface temperatures. These alternative technologies might appeal to permit applicants who might otherwise face burdensome and extremely expensive on-site carbon capture and sequestration technological requirements. Given the Trump Administration's recent moves to withdraw the Clean Power Plan, however, the prospects for EPA or state BACT determinations that require greenhouse gas emissions controls seem far less likely.

90 *See Heckler v. Chaney*, 470 U.S. 821 (1985). This grant of discretion assumes that Congress has not mandated enforcement for the particular violations at issue.

91 The paradigmatic example of this approach, of course, is *Massachusetts v. EPA*, which found that the Environmental Protection Agency had failed to adequately respond to a petition for rulemaking that would set emission standards for mobile sources under Title II of the federal Clean Air Act. *See Massachusetts v. EPA*, 549 U.S. 497 (2007).

92 *See* discussion, *supra*, at note 74. This decision is currently under appeal.

93 For example, the federal district court of the district of Oregon recently looked to the *Urgenda* decision in concluding that minors could rely on the public trust doctrine to argue that the state had failed to adequately control greenhouse gas emissions under its jurisdiction. *See Juliana v. United States*, Order and Findings & Recommendation, Civ. No. 6:15-cv-01517-TC at p. 11 (April 8, 2016).

94 *See Pelletier v. Her Majesty, the Queen* (filed March 10, 2016), T-431-16 (class action filed in Federal Court of Canada), www.geoengineeringwatch.org/documents/D%20Pelletier%20-%20Claim.pdf.

95 It should be noted, however, that many tort law actions arising under state law proceed nonetheless in federal courts under either diversity jurisdiction or federal statutes that authorize state law claims (e.g., the Federal Tort Claim Act).

96 T. Hester, *A New Front Blowing In State Law and the Future of Climate Change Public Nuisance Litigation*, 31 STAN. ENVTL. L. J. 101, 115–19 (2012).

97 RESTATEMENT (SECOND) OF TORTS § 821B; W. Page Keeton *et al.*, *PROSSER & KEETON ON THE LAW OF TORTS* § 86, at 616 (5th ed. 1984)- (discussion of the historical context of public nuisance claims).

98 D. Antolini, *Modernizing Public Nuisance: Solving the Paradox of the Special Injury Rule*, 28 ECOLOGY L.Q. 755 (2001).

99 Historically, cloud seeding or other attempts to enhance rainfall have not triggered state law trespass complaints because, in part, many states have statutes that bar trespass, negligence or other tort damage claims for properly permitted rain enhancement projects. *See* Hester, *supra* note 78, at 860.

100 Of course, a national government or agency that conducts a climate engineering research project or deployment would typically enjoy sovereign immunity against actions brought in its own courts or under its domestic laws (subject to any statutory waivers or other self-imposed restrictions). To the extent that an individual has received a permit from a national government to conduct a similar project, however, such authorizations do not typically provide a shield against tort liability claims unless provided by the underlying statute or by preemption or displacement of the underlying state law. *See, e.g., FPL Farming v. Envt'l Processing Systems, Inc.*, 351 S.W.3d 306, 310 (Tex. 2011) (a "permit is not a get out of tort free card").

101 Hester, *supra* note 78, at footnotes 27–32.
102 *See, e.g.*, J. May, *Climate Change, Constitutional Consignment, and the Political Question Doctrine*, 85 DENV. U. L. REV. 919 (2008); A. Thorpe, *Tort-Based Climate Change Litigation and the Political Question Doctrine*, 24 *J.* LAND USE & ENVTL. L. 79, 79 (Fall 2008).
103 Notably, while the Court found that the Clean Air Act displaced federal common law nuisance actions for greenhouse gas emissions, it explicitly declined to rule on whether the Clean Air Act also pre-empted state tort actions for climate change. *See* Hester, *supra* note 78, at footnote 32.
104 For example, Texas has barred nuisance and trespass actions based on the release of greenhouse gases if the facility operator is in substantial compliance with its environmental permits. *See* TEXAS WATER CODE § 7.257.
105 The issue of jurisdiction pertains primarily to private individuals. The ability of a domestic court to order a foreign sovereign to take action is highly questionable, and that court's powers would require that government's waiver of its sovereign immunity, either through its express consent or through actions within the forum nation that constitute waiver. For example, a government that engages in commercial activities within the United States that do not constitute state actions may subject itself to the jurisdiction of US federal courts under the Federal Sovereign Immunities Act. This liability, of course, does not constitute a grant of power to the federal court to issue injunctions against the sovereign, but US federal courts can issue contempt orders against sovereign defendants before them, as well as order the attachment of a sovereign's assets under the court's jurisdiction that can satisfy a judgment.
106 For example, both the *Kivalina* and the *Comer* lawsuits targeted numerous companies in the oil, gas, chemical, mining, and refining sectors because of their aggregate emissions of greenhouse gases.
107 Note, however, that climate engineering actions expressly seek to produce discernable effects in the environment, and consequently the statements of researchers or parties may unintentionally provide statements against interest that could support a lawsuit. *See* Hester, *supra* note 78, at 897.
108 *See id.* at 895. The US Supreme Court has indicated, however, that Congress may have a broader degree of power to define the concrete injury underlying constitutional standing than previously assumed. *See Spokeo, Inc. v. Robbins*, 578 S. Ct. — (2016).
109 *See* Hester, *supra* note 78, at 892.
110 *See id.*
111 For example, the testing of direct air capture of carbon dioxide and other greenhouse gases will likely not require a permit under US federal environmental statutes, although the operator might need to obtain permits for any impacts arising from the land use or corollary aspects of the direct air removal. For instance, a sizable direct air capture test or deployment might require an environmental impact statement, an authorization for any anticipated impacts on nearby protected species, or approvals to manage the captured CO_2 stream.
112 *See American Electric Power v. Connecticut*, 564 U.S. 410 (2011).
113 *See St. Bernard Parish v. United States*, 121 Fed. Cl. Ct. 687, 746 (2015) (holding the US Army Corps of Engineers liable for temporary takings caused by its failure

to prevent flooding after Hurricane Katrina). State cases have sought to hold state and local governmental agencies accountable for similar failures to anticipate foreseeable environmental effects of actions that they undertake or approve, such as causing increased flooding in downstream neighborhoods by failing to require detention ponds and other preventive steps. *See, e.g., Harris County Flood Control Dist. v. Kerr*, 499 S.W.3d 793 (2016) (dismissing homeowners' claims that the City failed to halt untrammeled upstream development that allegedly caused downstream flooding).

114 Such plaintiffs would have to show that they satisfy all of the prerequisites for the court to issue equitable relief of this nature. *See Weinberger v. Romero-Barcelo*, 456 U.S. 305 (1982).

115 For example, some federal environmental statutes expressly authorized declaratory judgments to guide allocation of future cleanup costs (after the parties have incurred at least some initial costs). 42 U.S.C. § 9613(g)(2); *US v. Occidental Chemical Corp.*, 200 F.3d 143, 153–54 (3d Cir. 1999).

116 To successfully argue that a climate engineering project has generated benefits which offset a damage claim, the defendants would likely have to show that the benefit accrued in the same nation or area that suffered the damage. For example, if India seeks redress for damages from a climate engineering project that generated benefits in Canada, the benefits to Canada would likely not provide a valid basis to reduce or reject the claim for damages suffered in India. From a larger perspective, the imposition of liability for damages to an individual for an action that benefits a larger portion of society as a whole raises fundamental issues about the allocation of tort liability and individual autonomy. In the United States, this debate frequently surfaces in judicial decisions to provide injunctive relief that protects an individual against trespass by pollutants or the designation of a nuisance for an activity that may benefit other persons. The paradigmatic decision discussing this principle in the United States is *U.S. v. Boomer Cement*, 26 N.Y.2d 219, 257 N.E.2d 870 (1970).

117 Domestic judges have limited competence and authority to substantively review judgments by foreign courts. Typically they have declined to enforce foreign judgments if the judgments resulted from fraud, duress, or corruption. *See Hilton v. Guyot*, 159 U.S. 113 (1895) (seminal case holding that US courts could decline to enforce foreign judgments because of fraud, prejudice, or lack of comity, including a failure by the other nation's courts to reciprocate on enforcement of judgments). *See also* V. Zeynalova, *The Law on Recognition and Enforcement of Foreign Judgments: Is It Broken and How Do We Fix It?*, 31 Berkeley J. Int'l L. 150 (2013).

118 H. Erichson, *The Chevron–Ecuador Dispute, Forum Non Conveniens, and the Problem of Ex Ante Inadequacy*, 1 Stan. J. Complex Litig. 417 (2013).

119 *See Chevron Corp. v. Donziger*, Civ. No. 14–0826 (2d Cir. 2016) (upholding a district court finding of fraud and an injunction against enforcement of an Ecuadorian judgment).

120 C. Perkel, *Canada in Court Over $9.5 Billion Owed to Ecuadorian Villagers* (Sept. 15, 2016), http://globalnews.ca/news/2943001/chevron-canada-in-court-over-9-5-billion-owed-to-ecuadorian-villages/.

121 *See* discussion *supra* at p. 232–233 (scope and application of the Space Liability Convention to SRM and other climate engineering technologies).
122 *See* Horton et al., *supra* note 20, at 232–33.
123 *See id.* at 233.
124 *See id.*
125 *See id.*
126 *See* International Convention on Civil Liability for Bunker Oil Pollution Damage art. 3, Mar. 23, 2001, 973 U.N.T.S. 3.
127 *See* Convention on Third Party Liability in the Field of Nuclear Energy, July 29, 1960, 956 U.N.T.S. 251; Vienna Convention on Civil Liability for Nuclear Damage, May 21, 1963, 1063 U.N.T.S. 265.
128 *See Survey on International Liability Regimes, supra* note 26, at 93–95, 109–14.
129 *See* Horton et al., supra note 20, at 235, n. 35.
130 *See Survey on International Liability Regimes, supra* note 26, at 106.
131 *See id.* at ¶ 295.
132 *See id.* at 106, ¶ 293 (criticisms of the 1984 Protocol to the 1969 Civil Liability Convention for imposing excessively low limits on liability for damages due to oil spills during shipping).
133 *See* Paragraph (b)(ii) and (iii), Article 3 of the Supplementary Convention to the Convention on Third Party Liability in the Field of Nuclear Energy, Jan. 31, 1963, 1041 U.N.T.S. 374.
134 *See, e.g.*, the Vienna Convention on Civil Liability for Nuclear Damage; the Convention on the Liability of Operators of Nuclear Ships; the Convention relating to Civil Liability in the Field of Maritime Carriage of Nuclear Material; and the Paris Convention.
135 Under Article 13 of the CRTD:

1. The carrier's liability shall be covered by insurance or other financial security, such as a bank guarantee, if the dangerous goods are carried in the territory of a State Party.
2. The insurance or other financial security shall cover the entire period of the carrier's liability under this Convention in the sums fixed by applying the limits of liability prescribed in article 9 and shall cover the liability of the person named in the certificate as carrier or, if that person is not the carrier as defined in article 1, paragraph 8, of such person as does incur liability under this Convention.
3. Any sums provided by insurance or by other financial security maintained in accordance with paragraph 1 of this article shall be available only for the satisfaction of claims under this Convention.

See Arts. 13 and 14, Convention on Civil Liability of Damage caused during Carriage of Dangerous Goods by Road, Rail and Inland Navigation Vessels (Oct. 10, 1969).
136 *See* Convention on Damage Caused by Foreign Aircraft to Third Parties on the Surface, Oct. 7, 1952, 310 U.N.T.S. 181 (also known as the Rome Convention).
137 *See* INT'L OIL POLLUTION COMPENSATION FUNDS, THE INTERNATIONAL REGIME FOR COMPENSATION FOR OIL POLLUTION DAMAGE: EXPLANATORY NOTE, 3–7 (2014).
138 *See* International Convention on Civil Liability for Oil Pollution Damage (in force June 19, 1975); International Convention on the Establishment of an International Fund for Compensation of Oil Pollution Damage (Oct. 16, 1978); International Convention on Civil Liability for Oil Pollution Damage (May 30,

1996); International Convention on the Establishment of an International Fund for Compensation for Oil Pollution Damage (May 30, 1996); Protocol of 2003 to the International Convention on the Establishment of an International Fund for Compensation for Oil Pollution Damage (March 3, 2005).

139 *See* L. Rutkow, *Balancing Consumer and Industry Interests in Public Health: The National Vaccine Injury Compensation Program and Its Influence During the Last Two Decades*, 111(3) PENN. STATE L. REV. 681 (2006).

140 For example, an environmental non-governmental advocacy group in Canada has repeatedly and persistently urged that all research and deployment of climate engineering should be banned until a comprehensive treaty is in place to govern development of disruptive novel technologies (including nanoscale materials and genetically modified organisms). *See* The ETC Group, *ETC Group Calls for a Ban on All Testing or Deployment of Geoengineering Technologies*, www.etcgroup.org/content/etc-group-calls-ban-all-testing-or-deployment-geoengineering-technologies-and-supports-move.

141 *See* J. Long, F. Loy, and M. G. Morgan, *Start Research on Climate Engineering*, 518 NATURE 29 (Feb. 5, 2015) (calling for a self-implementing threshold to allow small-scale field testing of climate engineering technologies).

142 D. Keith, A CASE FOR CLIMATE ENGINEERING at XX-XXI (The MIT Press, 2013) (discussion of the fundamental differences between SRM versus CDR and the need to distinguish the issues and policy demands of each technology).

143 The application of the precautionary principle has already spurred a large, and fast-growing, body of academic literature. Reynolds, Chapter 3 this volume, 3.7.3. While the precautionary principle may challenge some deployments or tests of climate engineering technologies, its application is less clear when the risk of inaction is greater than the comparable (albeit uncertain) risk of acting. *See id.*

144 Some researchers have preliminarily estimated that stratospheric aerosol injection could potentially cause damages equal to roughly 1 percent of gross domestic product. As a result, monetary damages incurred by particular states will vary according to the size of their economies (e.g., China could face damages of $100 billion, while Indonesia could incur damages up to $10 billion). *See* Horton et al., *supra* note 20, at 243.

145 For example, the broad-scale deployment of BECCS could have significant biodiversity impacts arising from deforestation or accelerated species loss. *See* discussion *supra* at I (text accompanying fn. 11).

6

Research Governance

Michael Burger and Justin Gundlach

There is a broad consensus on several basic points pertaining to the relationship between climate engineering research and its governance. First, governance should precede deployment of climate engineering technology[1] even if governance does not precede all research.[2] Second, governance will be prompted by research efforts.[3] Third, useful research is unlikely to advance steadily without adequate governance.[4] And fourth, neither climate engineering research nor institutions engaged in its governance currently exist to a substantial degree.[5] As the US Government Accountability Office (GAO) wrote in its 2010 review of the US federal government's role in climate engineering research efforts, "a general lack of significant efforts to pursue geoengineering is a contributing factor to why geoengineering governance has not been pursued further to date."[6] The consensus that surrounds these conclusions is remarkable, if only because it indicates a fundamental confusion over what comes first, research or governance.

The years since the GAO made the above statement have seen a handful of forays into climate engineering field research, a stack of new articles and reports recommending that climate engineering research proceed, others recommending approaches to its governance, multiple articulations of principles relevant to such governance, and few if any steps toward actual, purposive governance. Thus, now as in 2010, governance of climate engineering research is still nascent and inchoate. Notably, however, the intervening years saw the launch of "the first coordinated geoengineering research project supported by the National Key Basic Research Program of China,"[7] and a recommendation by the National Research Council that the US undertake its own climate engineering research program.[8]

This chapter approaches the complex topic of climate engineering research governance in four Parts. Part I describes the forms research has taken so far and those that are expected in the future. It also offers short summaries of five instances of climate engineering field research conducted since 2009. Part II considers the key issues and concerns that have prompted calls for governance and that have inspired sometimes heated debate of what it should involve. Part III discusses governance directly, including its goals and functions, as well as issues arising from implementation. This Part also surveys the institutional landscape and classifies bodies with potential jurisdiction into three groups: those that are currently serviceable for the governance of research into particular climate engineering technologies, those that seem capable of adapting to the task, and those that will need to be created anew in order to fill critical gaps. Part IV concludes.

6.1. WHAT IS CLIMATE ENGINEERING RESEARCH?

Climate engineering research has been characterized as "more than simply a scientific procedure: it is a socially constructed and contested phenomenon."[9] The importance of this point is underscored by the fact that knowledge of the climate will always be partial and imperfect, such that any decision to deploy climate engineering technologies will follow not from precise understanding and effective control but from a gamble – possibly a gamble informed by research but necessarily a gamble – that deployment will do more good than harm, at least to the parties participating in the decision.

To date, the bulk of research bearing the "geoengineering" label has involved comparing the outputs of computer models to historical data and conducting laboratory tests.[10] For instance, a number of models have drawn on data gathered in the aftermath of the 1991 eruption of Mount Pinatubo, which launched a large volume of particulates into the stratosphere and was followed by a period of reduced global temperatures and disruption to precipitation patterns.[11] Such modeling is generally understood to be the first step on any of the paths climate engineering research would take.[12] Subsequent steps on that path have been described as "laboratory" or "indoor" testing, followed by "field" or "outdoor" testing on small, medium, and large scales.[13] However, there is disagreement about whether field testing of some climate engineering technologies can be meaningfully distinguished from deployment for governance purposes.[14]

Although the issues discussed in this chapter relate to all climate engineering research activities, we devote particular attention to issues arising from field testing. The following brief summaries of field tests conducted since 2009 illustrate what the term "small-scale field testing" has described to date. These field tests, which have generally proven controversial, also provide a useful illustration of features that inform definitions and categories relevant to climate engineering research.

6.1.1. *Examples*

6.1.1.1. LOHAFEX

The LOHAFEX project was a collaboration between India's Council of Scientific Industrial Research and the German Alfred Wegener Institute for Polar and Ocean Research, first agreed to in 2007 and carried out in January 2009. *Loha* is the Hindi word for iron, and the project plan involved spreading 6 tons of iron sulfate dust on the surface of a patch of the Southern Ocean in a planktonic algae fertilization experiment – hence "FEX."[15] This action would test the hypothesis, described in more detail in Chapter 2 of this book, that ocean iron fertilization (OIF) can sequester CO_2 in the deep ocean by promoting algae blooms that absorb CO_2 from the ambient air before dying and taking that CO_2 with their remains below the ocean's surface level.[16] (The hypothesis was not novel,[17] and multiple small-scale field tests had explored it since 1993.[18]) The LOHAFEX investigators sought and received authorization for their experiment from the parties to the Convention on the Prevention of Marine Pollution by Dumping of Wastes and Other Matter (the London Convention) in October 2008,[19] and, after a two-week emergency delay, from the German Ministry for Education and Research.[20] But the authorization followed announcements of disapproval by the German Ministry of Environment and condemnation by a Canadian nongovernmental organization (NGO),[21] which both noted the OIF moratorium issued by the (nonbinding) UN Convention on Biological Diversity in May 2008 (discussed in Chapter 3 of this book).[22] On January 27, 2009 – the day after the investigators began depositing iron sulfate dust in the Southern Ocean – the German Ministry for Education and Research explained its approval of the project in a statement that referred to the test as an "oceanographic research experiment" and said that "fears of this being a step toward geoengineering are unjustified."[23] Though the Environment Ministry had not backed down,[24] the German government, citing the London Convention Scientific Group's January 16, 2009 Risk Assessment, endorsed the reasoning of the Ministry for Education and Research.[25] The experiment's findings contributed to mounting evidence that OIF is not an effective means of extracting ambient CO_2 from the atmosphere in climatically relevant amounts and sequestering it in the deep ocean.[26]

6.1.1.2. Aerosol Sulfate Injection in Russia

In 2009, about 200 miles south of Moscow, a Russian team led by Yuri Izrael of the Moscow Institute of Global Change conducted a field test to evaluate how much injecting sulfate aerosols into the troposphere would reduce incoming solar radiation.[27] The Russian military supplied the team with the helicopter and truck from which the sulfates would be released.[28] Based on preliminary modeling, the research team had concluded that the test would not cause any adverse environmental impacts and, because of its benign nature and small scale, should not be subject to the restrictions codified in the international ENMOD agreement – the 1977 Convention on the Prohibition of Military or Any Other Hostile Use of Environmental Modification Techniques (which is discussed in Chapter 3 of this book).[29] On this basis, the team did not announce the field test to any international organizations – or really anyone at all – before going ahead.[30] Two papers published by the team characterize the experiment as showing that the release of aerosol sulfates reduced insolation (i.e., the amount of solar radiation reaching the surface) to a degree consistent with what the team's models had predicted.[31] Few have sought to build on those results. One non-technical reason could be that Izrael is known as an advocate for solar radiation management (SRM), and was known as such before leading the field test.[32] Another could be that the secrecy with which the test was carried out reduced confidence in the reported findings.

6.1.1.3. SPICE

The Stratospheric Particle Injection for Climate Engineering (SPICE) project, funded in 2010 by the British government as one of a trio of climate engineering research initiatives,[33] ultimately provided more insights into research governance than into technical aspects of climate engineering. Plans for the project included study of the characteristics of various aerosol particles, delivery vehicles and systems, and further modeling.[34] The planned delivery vehicle field test involved attaching a hose to a balloon, floating the balloon 1 kilometer off the ground, pumping water from the ground up to the nozzle of the hose, and measuring various results, including what happened to the water sprayed from the hose and how the hose and other equipment behaved throughout. Two different university ethics committees reviewed and approved the plan.[35] This modest test was meant to inform further plans for floating a larger balloon 20–25 kilometers up and spraying a sulfate aerosol into the air. However, the planned field test fell apart after the publication by some research team members of applications for patents on pump components that would be used in the test.[36] There was a forceful negative reaction from several NGOs and some members of the public,[37] apparently out of concerns that the experiment was a first step onto a slippery slope that would end in climate engineering driven by military and commercial interests.[38] After one of the agencies funding SPICE announced in September 2011 the decision to delay the test, the principal investigator finally cancelled it in April 2012.[39]

6.1.1.4. E-PEACE

The Eastern Pacific Emitted Aerosol Cloud Experiment, or E-PEACE was conducted in July and August 2011 off the California coast with funding from the National Science Foundation and the Office of Naval Research. It sought to better understand the relationships among aerosols that marine vessels emit, characteristics and behavior of marine clouds, and those clouds' albedo.[40] Marine cloud "brightening" has been on the list of theoretical climate engineering options since 1990 when Jonathan Latham first began exploring how to respond to global warming by making deliberate use of the cloud-changing effects of emissions from marine vessels.[41] A host of research projects since then, including efforts to build wind-powered vessels and energy-efficient spraying apparatus,[42] have given shape to Latham's initial vision. E-PEACE involved coordinated efforts to emit smoke (from a ship) and salt aerosols (from aircraft) into cloud formations, and to measure the effects on cloud size and albedo.[43] Notably, however, according to its principal investigator and the agencies that funded it, E-PEACE was not expressly a climate engineering experiment, but merely a scientific experiment devised to answer narrow questions about cloud perturbation.[44] Given the clear relevance of this sort of inquiry to climate engineering via marine cloud brightening, however, commentators quickly observed that "we're seeing … research that could give geoengineering answers, but isn't labelled as such."[45] Issues relating to such "dual-purpose" research are discussed in section 6.2.2 below.

6.1.1.5. The Haida Salmon Restoration Corporation

In the summer of 2012, about 180 miles off of Canada's west coast, the Haida Salmon Restoration Corporation, a for-profit venture led by an American entrepreneur and supported financially by a Haida Nation village reliant on the regional salmon fishery, conducted a controversial OIF field test involving 100 tons of iron-rich dust.[46] Whereas LOHAFEX had led to conflicts among authorities with overlapping oversight responsibilities, the company undertaking the Haida test did not seek permissions from any authority at all – notwithstanding allegedly pertinent Canadian environmental regulations and Canada's ratification of the London Convention and London Protocol.[47] A second feature propelled the controversy still further: The company had planned to repay its Haida Nation investors by selling carbon emissions credits for the carbon capture and sequestration it expected to accomplish via OIF.[48] Condemnations of the "rogue geoengineering experiment" noted that the company had dumped five times as much material as any previous OIF test, and had performed so little data collection that its carbon-related results were difficult to discern.[49] Canada's environmental agency investigated whether the test violated international or Canadian law, though several commentators have apparently concluded that it did not violate international law,[50] and others have noted that Canada may have failed to enact or enforce laws called for by international agreements.[51] In any case, the test did seem to accomplish its stated primary purpose of promoting the phytoplankton that would support a larger salmon fishery.[52] Notwithstanding the controversy over the climate engineering component of the test, the company's website continues to list arguments and support for the idea that OIF can sequester CO_2 in the deep ocean.[53]

6.1.2. *Defining Climate Engineering Research*

Defining climate engineering research requires the articulation of three things: first, what climate engineering is; next, what distinguishes research that counts as climate engineering research from research that does not; and last, what distinguishes research from deployment of climate engineering technologies. This subsection discusses each of these in turn.

As also discussed in Chapter 1 of this book, the Royal Society and the Intergovernmental Panel on Climate Change (IPCC) have defined climate engineering as the "deliberate large-scale manipulation of the planetary environment to counteract anthropogenic climate change."[54] By this definition, geoengineering is intentional – incidental large-scale alterations of the climate do not count – and only occurs when the alterations are undertaken on a large scale, not a small one (though there are some discussions of regional-level climate engineering).[55] In addition, by this widely adopted definition,[56] some forms of climate engineering that capture and sequester greenhouse gases so closely resemble forms of mitigation that the British House of Commons Science and Technology Committee identifies carbon capture and storage technologies as "mitigation" if they are coupled with a power-generating facility but as climate engineering if they operate independently.[57] Other forms, such as brightening elements of the landscape to increase albedo, overlap with adaptation measures intended to reduce temperatures in urban heat islands.[58]

A subtle but important further aspect of what climate engineering encompasses relates to the particular activity that qualifies. That is, because legal decisions are likely to examine this issue closely, it is important to note that climate engineering is not a category comprised solely of technologies, nor solely of activities, but of what this chapter refers to as "approaches" or "techniques," meaning *combinations* of technologies and purposive activities.[59] Thus, the capture and sequestration of carbon using biochar cannot be reduced to just the apparatus nor to just the actions involved in, say, the use of bioenergy with carbon capture and storage (BECCS).[60] (Chapter 2 of this volume describes the various climate engineering approaches that have been hypothesized or explored to date.[61])

When should research efforts get a "geoengineering" or "climate engineering" label? As with the definition of geoengineering above, the current approach relies heavily on the intent of the researcher.[62] The Haida Salmon Restoration Corporation's fertilization project demonstrates both that relying on intent can invite researchers to dodge the label,[63] but also that intent can be inferred even if researchers do not announce that their work focuses on altering the climate. Haida Salmon and the E-PEACE project also provide examples of dual-use research that, whether or not it informs future climate engineering efforts, clearly also pertains to an inquiry about non-climate issues. The list of research topics that qualify as dual use is long, and, much as climate engineering overlaps with mitigation and adaptation, the list of topics that might be labeled "geoengineering research" overlaps substantially with more general atmospheric and climate research.[64] Thus, the ARM Cloud Aerosol Precipitation Experiment (ACAPEX), funded by the US Department of Energy, is not currently labeled as geoengineering research,[65] even though a future geoengineering effort would likely draw on what it discovers about atmospheric "rivers" and the effects on precipitation of aerosol particulates from various sources in marine cloud formations. Recognizing that labeling such experiments as "geoengineering research" might prove unwelcome and cumbersome, the British government stated in a 2010 report that "weather techniques such as cloud seeding should not be included within the definition of geoengineering used for the purposes of activities designed to effect a change in the global climate with the aim of minimising or reversing anthropogenic climate change."[66]

What distinguishes research from deployment?[67] Several international agreements concerned with preventing transboundary environmental harm contemplate such a distinction,[68] and the London Convention/London Protocol's Assessment Framework sets out a process for distinguishing prohibited ocean dumping from scientific research, including research into the efficacy of OIF.[69] Some have suggested that technical thresholds, such as the amount of solar radiation deflected by a given experiment, might provide the needed specification.[70] However, particularly for stratospheric or tropospheric aerosol injection, the impacts of which are poorly understood,[71] any specification would be difficult to agree on or implement.[72] Thus it would be very difficult to say when governance should transfer from institutions and authorities responsible for research oversight to those responsible for overseeing deployment.[73] In addition to this difficulty, there is also a deeper problem of how to maintain effective control over new technologies, dubbed the "Collinridge Dilemma."[74] This Dilemma attends any new and potentially powerful technology because (a) that technology's role and impacts are difficult to predict before its deployment, but (b) once it is deployed, it can be difficult to alter the role it plays, notwithstanding its impacts.[75] Restated, even if small-scale research fails to resolve key uncertainties or control foreseeable dangers, "[r]esearch may generate its own momentum and create a constituency in favor of large-scale research and even deployment."[76]

6.1.3. *Categories of Climate Engineering Research*

As many commentators have observed, any effort to implement governance of climate engineering research cannot take a one-size-fits-all approach.[77] The range of activities that qualify as "geoengineering" or "climate engineering" is simply too wide to be usefully addressed by a uniform approach to governance. Imposing governance on different forms of climate engineering therefore requires specifying their differences and organizing them into categories useful for the purpose of governance.

Most efforts to categorize climate engineering approaches have built on the separation of carbon dioxide removal (CDR), which "address[es] the root cause of climate change by removing greenhouse gases from the atmosphere," from SRM, which "attempt[s] to offset effects of increased greenhouse gas concentrations by causing the Earth to absorb less solar radiation."[78] This division aligns with what others have called "remediation" and "intervention" approaches.[79] Both "have the ultimate aim of reducing global temperatures, but there are major differences in their modes of action, the timescales over which they are effective, temperature effects and other consequences, so that they are generally best considered separately."[80]

As explained by the US National Research Council (NRC),[81] separate consideration of CDR/remediation and SRM/intervention makes sense for a number of reasons. First, whereas CDR is a "potentially viable option," SRM comprises a "more speculative family of approaches."[82] Second, "CDR strategies … are generally of lower risk [than SRM] and of almost certain benefit."[83] Third, in contrast to SRM, "it is almost inevitable that some CDR will be needed in the long term" to offset emissions and then to shorten the duration of planetary recovery to preindustrial atmospheric concentrations of greenhouse gases (GHGs).[84] China's planned climate engineering research program has also adopted the basic categories of CDR and SRM.[85]

One alternative categorization places climate engineering in the broader context of all policy responses to anthropogenic climate change,[86] reconfiguring climate change responses into five categories: anthropogenic emissions reductions, climate change adaptation measures, domestic removal of atmospheric GHGs, transboundary removal of atmospheric GHGs, and targeted climate modification.[87] This alternative formulation reframes the existing categories chiefly by distinguishing between "transboundary" GHG removals like OIF in the open ocean, which directly affect the global commons, and "domestic" removals, like afforestation within a single nation's jurisdiction.

Whether building on the CDR/SRM categories or on the alternative configuration, analysts and commentators have employed further subcategories that take into account some or all of the following features of a given approach: scale, locus (e.g., indoors or outdoors, intra- or transboundary), temporality, utility (e.g., scientific versus commercial), and risk.[88] Though different categorizations employ different terms, they get at the same underlying issues. For instance, references to the "cost" and "effectiveness" of an approach[89] address the same point as references to its "leverage"[90] – i.e., its cost-effectiveness. "Reversibility" and "permanence of intended effect" are also synonyms, as are "timeliness" and "rapidity." Some alternative terminologies overlap but are not cleanly interchangeable: "Encapsulation" and "safety" are not the same as "scale of action," "scale of intended effects," "effect on the global commons," or "transboundary side effects," but they focus on the same features and implications of climate engineering research approaches.

As described further in Section 6.3, these categories and subcategories are critical for research governance because they help to determine questions of jurisdiction, the nature of oversight, and the intensity of oversight, among others.

6.2. ISSUES AND CONCERNS

The prospect of climate engineering and its research generate a number of governance-related concerns, including "moral hazard," "forum shopping," a slippery slope from research to deployment, irrational "lock-in" of a particular technology, the potential for a "governance trap," and disruption to international relations and particularly to the Paris Climate Agreement. Each is discussed in turn.

6.2.1. *Moral Hazard*

One of the concerns most frequently expressed about climate engineering relates to "moral hazard," a term common to insurance and economics that describes scenarios in which one party takes a risk knowing that she is protected from the cost of that risk, which will accrue to other parties.[91] There are two ways in particular in which "the decision to 'insure' via geoengineering may influence the conduct of the 'insured.'"[92] First, governments, industry, and the public – misunderstanding climate engineering to be an effective substitute for reducing GHG emissions – might cease to invest in the expensive and politically difficult tasks of mitigation.[93] For example, the world's largest fossil fuel companies might undertake an enormous climate engineering research and deployment effort in order to "offset" persistent GHG emissions and thereby protect their revenues from the consequences of effective mitigation efforts.[94] Put at ease by early results of these efforts, voters' and governments' demands for mitigation might wane.[95] Alternatively, governments of wealthy nations might collaborate to pursue, over the objections of poorer nations, stratospheric aerosol injection, even though doing so could foreseeably result in disruption of the monsoon relied upon by India, Bangladesh, and other nations for agricultural productivity.[96]

The second form of moral hazard could arise from ignorance – or at least a greater degree of uncertainty. In this scenario, a lack of research would leave the risks of deploying a technology obscure, such that an optimistic government or other entity deploys it in spite of that technology's uncertain net effects within and especially outside that country's borders. This latter version of the moral hazard concern aligns closely with more general concerns about climate engineering approaches like stratospheric aerosol injection and large-scale marine cloud brightening, which are thought to be potentially cost-effective and whose effects would certainly be felt across international boundaries. One study put it this way: "the main problem is establishing legitimate collective control over an activity that some might try to do unilaterally without prior consultation or international risk assessment."[97]

6.2.2. *Forum Shopping*

The entity seeking to research or deploy climate engineering technology in the foregoing scenario could be either a government or a private actor seeking to take advantage of the lax oversight imposed by a government in a given jurisdiction.[98] Such "forum shopping" could arise in several forms. One that has already occurred was the Russian government's sponsorship of a favored scientist's sulfate aerosol injection experiment with an inappropriately nonchalant attitude to compliance with international agreements governing weather modification.[99] Another that could occur is an experimenter with corporate sponsorship seeking a jurisdiction that does not require environmental review so long as no government funding is involved and no permits are required for the activity under existing law.[100] This might be an apt description of the Haida Corporation's 2012 OIF deployment, which may well not have violated Canadian law,[101] and it describes scenarios conceivable in other jurisdictions, including in international waters using ships sailing under a "flag of convenience."[102] This potential for forum shopping is often the reason why some have recommended prohibiting any climate engineering research by for-profit entities without either a public partner or public oversight.[103]

6.2.3. *The Slippery Slope and Lock-in Effect*

Concerns about a "slippery slope" and "lock-in" both amount to versions of the Collinridge Dilemma, which arises because research into a powerful but poorly understood technology may generate momentum that propels it to imprudent deployment. A characteristic version of the "slippery slope" envisions a research program funded and structured not so much to understand a technology's potential effects as to develop that technology.[104] Such a program might be guided by a definition of success that assumes eventual deployment, or one that proceeds unfettered by the need to justify go/no-go decisions based on something other than operational capability.[105] Whatever form it takes, the key feature of a slippery slope scenario is the *absence* of external, non-technical feedback capable of steering the progress of technological development.[106]

Discussions of "lock-in," or "resistan[ce] to change even if negative impacts were later discovered," are concerned with irrational commitments to adoption after an initial experiment or small-scale deployment.[107] Thus the most obvious form of lock-in for SRM technologies is the "termination problem" that would result from successfully preventing further increases in the Earth's average temperature only so long as deployment of a particular technology continues: Cessation of that deployment would lead to a rapid and dangerous temperature rise.[108] This example is one of "technical" as opposed to "social" lock-in.[109] The latter focuses on the constituency that could be expected to coalesce around maintenance of a particular climate engineering approach, either because of gains to be had from the large-scale investments required for its research or deployment, or because that approach enables continued profitability for an emissions-intensive industry.[110] Such a constituency would be invested in overlooking evidence that the benefits of the approach do not exceed its costs.

As Long and Scott have observed, with somewhat forced alliteration, there are at least four potential motives for this social form of lock-in: fortune, fear, fame, and fanaticism.[111] "Fortune" sums up the perspective of a group or entity that stands to lose from mitigation and to gain, or continue not losing, from use of climate engineering as a mitigation substitute. "Fear" refers to the reticence of an individual or a funder to publish negative results[112] – a reticence that, if widespread, would bias scientific and public understanding of the climate engineering approaches at issue.[113] "Fame" recognizes that individuals might lose perspective and give undue priority to public recognition. "Fanaticism" acknowledges that any ideology, whether it comes to favor or oppose a particular path of climate engineering research or deployment,[114] would ignore facts crucial to a rational analysis of the prospects of approaches to climate engineering.

6.2.4. *The Governance Trap*

The "governance trap" occurs if existing institutions oriented to some other purpose, but with some measure of jurisdiction over climate engineering research activities, come to steer the governance of climate engineering research – and possibly do so badly. For example, the Montreal Protocol could apply to the dispersion of sulfate aerosols into the stratosphere. However, the Montreal Protocol is a mechanism set up to phase out certain substances that contribute to ozone depletion; it is not designed to assess the potential impacts of field tests for new technologies.[115] In addition to an institutional mismatch that inappropriately shuts down research, the trap could manifest in a situation where the authority that claims jurisdiction over climate engineering research essentially fails to govern, requiring merely ministerial compliance with existing regulations in a way that does not stymie imprudent research pathways. Alternatively, the authority might apply regulatory techniques that are well suited to its core mission but not to climate engineering. Thus a governance trap is usefully understood as poor or non-governance that increases the likelihood of such scenarios as a slippery slope, a lock-in scenario, premature termination of justifiable research, use of inappropriate controls, or failure to exercise needed controls.

For climate engineering research, perhaps the most likely "trap" lies in the law governing intellectual property rights.[116] Since 2010 the US Patent and Trademark Office has witnessed a climate engineering "patent land-grab," with a focus on technologies related to CDR approaches.[117] This land-grab – and patenting more generally – could derail an optimal climate engineering research program.[118] The authors of the Oxford Principles of Geoengineering Research identify another problematic implication of patent law in this context:

> The ability to obtain patents on geoengineering technique [*sic*] could create a culture of secrecy and may lead to the concealment of negative results. This has been observed in the pharmaceutical industry, where negative research results are deliberately concealed. This is doubly damaging – firstly, the negative consequences of a geoengineering technique could be far more wide-ranging than from a drug trial, and secondly, the concealment of negative results could lead to a public backlash against all geoengineering research and research scientists.[119]

Thus the normal operation of patent law, in which actors are arguably incentivized to conceal adverse findings so as not to impact the commercial potential of their products, could impair a program of climate engineering research governance that seeks to foster rapid improvements in the understanding of climate engineering.[120]

Intellectual property may also present barriers in the international realm. If one or more national governments seek to impose special requirements on intellectual property related to climate engineering research, they will have to vie with the Agreement on Trade-Related Aspects of Intellectual Property Rights (TRIPs).[121] Members of the World Trade Organization are bound by the provisions of TRIPs, which strengthen the ability of intellectual property owners to enforce their rights across international borders.[122] Although TRIPs has given rise to notably few enforcement actions,[123] and although its provisions could be especially difficult to enforce in the climate engineering research context, it nonetheless presents a formal and functional hurdle that national governments cannot ignore should they create exceptions to the treatment of climate engineering-related intellectual property.[124]

6.2.5. *Disruption of International Relations*

Because of the potential enormity of climate engineering's transboundary consequences, its research is inevitably a charged issue for international relations.[125] Potential disruptions incidental to disagreement over the conduct of research or its governance could take a number of diverse forms.[126] A full discussion of imagined paths leading from research to international strife is beyond the scope of this chapter, but we acknowledge the potential for rifts to open between states over climate engineering research and thus the importance of reducing that risk by devising internationally applicable norms, mechanisms, and institutions for governance.[127]

6.3. GOVERNANCE: WHAT? WHEN? AND BY WHOM?

Envisioning governance of climate engineering research leads to three broad categories of possible approaches: a ban (typically on both research and deployment);[128] unilateral, uncoordinated, and potentially variable approaches by national governments;[129] or an effort on the part of actors at multiple levels – ranging from research facilities to international bodies – to monitor, coordinate, steer, and limit research in an agreed fashion.[130] Many have warned against pursuing the first two of these, and – although calls for a ban will certainly persist – discussion in the scientific and policymaking communities has effectively become a debate on how to govern climate engineering research – and possibly also eventual deployment – through a combination of international and national institutions.

It is useful to frame this ongoing discussion as an effort to determine how best to apply the prevention and precautionary principles[131] – features of international law[132] and of many countries' domestic laws[133] – in the context of climate engineering and its research. Some have said that prevention/precaution in the context of climate engineering research is straightforward and means simply not meddling with the global climate.[134] Others have said that anthropogenic climate change makes the application of these principles ambivalent: Given what we understand of the rapidly changing state of the climate, doesn't the greater risk lie in failing to explore, understand, and *possibly* employ climate engineering techniques?[135] As one commentator put it: "ignoring geoengineering today, and only considering it when all else has failed, is a recipe for bad, politics-led decision-making."[136] This chapter understands the basic goal of climate engineering research to be *the reconciliation of these interpretations of prevention and precaution by enabling useful and informative research to proceed – or be terminated – safely, while preventing the conduct of research that is not useful, not safe, or otherwise undesirable*.[137]

This section examines the prospective features of climate engineering research governance. It divides that examination into three subparts. The first provides a definition of governance and discusses the widely accepted Oxford/Asilomar principles as well as the functions that any climate engineering research governance institution would perform. The second examines the question of when and how particular climate engineering governance approaches should start to apply to research efforts. The third addresses the question of who has the authority and the capacity to administer a governance regime, identifies existing institutions, and highlights the gaps that would need to be filled.

6.3.1. *What: Definition, Principles, and Functions*

As Professor Daniel Bodansky has observed, "the essence of governance is to make decisions for a collective."[138] The SRM Governance Initiative (SRMGI), which the Royal Society, the Environmental Defense Fund, and the World Academy of Sciences established in 2010, has proposed the following definition for research governance: "the resources, information, expertise, and methods needed for the control of an activity in order to advance the potential societal benefits provided by SRM, while managing associated risks."[139] This definition, in keeping with Bodansky's observation, encompasses all key aspects of decision-making about SRM, including both "hard" governance, meaning prohibitions and forms of direct governmental control over experiments, as well as "soft" governance, meaning the allocation of financing and requirements for reporting and transparency. In adopting this definition, we note that it is compatible with the admonitions of the NRC that "'[g]overnance' is not a synonym for 'regulation,'"[140] and of the German Environment Agency that "governance encompasses more than binding legal rules."[141]

A governance system ought to be attached to a sound set of principles. Although there will likely be further statements about what should inform climate engineering research governance, the five principles articulated by participants in the 2010 Asilomar International Conference on Climate Intervention Technologies[142] have been especially influential on subsequent efforts, including by governments, to articulate increasingly concrete recommendations for climate engineering research governance.[143] In addition to codifying the consensus views of the scientific community, these principles also reflect the broadly similar Oxford Principles[144] which were published in 2009 and were endorsed by the British government in 2010.[145] The Asilomar Principles are as follows:

1. Climate engineering research should be aimed at promoting the collective benefit of humankind and the environment;
2. Governments must clarify responsibilities for, and, when necessary, create new mechanisms for the governance and oversight of large-scale climate engineering research activities;
3. Climate engineering research should be conducted openly and cooperatively, preferably within a framework that has broad international support;

4. Iterative, independent technical assessments of research progress will be required to inform the public and policymakers; and
5. Public participation and consultation in research planning and oversight, assessments, and development of decision-making mechanisms and processes must be provided.

While these principles leave much room for further specification, they distill several basic priorities: First, Principle 1 makes clear that the aim of research is greater understanding, not greater capacity to deploy one or more forms of climate engineering. Thus research guided by this principle could be considered successful not only if it assisted in deployment, but also if it instead supported the conclusion that deployment would be unhelpful or harmful. Second, Principle 2 recognizes that national governments are primarily responsible for climate engineering research governance but also that they will fail in that responsibility if they do not act to establish or actively support the establishment of governance mechanisms. Third, Principle 3 highlights that the mechanisms contemplated by Principle 2 should enable international transparency, and, ideally, be the result of international consultation and agreement. Fourth, Principle 4, by calling for "independent technical assessments," effectively proposes that no research findings should be treated as sufficient to justify policy decisions unless those findings have withstood the scrutiny of review by experts working independently of the original research team. Last, Principle 5 calls for "public participation and consultation." This marks a rare departure from the Oxford Principles, which call for notice, consultation, "and ideally … prior informed consent of those affected."[146] Thus, taken together, Asilomar Principle 5 and Oxford Principle 2 establish both a minimum of public consultation and an ideal of consent. An approach that nears the ideal is akin to the concept of "social license," which was developed in relation to extractive industry activities[147] and has since been discussed in relation to climate engineering as well.[148]

Implementation of the basic Oxford/Asilomar principles will entail performing familiar functions in potentially novel ways.[149] Those functions, elaborated below, include: authorizing and financing proposed and ongoing research; identifying and resolving potential conflicts of interest; establishing and enforcing requirements for disclosure and dissemination of plans, data, and analyses; conducting environmental impact assessments in advance of experiments; providing for public consultation and engagement; managing intellectual property rights; and defining, assigning, and enforcing liability for harms arising from research.

6.3.1.1. Funding and Authorizing Research

While there is nothing unusual about public or private entities funding research, there is something unusual (but not unprecedented) about restrictions, whether through ethical norms or legal strictures, on not just *how* a research agenda may be pursued but also on who may support it and whether it may be pursued at all. Thus climate engineering research governance, insofar as it seeks to apply such limits, will be in contentious territory – both ethically[150] and legally[151] – alongside the governance of research into technologies like nuclear power, biotechnology for medicine and agriculture, and nanotechnology.[152] On the other hand, funding sources shape research paths,[153] and several concerns noted in Part II of this chapter would seem to justify imposing oversight on some forms of research and some sources of research funding. Thus any governance regime will need to develop and articulate answers to difficult questions, such as:

- how to distinguish "climate engineering research" from non-climate engineering research;
- whether to impose special requirements on privately funded geoengineering research – for instance, reporting of plans, data, and results analysis to a publicly available registry[154] – in order to prevent efforts to end-run restrictions, such as public records access, that would accompany the same research if funded publicly;
- how to reconcile and potentially integrate governmental oversight with self-regulation by the scientific community, such as the arrangement that grew out of the 1975 Asilomar International Conference on Recombinant DNA Molecules;[155]
- whether to prohibit particular research efforts, either by articulating categories of research that are "off limits," or as the result of a process that considers proposals ad hoc;
- how to justify and establish legal predicates for any of the foregoing types of restrictions.

6.3.1.2. Identifying and Resolving Potential Conflicts

Research governance in general has concerned itself in part with rooting out conflicts of interest to protect the integrity of research for at least several decades.[156] The standard procedural approach to this threat to research integrity entails disclosure of financial and other interests to a supervisory body and/or to the public.[157] When a US federal agency seeks expert advice on a particular topic, the heightened requirements of the Federal Advisory

Committee Act may apply.[158] Nonetheless, conflicts review in several academic and professional contexts continues to spark disputes over the adequacy of basic definitions and procedures[159] in part because factors like publication bias have persisted – even flourished – notwithstanding applications of the measures listed above.[160]

The pernicious potential of conflicts – actual or perceived[161] – in the fraught context of climate engineering argues for employing two further measures when dealing with consequential research experiments or campaigns. First, a steering committee could be formed that includes representation from ethical perspectives as well as scientific ones and that would be responsible for the review of proposed research and possibly also for decisions about funding.[162] And second, a general expectation could be created that all field testing will employ a "blue team/red team" model that assigns related tasks but different goals to each team: blue to test a hypothesis or the effectiveness of a given apparatus or technique, and red to identify weaknesses, perhaps by simply critiquing the first team's approach, or by assessing the levels of cost and environmental impact related to the test's subject.[163] Actually employing either a steering committee or rival teams model, or both (they would be not only compatible, but complementary), would require specifying a long list of particulars, including the source and scope of the steering committee's authority over research and the process for determining goals and membership for the "red team" in any experiment.

6.3.1.3. Disclosure Requirements for Research Plans, Data, and Results

Existing disclosure requirements for scientists and engineers who are engaged in research and testing vary widely and depend on the researchers' affiliation, funding, field of study, and object of study. Such variability can give rise, in the field of climate engineering, to a lack of transparency, public distrust, or needless additional environmental impacts from duplicative research efforts, presenting problems for the governance of climate engineering research. One logical response would include "open access to SRM knowledge,"[164] or "a commitment to make public the existence of all SRM research activities" by "providing public notification of proposed field experiments and providing decisionmakers and the public with full access to the results of the research … (preferably through publication in refereed journals)."[165] Open access along these lines could help build societal trust in climate engineering.[166] Yet, depending on how much detail is publicized in advance of an experiment, it could also make it easy for opponents of research to attempt to disrupt it, so under some circumstances the exact time and place of experiments might not be disclosed. Scientists and policymakers can rely on research funding applications, publications, and similar media for sharing information about experiments' designs, protocols, and results.[167] Public disclosure of experiments also would not substitute for the processes required to effectively engage the public in deliberations about SRM research as it relates to climate change management more generally.[168]

In sum, there is broad agreement that climate engineering research governance must provide for transparency, both among researchers themselves and between researchers, policymakers, and the public. Further, there is also broad agreement that this function is best performed by coordinating among multiple mechanisms – some new, some extant – rather than designating a single sluice as the channel through which all research information must flow. Whether or not compliance with the norm that all proposals, data, and analyses be made accessible is initially voluntary or immediately mandatory,[169] the lack of a unitary sluice for information cannot be taken as license to withhold or obscure the details of research efforts.

6.3.1.4. Environmental Impact Assessments

As discussed in detail in Chapter 3 of this book, in the USA the National Environmental Policy Act requires a federal agency to assess environmental impacts before it permits, licenses, or funds any action significantly affecting the human environment.[170] Environmental impact statements (EISs) must be made public and allow for public comment in the administrative record.[171] The agency must consider the final version of that EIS as it makes decisions about the action.[172] Similar rules govern actions in the European Union and other jurisdictions,[173] and international law requires environmental assessment where environmental impacts could be felt across international boundaries.[174] Thus EISs are a readymade tool for a program of climate engineering research governance that seeks to adhere to the principle of public consultation discussed above. Reflecting that suitability, the London Convention/London Protocol's Assessment Framework for OIF experiments requires that any instance of OIF conducted for scientific research purposes be preceded by an environmental assessment process involving detailed disclosures of the experiment's goals, parameters, and expected effects to countries in the region of the proposed experiment.[175]

However, some aspects of EIA practice do not well fit the purpose of assessing the environmental impacts of climate engineering research. For one, the impacts of individual field tests may not rise to the level of significance necessary to trigger detailed analysis, even if some or all of those tests' impacts are considered cumulatively.[176] Similarly, should a class of experiment be exempted from impact assessments based on a categorical exclusion – such as the ones that the US Department of Energy and National Science Foundation frequently issue for weather-related research[177] – the law devised to ensure scrutiny and transparency could actually supply a means of *avoiding* impact assessment and public participation. Private funding, should it cause an experiment or research campaign to avoid the trigger of governmental involvement, could provide another potential means of avoiding the EIA process, barring a trigger based on other laws, such as pollution restrictions.[178]

Just as gaps in coverage by NEPA and similar laws might allow some research to go inappropriately unexamined, it is also possible that compliance with these laws could encumber a particular experiment with costs and procedural burdens unwarranted by the risks and uncertainties that attend it, and so inhibit the advancement of knowledge.[179]

Issuance of a programmatic EIS (or the equivalent for non-US jurisdictions) could ameliorate at least some of these gaps and mismatches by crafting a tiered review structure that takes into account the unusual purpose of climate engineering experiments, pays close attention to the potential accretion of cumulative impacts over the course of multiple experiments, and digests research proposals more quickly than could be expected if each experiment were reviewed as a unique action.[180] In addition to providing for a coherent assessment process in advance of experimentation, a programmatic EIS would make at least some interaction among scientists, regulators, political authorities, and the public mandatory in advance of field testing. This stricture might inspire discontent within the research community. It also would *not* provide a complete means of integrating scientific, political, and ethical priorities. Rather, it would likely serve as a resource on which to draw and a platform on which to base additional forms of review and public engagement.

6.3.1.5. Public Engagement

Just as "the consent of those who are affected by climate engineering research and deployment can confer legitimacy on the research and deployment or on the institutions in charge of geoengineering decision-making,"[181] a lack of consent – or even a perceived lack – can deprive geoengineering research of social license.[182] Although public consent to climate engineering research is difficult even to define,[183] much less to garner, some measure of consent – whether manifest through reasoned public approval by a government agency or a more direct means – is important to a viable program of climate engineering research in liberal democracies, and effective climate engineering research governance should entail public engagement.[184] However, as the experiences of researchers, government officials, and the public with climate engineering experiments such as SPICE and the Haida Corporation's fertilization have demonstrated, it can be difficult to engage effectively with the public because the myriad concerns they express tend to focus not only on the risk posed by an experiment but also on its motives, the values it implies, and its implications for other priorities.[185] Such reactions are consistent with the characterization of climate engineering as not simply a technical capacity but "a socially constructed and contested phenomenon."[186] Because public reaction to the proposal or conduct of early climate engineering field experiments could, in the absence of established and tested governance mechanisms or institutions, function as a referendum on such experimentation,[187] the stakes of engaging the public well or badly are high.[188]

6.3.1.6. Managing Intellectual Property Rights

The usual tension between private and public interests in intellectual property is greatly heightened in the climate engineering context, where private research funding currently rivals public funding in scale,[189] and where control of a technology's patent rights could mean partial control of the climate.[190] This tension led the authors of the Oxford Principles to clarify how their call for regulation of climate engineering technologies as "a public good" would affect intellectual property rights:

> There should … be a presumption against exclusive control of geoengineering technology by private individuals or corporations … [W]hilst [geoengineering's characterization as a public good] does not mean that there can be no intellectual property in geoengineering, it highlights that there might be a need for restrictions to ensure fair access to the benefits of geoengineering research.[191]

To answer the concerns informing this statement and those noted in Part II above, climate engineering research governance may need to steer around or wholly displace the normal operation of intellectual property law when it encounters climate engineering technologies and processes.

No one is publishing arguments in favor of a business-as-usual approach to intellectual property in the climate engineering context. (This is not surprising; most patent advocates instead quietly focus on building portfolios of claims before engaging in public debate.) Though proposals for intervention vary in their details, they share the basic premise that climate engineering research governance should limit the ability of individual private interests to exclude others from useful knowledge and know-how.[192] Under one such proposal, the US Patent and Trademark Office would screen for climate engineering-related applications and place them in a compulsory patent pool.[193] Another proposal envisions crafting multiple, coordinated intellectual property mini-regimes, each specific to a particular climate engineering approach.[194] Yet a third proposal takes the US approach to nuclear energy patents as a model for a *sui generis* climate engineering intellectual property regime,[195] and makes four specific recommendations with this model in mind.[196] A final proposal points to the suite of intellectual property policies employed by the European Organization for Nuclear Research, which successfully supports an "open science" model that does not eliminate intellectual property rights but privileges public dissemination if ever those two priorities collide.[197]

6.3.1.7. Liability

Because the effects of most approaches to climate engineering would be felt unevenly across geographies, populations, and generations, any climate engineering deployment can be expected to impose costs and deliver benefits disparately.[198] Having looked to other contexts in which human activity tends to yield similarly disparate outcomes, several authors have proposed regimes for defining, assigning, and enforcing liability among those who cause and those who are affected by the activity at issue.[199] Several factors present steep challenges to the successful operation of any such regime for climate engineering deployments, however,[200] chief among them the problem of determining causation.[201] As liability issues arising from climate engineering research activities do not differ in material ways from those arising from deployment (discussed in Chapter 5 of this volume), there is every reason to try to develop answers to all major liability questions in advance of significant research efforts.

6.3.2. *When*

At what point should activities that qualify as "climate engineering research" receive heightened or particular oversight? Should it be as soon as the research gets the "climate engineering" label, such that even computer modeling activity could come within the compass of oversight? Or should it be only after the potential impacts of a given experiment are determined to be likely to exceed a particular impact threshold? Should moratoria apply until such time as key stakeholders agree to establish governance structures?

Answers to the foregoing questions vary for different approaches to climate engineering. Less is at stake in determining the phases of research for climate engineering approaches like direct air capture or afforestation whose potential adverse effects could be contained. As the National Research Council has observed, there is little doubt that one or more forms of CDR would be net beneficial and important for the purpose of restoring and maintaining the Earth's carbon balance,[202] and as a result decisions about whether to deploy these approaches "will be largely based on cost and scalability" and less on the improved understanding of the risks presented by adverse side effects.[203] By contrast, approaches like stratospheric aerosol injection and marine cloud brightening have the potential to cause unknown amounts of incidental harm through complex and little understood mechanisms.[204] This disparity makes it important for experiments that must create the risk of harm to do so minimally, which in turn means structuring research phases with care. Accordingly, this section first discusses the phases of research particular to approaches whose potential adverse effects could not readily be contained (stratospheric aerosol injection, marine cloud brightening, OIF, and ocean alkalization) before turning to timing-related features of governance that potentially apply to all climate engineering approaches.

6.3.2.1. The Phases of SRM and OIF Research

The phases through which research into SRM technology would likely proceed are as follows:

- Laboratory – useful to develop understanding of efficacies and risks well represented by laboratory experiments or computer or physical models.
- Technology development – focused on hardware development and operations; no chemical processes.
- Process study – micro-scale analysis of physical, chemical and, radiative processes; not going beyond the scale of natural perturbations.
- Scaling test – conducted at the mesoscale level of 1 to 1,000 km^2; intended to validate models and assess how processes may vary across scales.
- Climate response test – designed to elicit a large-scale climate response.[205]

The first of these phases entails no environmental impacts; the second and third, if they entail any impacts, do not pose significant ones.[206] To illustrate the nature and utility of the third phase, one report offers the example of studying "the possible loss of stratospheric ozone in response to aerosol injection":

> The large-scale ozone response depends first on the small-scale physical and chemical interactions that determine how the chemistry of an air parcel evolves, and second on the large-scale atmospheric dynamics that transport

constituents within the stratosphere. Most (but not all) of the uncertainty in predicting the response of ozone to injection of a novel kind of aerosol stems from uncertainty in small-scale processes so it is possible for small-scale experiments to reduce uncertainty in large-scale predictive models.[207]

Thus, although impacts would vary among experiments undertaken to explore each of four climate engineering approaches considered here (i.e., those whose adverse impacts cannot readily be contained), it is generally true that only the commencement of field tests at the fourth phase promises non-negligible environmental impacts.[208]

6.3.2.2. The Phases of Climate Engineering Research, in General

This subsection discusses thresholds, moratoria, and the "allowed zone" for climate engineering research. These mechanisms are all potentially useful to the governance of research of any climate engineering approach.

(1) **Thresholds**. Three broad categories of threshold pertain here: one relating to the scale of a given experiment, another relating to the state of the global climate, and the third relating to the existence and nature of climate engineering governance institutions.

One proposal for scalar thresholds establishes numeric criteria for individual SRM experiments based on how much all SRM experiments in a given year would dim the Sun (measured in watts per square meter or Wm^{-2}), where they would do so, and for how long.[209] This threshold would pertain to the annual average effects of all SRM experiments, thereby limiting cumulative, worldwide "radiative forcing perturbation" in a given year.[210] In contrast to this ostensibly precise numeric dividing line,[211] the scalar threshold for OIF experiments exempted from the London Convention/London Protocol's prohibition on "dumping" takes the form of an environmental impact assessment without a predetermined numeric component.[212]

A second approach to establishing research thresholds would set limits based on the rate and degree to which the climate is changing. This approach would mark the tipping point at which the danger of failing to intervene in the global climate is felt to match or outweigh the danger of previously impermissible climate engineering experimentation or deployments. Because this threshold could take many forms, all of them uncertain, no one has yet suggested that national governments or the scientific or international communities should articulate it precisely.

The third type of threshold is less literal, and would be satisfied by the establishment and operation of governance mechanisms that satisfy particular criteria.[213] The draft Berlin Declaration – a statement about SRM that was presented, debated, but never finalized or agreed to by attendees at the 2014 Climate Engineering Conference[214] – articulates such a threshold in this way: "the establishment of an open and transparent review process … At a minimum, such a review process should involve prior disclosure of research plans."[215] The London Convention/London Protocol's imposition of the Assessment Framework satisfies this threshold for OIF.

As these types and examples reflect, thresholds could – and in the context of OIF do – describe the boundary within which climate engineering research field tests are agreed to be safe. What the foregoing also makes clear is that thresholds can accomplish this goal only when they are combined with other governance mechanisms.

(2) Moratoria – full or partial. A moratorium differs from a ban in that a moratorium is temporary.[216] Nonetheless, it creates risks akin to those that attend outright permanent bans on climate engineering research – chiefly by inhibiting scientists who are likely to conduct research in a responsible manner,[217] but not necessarily those that are more likely to do so irresponsibly.[218] Thus most proposals for moratoria on one or more forms of climate engineering research call for *partial* moratoria,[219] or what the German Environment Agency calls a "general prohibition with exemptions."[220] The London Convention/London Protocol (LC/LP) govern OIF research and deployment in this way, albeit in a legally nonbinding fashion, by maintaining a moratorium on all OIF unless it satisfies the LC/LP Assessment Framework. The same outcome could be accomplished for stratospheric aerosol injection and marine cloud brightening by placing a moratorium on any experiment or experiments that would have a cumulative worldwide diminishment of insolation above a threshold amount.[221]

In addition to these actual and proposed moratoria which focus on the effects of climate engineering and its research, others, including the NRC and the authors of the draft Berlin Declaration, have proposed moratoria on SRM field testing pending the establishment of an international research governance architecture.[222]

(3) An "allowed zone." Integrating thresholds and moratoria in a way that is sensitive to concerns about needlessly impeding useful research leads to what has been called an "allowed zone."[223] Much like a Categorical Exclusion under NEPA, that zone would pertain to "small experiments that operate within agreed, safe limits"[224] and would exempt researchers from facing more burdensome requirements that might accompany experiments which exceed those limits.[225] Importantly, it would not constitute an exemption from all governance measures, and indeed its existence would rely on an encompassing governance scheme that imposes at least reporting requirements on research into one or more climate engineering approaches.[226]

Thus an "allowed zone" need not give rise to concerns that imposing an artificial distinction between research and deployment could invite a slippery slope or lock-in scenario. Rather, it fits nicely with a stepwise approach to climate engineering research governance that understands preliminary model-based and small-scale field testing phases to be part of, rather than distinct from, subsequent phases of larger-scale testing and potential deployment.[227] In a 2014 paper, Professor David Keith and his coauthors depict this sort of process schematically for SRM research (see Figure 6.1), and show how it would incorporate predetermined decision points and criteria for proceeding

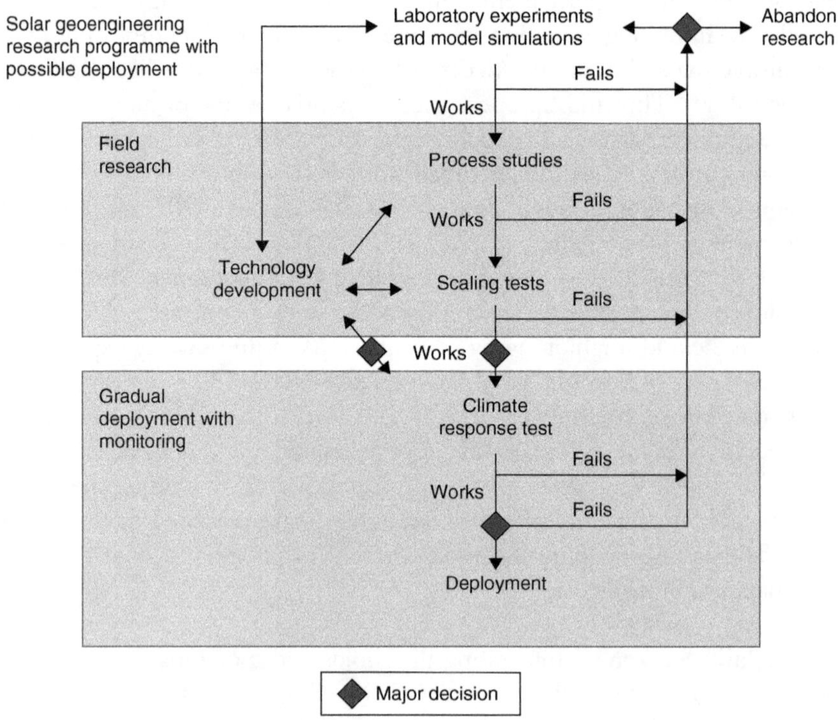

FIGURE 6.1 Research program paths and decisions

Taken from David W. Keith et al., *Field Experiments on Solar Geoengineering: Report of a Workshop Exploring a Representative Research Portfolio*, 372 *Phil. Transactions Royal Soc'y* 1, 5 (December 2014)

from laboratory testing to field testing, and from field testing to "gradual deployment with monitoring."[228]

By incorporating perspectives informed by political as well as scientific priorities into decision points, such a process would also ensure that those decisions are legitimate as well as responsible, and that the process as a whole is designed to shut down particular lines of inquiry on either basis.

6.3.3. *By Whom*

Climate engineering techniques – such as direct air capture and stratospheric aerosol injection, to name just two examples – are too unlike each other in terms of goals, containment of environment impacts, safety, scope of effect, and leverage for their research and deployment to be governed in the same way.[229] Accordingly, governance of climate engineering research is inevitably going to be "clumsy" because it will evolve toward oversight through an untidy compilation of overlapping authorities,[230] including international bodies, national governments, and the scientific community.[231] The Institute for Advanced Sustainability Studies (IASS) commentary on its proposed Code of Conduct elaborates on this vision:

> In terms of instrument choice, there are widespread calls for a flexible governance framework for research activities that interacts at multiple levels. There is also a need for an instrument that reaches beyond the traditional sphere of international law, in which States remain the principal actors, to involve other sectors of society, including intergovernmental and nongovernmental organisations, companies, and scientific institutions, academies and individual scientists in order to respond to the transnational demands of climate governance.[232]

As the IASS notes, the Food and Agriculture Organization (FAO) Code of Conduct for Responsible Fisheries may serve as a useful model for governing climate engineering research at this early stage. This is so because the FAO's code:

- was negotiated by states, but is directed at state and nonstate actors, including members of the scientific community;
- sought to provide a harmonizing instrument that would enhance governance, in part by facilitating the implementation of relevant existing agreements; and
- recognized that, as it was a nonbinding guidance document, it would foster subsequent iterations of governance rather than being the last word on fisheries governance.[233]

These points suit the present moment in climate engineering research governance, and may inform efforts to stitch together a comprehensive governance framework.

Such a framework must begin by matching up existing institutions to proposed climate engineering approaches based on those institutions' jurisdictions and competencies.[234] This match yields three groups of institutions that could oversee climate engineering governance: (1) existing institutions, (2) adapted versions of existing institutions, and (3) new institutions for spaces where regulatory gaps exist.[235]

6.3.3.1. Governance by Existing Institutions

Research into boosting albedo using desert reflectors or other land-based forms of CDR can likely be governed adequately by existing scientific and governmental institutions.[236] Though these approaches vary in significant ways, on a research (as opposed to deployment) scale they all share high degrees of encapsulation and reversibility, low degrees of leverage or cost-effectiveness, and promise limited (if any) transboundary impacts.[237] As such, experimentation involving small or even mid-scale field testing will ordinarily not implicate international legal requirements related to transboundary harms.[238] Thus national governments and the scientific community are well positioned to monitor and coordinate such research, and to resolve disputes arising from its economic and environmental effects. This is not to say that national governments have grasped the nettle of climate engineering research governance (they have not),[239] but only that they are able to do so in a way that addresses the concerns and performs the functions described in Parts II and III.A above.

6.3.3.2. Governance by Adapted Institutions

The recent history of governance of research into OIF provides a helpful illustration of how adaptation by existing institutions can take place. It also provides something of a blueprint for possible adaptations for governing research into other climate engineering approaches, such as marine cloud brightening.[240]

In June 2007 Planktos, Inc., a for-profit entity seeking revenue from carbon offsets, planned to discharge 100 tons of iron dust into waters near the Galapagos Islands.[241] The plan, which was never carried out, spurred the Scientific Group of the LC/LP to issue a Statement of Concern that called for the Contracting Parties to address OIF at their next meeting.[242] In 2008 the parties resolved that "ocean fertilization" is subject to the LC/LP and that, "given the present state of knowledge, ocean fertilization activities other than legitimate scientific research should not be allowed."[243] That resolution also called for the Scientific Group to develop an assessment process for proposed projects. In October 2010 the parties issued the Assessment Framework for Scientific Research Involving Ocean Fertilization.[244]

In 2010, even while the LC/LP Scientific Group issued the OIF Assessment Framework and the parties to the Convention on Biological Diversity incorporated it by reference,[245] several members of the United Nations Educational, Scientific and Cultural Organization (UNESCO)'s Intergovernmental Oceanographic Commission (IOC) resisted a proposal to give UNESCO oversight over OIF.[246] Those members – including the USA, Britain, and India – prevailed in their argument that all salient aspects of OIF research governance were being overseen adequately pursuant to the LC/LP, and the IOC's input should be scientific only.[247] As subsequent experience with the Haida Corporation and Canadian government in 2012 demonstrated,[248] however, this consolidation of jurisdictional authority over OIF and related research still relied entirely on national governments to monitor and enforce compliance with the LC/LP – a legally nonbinding regime.[249]

This series of steps sorted out which international body would take responsibility for guiding decisions about OIF research, educated national governments about their role in implementing that guidance, and also informed the NGO and scientific communities about the legal aspects of OIF research governance. In sum, while the unfolding of these steps appears somewhat messy in retrospect, it is also clear that existing national and international institutions adapted themselves quickly – or at least not slowly – to fill a gap in governance.

6.3.3.3. Governance Gaps for which New Institutions Are Needed

To be effective, the governance of climate engineering research that has potentially significant transboundary impacts must restrict, encourage, and steer various activities, and must do so in a way that provides scientific certainty, political and international legitimacy, and mechanisms for dispute resolution.[250] No national governments or international bodies have as yet equipped themselves to handle this composite task. As discussed at length in Chapter 4 of this volume, several US laws would likely require the review and permitting of stratospheric aerosol injection activities that occur within US jurisdiction. The same is true of laws in Germany[251] and Britain,[252] and presumably in many other jurisdictions as well. But multiple pieces must fit together to govern climate engineering research effectively. Relying on inchoate and mismatched protections such as national laws currently provide will not address all of the concerns that surround climate engineering research.

Similarly, while several international agreements would seem to be candidates for filling the governance gap created by those activities' transboundary atmospheric effects, multiple analyses have considered and rejected them all.[253] For instance, the 1987 Montreal Protocol on Substances that Deplete the Ozone Layer and its parent agreement, the Vienna Convention on the Protection of the Ozone Layer, *could* provide regulatory oversight,[254] but these agreements concern only the restoration and protection of the ozone layer and seek to facilitate the substitution of mature single-use technologies, and they do not address the enormous question of whether and how to intervene in the operation of the climate.[255] Similarly, the 1984 and 1995 Protocols to the 1979 Convention on Long-Range Transboundary Air Pollution, which impose binding limits on sulfur dioxide emissions and so already in effect govern sulfate aerosol injection in some contexts, are clearly a mismatch for climate engineering research governance: The Convention and its Protocols are oriented to protect "against air pollution," seek to "gradually reduce and prevent air pollution including long-range transboundary air pollution,"[256] and impose thresholds far above the levels contemplated by even mezo-scale field testing.[257] A third example, the UNFCCC, is in many respects ideally suited to perform climate engineering research governance functions: It is climate-oriented, established, capable of addressing scientific and political issues, and has nearly universal participation among states. However, several analyses reject it because it was created chiefly to pursue climate change *mitigation*[258] and it carries out its mandates through a time-consuming consensus-based process.[259] Finally, the Outer Space Treaty of 1967 has a narrow focus and no dedicated secretariat or other institutional infrastructure.[260]

SRMGI, established in 2010 by the Royal Society, Environmental Defense Fund, and the World Academy of Sciences to begin a discussion about "effective governance arrangements for potentially risky research,"[261] has arguably done just that. However, that discussion has not yet given rise to purpose-built governance institutions at the national or international level.

National governments have established several possible models for gap-filling institutions, capable of providing for international coordination and oversight of the development of other new and consequential technologies, such as nuclear power development,[262] nuclear waste disposal,[263] and, most recently, genetically modified organisms.[264] To date, however, and notwithstanding increasingly frequent calls to do so,[265] national governments and international bodies have, at best, only begun to create effective climate engineering governance norms, mechanisms, and institutions for those climate engineering approaches that are most likely to be both effective and dangerous.

6.4. CONCLUSION

There is broad agreement that climate engineering research must be governed, yet this consensus has failed to spur substantial efforts by national or international actors to create and administer appropriate forms of governance. London Convention/London Protocol Resolutions on ocean iron fertilization research represent a partial exception to this point – and they would be wholly exceptional if national governments would act to implement those resolutions' key provisions. The lack of governance has variable effects: It stymies some research efforts while failing to touch others. But climate engineering proposals cannot simply be ignored. Rather, as the effects of climate change grow increasingly severe, and as the need to find non-conventional means to address global greenhouse gas concentrations grows pressingly urgent, it seems quite likely that climate engineering research – including field tests that verge on deployment – will also become more prevalent. The lack of an effective governance regime could leave the climate engineering research community, and the societies potentially affected by its decisions, susceptible to the various concerns noted in Section 6.2 above, including moral hazard, forum shopping, slippery slopes, governance traps and international tensions, or even conflicts among nations. Accordingly, there is a very real and increasingly urgent need to answer the key questions surrounding governance: What qualifies as climate engineering research that subject to governance, at what point do governance requirements kick in, what substantive rules should apply, and who should do the governing? Though this chapter provides no definitive answers to these questions, it has sought to map the terrain for researchers and decision-makers and to explain the importance of moving ahead.

NOTES

1 Committee on Geoengineering Climate, National Research Council, Climate Intervention: Reflecting Sunlight to Cool Earth (2015) [hereinafter NRC, Reflecting Sunlight]; Stefan Schäfer et al., The European Transdisciplinary Assessment of Climate Engineering (EuTRACE): Removing Greenhouse Gases from the Atmosphere and Reflecting Sunlight away from Earth 111 (2015), bit.ly/2oJsjB8; John Shepherd et al., Royal Society, Geoegineering the Climate: Science, Governance and Uncertainty (2009) [hereinafter Royal Society]; *see also Geoengineering I: Assessing the Implications of Large-Scale Climate Intervention: Hearing Before the H. Comm. on Sci. & Tech.*, 111th Cong. 83 (Statement of Chairman Gordon: "I

think we can submit unanimously that this panel would say that there should be no deployment, only research. I don't think you are going to find anybody that is going to disagree with that").

2 *Compare* Edward A. Parson & David W. Keith, *End the Deadlock on Governance of Geoengineering Research,* 339 *Science* 1279 (Mar. 15, 2013) (proposing technical threshold for categories of field research to be subject to or exempt from governance), *with* Clive Hamilton, Geoengineering: Governance Before Research Please, Our World-United Nations University (Sept. 27, 2013), http://bit.ly/1NXs697, *and* Editorial: A *Charter for Geoengineering,* 485 *Nature* 415 (May 2012), bit.ly/1mFihma ("Geoengineers should … come together and draft detailed, practical actions that need to be taken to advance governance in the field").

3 *See, e.g.*, Government Response to the House of Commons Science and Technology Committee 5th Report of Session 2009–10: The Regulation of Geoengineering 4 (Sept. 2010) ("The Government considers it too early to be able to establish appropriate regulatory frameworks for geoengineering research or deployment on a comprehensive basis without a clear view of what needs to be regulated and how"); David G. Victor, *On the Regulation of Geoengineering,* 24 *Oxford Review of Economic Policy* 322, 323–24 (2008).

4 *See* Editorial: A *Charter for Geoengineering, supra* note 2 ("More troubling is the lack of an overarching governance framework … Geoengineers should keep trying. They should come together and draft detailed, practical actions that need to be taken to advance governance in the field"); Parson & Keith, *supra* note 2, at 1278; Daniel Bodansky, *The Who, What, and Wherefore of Geoengineering Governance,* 121 *Climatic Change* 539, 546 (Dec. 2013) ("The rationale for international governance in this scenario is to encourage national spending, develop cost-sharing arrangements, and incentivize private investment").

5 Piers Forster, *Not Enough Time for Geoengineering to Work?*, Bulletin of the Atomic Scientists, Feb. 2, 2015, http://bit.ly/1NRN2Ox ("None of the proposed technologies really exist on anything other than paper"); David W. Keith et al., *Field Experiments on Solar Geoengineering: Report of a Workshop Exploring a Representative Research Portfolio,* 372 *Philosophical Transactions of Royal Society* 175 *(Dec. 2014)* (setting out taxonomy of proposed research projects). The article generally credited as incipient of broader discussions of geoengineering was published in 2006. *See* Paul J. Crutzen, *Albedo Enhancement by Stratospheric Sulfur Injections,* 77 *Climatic Change* 211 (2006).

6 GAO, GAO-10–903, Climate Change: A Coordinated Strategy Could Focus Federal Geoengineering Research and Inform Governance Efforts 34 (2010); *see also* Jane C. S. Long et al., *Start Research on Climate Engineering,* 518 *Nature* 29 (Feb. 5, 2015) ("governance and experimentation must co-evolve").

7 Cao Long et al., *Geoengineering: Basic Science and Ongoing Research Efforts in China,* 6 *Advances in Climate Change Research* 188, 193 (2015), http://bit.ly/22wOUmr.

8 NRC, Reflecting Sunlight, *supra* note 1, at 6 (recommending investment into R&D for carbon removal and sequestration), 10 (recommending launch of albedo modification research program).

9 Rob Bellamy, *Safety First! Framing and Governing Climate Geoengineering Experimentation,* Climate Geoengineering Governance Working Paper No. 14, at 33 (2014).

10 *See* Kelsi Bracmort & Richard K. Lattanzio, Cong. Res. Serv., Geoengineering: Governance and Technology Policy 3 (Nov. 2013) ("Little research has been done

on most geoengineering methods, and no major directed research programs are in place. Peer reviewed literature is scant"); *see also* Committee on Geoengineering Climate, National Research Council, Climate Intervention: Carbon Dioxide Removal and Reliable Sequestration 86 (2015) [hereinafter NRC, Carbon Removal] (indicating that nearly all CDR technologies exist only in pre-prototype or prototype form).

11 *See, e.g.*, Kevin E. Trenberth & Aiguo Dai, *Effects of Mount Pinatubo Volcanic Eruption on the Hydrological Cycle as an Analog of Geoengineering*, 34 *Geophysical Research Letters* 1 (Aug. 2007).

12 *See, e.g.*, Ben Kravitz et al., *The Geoengineering Model Intercomparison Project (GeoMIP)*, 12 *Atmospheric Science Letters* 162 (2011) (reporting results of comparison of how each of several climate models predicts effects of stratospheric geoengineering with sulfate aerosols).

13 *See* NRC, Reflecting Sunlight, *supra* note 1, at 188 (recommending that, for some approaches, "[s]mall-scale field experiments" will help reduce uncertainty, verify modeling, and validate theory); Bellamy, *supra* note 9, at 13–14 (reporting research steps anticipated by several authors: non-field experiments, field experiments, and "de facto experiments in deployment"); Jane C. S. Long et al., Bipartisan Pol'y Ctr., Task Force on Climate Remediation Research 29 (2012) [hereinafter "BPC Task Force"] (anticipating "large-scale field tests" will occur "at some point"); Solar Radiation Management Governance Initiative, Solar Radiation Management: The Governance of Research 45–53 (2011) (listing steps or successive "categories" of research activity).

14 *See* Edward A. Parson & Lia N. Ernst, *International Governance of Climate Engineering*, 14 *Theoretical Inquiries in Law* 307, 329 (2013) ("there will not be a clean boundary between an early period of 'scientific' governance and some later period of 'operational' governance"); M. Granger Morgan and Katharine Ricke, Cooling the Earth through Solar Radiation Management: The Need for Research and an Approach to its Governance, 19 (2010) ("one of the first objectives of an SRM research programme should be to give more precise meaning to the phrase 'modest low-level [field testing]'"); Royal Society, *supra* note 1, at 41 ("In some cases … it is not clear that field trials can usefully be conducted on a limited scale, or without appreciable and widespread environmental impacts").

15 Press Release, Alfred-Wegener-Institute for Polar and Marine Research, The Indo-German Iron Fertilization Experiment LOHAFEX (Jan. 13, 2009), bit.ly/23JLBZN.

16 NRC, Carbon Removal, *supra* note 10, at 58.

17 *See, e.g.*, Fortunat Joos & Ulrich Siegenthaler, *Possible Effects of Iron Fertilization in the Southern Ocean on Atmospheric CO_2 Concentration*, 5 *Global Biogeochemical Cycles* 135 (1991).

18 Phillip Williamson et al., *Ocean Fertilization for Geoengineering: A Review of Effectiveness, Environmental Impacts and Emerging Governance*, 90 *Process Safety & Environmental Protection* 475 (Nov. 2012) (listing 13 ocean fertilization field experiments conducted between 1993 and 2009: IronEx I (1993); IronEx II (1995); SOIREE (Southern Ocean Iron Enrichment Experiment, 1999); EisenEx (2000); SEEDS I (Subarctic Pacific Iron Experiment for Ecosystem Dynamics Studies, 2001); SOFeX North and SOFeX South (Southern Ocean Iron Experiment, 2002); SERIES (Subarctic Ecosystem Response to Iron Enrichment Study, 2002);

CY CLOPS (Cycling of Phosphorus in Eastern Mediterranean, 2002); EIFEX, European Iron Fertilization Study, 2004); SEEDS II (2004); SAGE (SOLAS Air–Sea Gas Exchange experiment, 2004); FEEP (2004); and LOHAFEX (2009)).

19 Resolution LC-LP.1 (2008); *see also* Scientific Group of London Convention and London Protocol, Int'l Maritime Org., Statement of Concern Regarding Iron Fertilization of the Oceans to Sequester CO_2 at 1, IMO Ref. T5/5.01, LC-LP.1/Circ. (14) (July 13, 2007).

20 Press Release, Statement von Bundesforschungsministerin Annette Schavan zum deutsch-indischen LOHAFEX-Experiment im Südatlantik, Jan. 27, 2009, https://perma.cc/XZ6T-FFA7.

21 *See* UN Convention on Biological Diversity Decision XI/16 "Ocean Fertilization" (May 2008) (calling for a moratorium on OIF).

22 Mason Inman, *Planning for Plan B*, 4 *Nature Climate Change* 7, 7–8 (2010); *see also* ETC Group, *LOHAFEX Update: Throwing Precaution (and Iron) to the Wind (and Waves)*, Jan. 28, 2009, bit.ly/1SYVH4W.

23 Statement von Bundesforschungsministerin Annette Schavan zum deutsch-indischen LOHAFEX-Experiment im Südatlantik, Jan. 27, 2009 (author's translation).

24 German Ministry for Environment, Press Release No. 025/09: Federal Environment Ministry Regrets Approval by Federal Research Ministry of Iron Enrichment Experiment (Jan. 26, 2009), https://perma.cc/ZCK6-NXL7 ("takes note with regret of the decision by the Federal Research Ministry to approve the LOHAFEX experiment").

25 Deutscher Bundestag, 16. Wahlperiod, Antwort der Bundesregierung auf die Kleine Anfrage der Abgeordneten Eva Bulling-Schröter, Lutz Heilmann, Hans-Kurt Hill und der Fraktion Die Linke, Drucksache 16/11689: Polarstern-Expedition und Algendüngung, Feb. 9, 2009, http://bit.ly/1IdIfXF (responding to a parliamentary query about the decision to authorize the conduct of an experiment).

26 Catherine Brahic, *Hungry Shrimp Eat Climate Change Experiment*, *New Scientist*, Mar. 25 2009, http://bit.ly/1RtZVkw ("Early results from the latest field experiment suggest the technique will fail"); *see also* NRC, Carbon Removal, *supra* note 10, at 63 ("Given these limitations and unknowns, the committee concludes that the risks and costs [of OIF] currently outweigh the benefits"), 109 ("there is a near consensus that at climatically relevant levels of deployment potential risks [of OIF] outweigh potential benefits").

27 Yuri A. Izrael et al., *Field Studies of a Geo-Engineering Method of Maintaining a Modern Climate with Aerosol Particles*, 34 *Russian Meteorology & Hydrology* 635 (2009); Yuri A. Izrael et al., *Field Experiment on Studying Solar Radiation Passing through Aerosol Layers*, 34 *Russian Meteorology & Hydrology* 265 (2009).

28 Eli Kintisch, *Exclusive Excerpt: Hack the Planet*, WIRED, Mar. 24, 2010, http://bit.ly/1QUFMm6.

29 Aaron Welch et al., *Climate Engineering: The Way Forward?*, 2 *Environmental Device* 57, 63 (2012).

30 *Id.*

31 *See* note 27, *supra*.

32 Kintisch, *supra* note 28; V. P. Meleshko et al., *Is Aerosol Scattering in the Stratosphere a Safety Technology Preventing Global Warming?*, 35 *Russian Meteorology &*

Hydrology 433, 434 (2010) (citing Yuri A. Izrael, *An Efficient Way to Regulate the Global Climate Is the Main Objective of the Solution of the Climate Problem*, No. 10 *Russian Meteorology & Hydrology* (2005)).

33 The other two are the Integrated Assessment of Geoengineering Proposals, http://iagp.ac.uk/, and the Climate Geoengineering Governance project, Geoengineering Governance Research, *About the Project*, http://bit.ly/1RXfyjI.

34 SPICE Project, *Aims and Background*, http://bit.ly/1Yw9vCs.

35 *See* Jack Stilgoe, *Experiment Earth* 1 (Routledge, 2015).

36 Erin Hale, *Geoengineering Experiment Cancelled Due to Perceived Conflict of Interest*, *The Guardian*, May 16, 2012, http://bit.ly/1OiY95e.

37 Letter from ETC et al. to Chris Huhne, Secretary of State for Energy and Climate Change, Sept. 26, 2011, http://bit.ly/1Ik9EHf.

38 *Id.*; *see also* Stilgoe, *supra* note 35, at 139; Nick Pidgeon et al., *Deliberating Stratospheric Aerosols for Climate Geoengineering and the SPICE Project*, 3 *Nature Climate Change* 451, 455 (2013).

39 Phil Macnaghten and Richard Owen, *Good Governance for Geoengineering*, 479 *Nature* 293 (2011); *see also* SPICE, Work Package 2: Delivery Methods – Putting Particles up into the Stratosphere, http://bit.ly/1Yq3BTs ("The design produced by Work Package 2 will not now be tested outside … The decision to call-off the test-bed was made by the project management team because external delays to the project had reduced the time available for performing adequate stakeholder engagement prior to any test-bed activities. As such it was felt inappropriate to continue with a test-bed without such stakeholder engagement").

40 Lynn M. Russell, *Offsetting Climate Change by Engineering Air Pollution to Brighten Clouds*, The Bridge: Frontiers of Engineering, Winter 2012, at 10 ("The E-PEACE results provide a proof of concept that cloud brightening to reduce global mean warming is possible, with existing, decades-old technology, for some cloud conditions").

41 J. Latham, *Control of Global Warming?*, 347 *Nature* 339 (1990); *see also* J. Latham et al., *Marine Cloud Brightening*, 370 *Philosophical Transactions of Royal Society* A 4217 (2012); Christopher Mims, *"Albedo Yachts" and Marine Clouds: A Cure for Climate Change? A Deep Dive into One of the Least Scary Geoengineering Schemes to Control Global Warming*, *Scientific American*, Oct. 21, 2009, http://bit.ly/1OXBKEV; S. Twomey, *Influence of Pollution on the Short-Wave Albedo of Clouds*, 34 *Journal of the Atmospheric Sciences* 1149 (1977).

42 *See* Gary Cooper et al., *Preliminary Results for Salt Aerosol Production Intended for Marine Cloud Brightening, Using Effervescent Spray Atomization*, *Philosophical Transactions of the Royal Society* A 372: 20140055, at 1, 9–10 (2014), http://dx.doi.org/10.1098/rsta.2014.0055 ("existing snow making equipment can be adapted to launch the nuclei 60–100m into the air, requiring approximately 20kW of additional power"); Robert Wood & Thomas P. Ackerman, *Defining Success and Limits of Field Experiments to Test Geoengineering by Marine Cloud Brightening*, 121 *Climatic Change* 459 (2013); Stephen Salter et al., *Sea-Going Hardware for the Cloud Albedo Method of Reversing Global Warming*, 366 *Philosophical Transactions of the Royal Society* A 3989 (2008), http://bit.ly/1OwINtX.

43 Lynn M. Russell et al., *Eastern Pacific Emitted Aerosol Cloud Experiment*, *American Meteorological Society* 709, 713–14 (May 2013).

44 *See id.*

45 Richard Black, *Geoengineering: Risks and Benefits*, *BBC News*, Aug. 24, 2012, http://bbc.in/1NOWoto; Jack Stilgoe et al., *Public Engagement with Biotechnologies Offers Lessons for the Governance of Geoengineering Research and Beyond*, 11 *PLoS Biology* e1001707 (Nov. 2013), http://1.usa.gov/1kqUu01 ("The E-PEACE project has clear relevance to geoengineering research but has not labelled itself as a geoengineering research project, so has avoided some of the attention attracted by SPICE"); *see also* R. Wood et al., *The VAMOS Ocean–Cloud–Atmosphere–Land Study Regional Experiment (VOCALS-REx): Goals, Platforms, and Field Operations*, 11 *Atmospheric Chemistry and Physics* 627 (2011), http://bit.ly/22xfGek (similarly omitting mention of geoengineering from characterization experiment's motive or results).

46 Jeff Tollefson, *Ocean-Fertilization Project Off Canada Sparks Furor*, 490 *Nature* 458 (Oct. 2012).

47 *Id.*; Statement of Concern Regarding the Iron Fertilization in Ocean Waters West of Canada, LC 34/15 (2012), Annex 7, paras. 1–2 ("Parties to the London Convention and London Protocol (LC/LP) express grave concern regarding the deliberate ocean fertilization activity that was recently reported to have been carried out in July of 2012 in waters off the Canadian west coast").

48 Tolefsson, *supra* note 46, at 458.

49 *Id.* (quoting the LOHAFEX principle investigator as saying: "I'm not going to condemn it offhand, but this is just not the way to do this experiment").

50 *Compare* Parson & Keith, *supra* note 2, at 1278 ("the project was apparently done without knowledge of Canadian authorities, yet violated no international law"), *with* Robert F. Service, *Legal? Perhaps. But Controversial Fertilization Experiment May Produce Little Science*, *Science*, October 23, 2012 ("while the London Protocol is binding, it only applies to the release of material intended to be dumped as waste, not released as part of a scientific experiment"), *and* Dene Moore, *Ocean Fertilization Experiment Loses in B.C. Court; Charges Now Likely*, *The Globe & Mail*, February 3, 2014, http://bit.ly/1OhRovX.

51 Grant Wilson, *Murky Waters: Ambiguous International Law for Ocean Fertilization and Other Geoengineering*, 49 *Texas International Law Journal* 507, 544 (2014) ("determining whether Canada failed at its duty to enforce the LP is difficult").

52 *See* Peng Xiu et al., *Satellite Bio-Optical and Altimeter Comparisons of Phytoplankton Blooms Induced by Natural and Artificial Iron Addition in the Gulf of Alaska*, 145 *Remote Sensing of Environment* 38 (2014); Sonia D. Batten & James F. R. Gower, *Did the Iron Fertilization Near Haida Gwaii in 2012 Affect the Pelagic Lower Trophic Level Ecosystem?*, 36 *Journal of Plankton Research* 925 (2014).

53 Haida Salmon Restoration Corporation, *References and Citations*, http://bit.ly/2aqhv7R. The company has also, however, fired and brought suit against Russ George, who spearheaded the controversial experiment. *See Geo-engineering Guru Misled Haida Group, Documents Allege*, *Epoch Times*, Feb. 26, 2014, bit.ly/1W9a70W. For further information on legal proceedings and investigations concerning this experiment, *see* Joshua Learn, Geoengineering: Legal Mess Hampers Understanding of a Major CO_2 Sequestration Test, *Climate Wire*, Nov. 13, 2014, http://bit.ly/29nIWPh.

54 Royal Society, *supra* note 1; *see also* Climate Change 2013: The Physical Science Basis, Working Group I Contribution to the Fifth Assessment Report of the Intergovernmental

Panel on Climate Change 29 (Sept. 2013) ("Methods that aim to deliberately alter the climate system to counter climate change"); *see also* Wilfried Rickels et al., Kiel Earth Institute, Large-Scale Intentional Interventions into the Climate System? Assessing the Climate Engineering Debate; Scoping Report Conducted on Behalf of the German Federal Ministry of Education and Research (2011).

55 *See* Olivier Boucher et al., *Rethinking Climate Engineering Categorization in the Context of Climate Change Mitigation and Adaptation*, 5 *WIREs Climate Change* 23, 27 (2013) ("Planting trees on a small plot of land or whitening roofs in a small urban area do not constitute [climate engineering] as their impact on the climate system will be negligible … [Climate engineering] can be considered to start when there is a measurable climate impact at the regional or global scale"). Notably, the definition used by the National Academies of Science does not include a scalar criterion, and so encompasses all "intentional efforts to remove carbon dioxide from the atmosphere" and "to increase the amount of sunlight that is scattered or reflected back to space." NRC, Carbon Removal, *supra* note 10, at 2 (Box S.1). By contrast, the IPCC and the Umweltbundesamt (German Federal Environment Agency) emphasize the relevance of scale to its preferred definition. *See* IPCC Working Group III Technical Support Unit, Meeting Report of the IPCC Expert Meeting on Geoengineering 2 (Ottar Edenhofer et al., eds. 2012), bit.ly/1l5qICW ("scale and intent are of central importance"); Ralph Bodle, Sebastian Oberthür et al., Umweltbundesamt, Options and Proposals for the International Governance of Geoengineering 45–47 (2014) [hereinafter UBA], bit.ly/1RMhew5.

56 *See, e.g.*, Anna-Maria Hubert & David Reichwein, Inst. for Advanced Sustainability Studies, An Exploration of a Code of Conduct for Responsible Scientific Research Involving Geoengineering 18 (May 2015) [hereinafter IASS Code of Conduct] (proposing a definition for inclusion in the Code of Conduct for Responsible Scientific Research Involving Geoengineering that elaborates slightly on the Royal Society's definition).

57 House of Commons (UK) Science and Technology Committee, *The Regulation of Geoengineering*, Fifth Report of Session 2009–10, at 20, Mar. 18, 2010, http://bit.ly/1j7aLvE; BPC Task Force, *supra* note 13, at 7 (same); *see also* Boucher et al., *supra* note 55, at 23 (noting overlap and suggesting clarification through proposed categories and subcategories).

58 Boucher et al., *supra* note 55, at 26–27, 32 (noting overlaps and suggesting clarification through proposed categories and subcategories).

59 *See* UBA, *supra* note 55, at 42–44; Phil Macnaghten & Bronislaw Szerszynski, *Living the Global Social Experiment: an Analysis of Public Discourse on Solar Radiation Management and Its Implications for Governance*, 23 *Global Environmental Change* 465, 466 (2013) ("unlike many technoscientific issues, the distinctiveness of [SRM] does not lie in the use of novel technologies with new properties: the actual interventions themselves typically involve mundane technologies … albeit deployed at a very large scale. Its novelty rather lies in the intention to use these technologies to establish a radically new relationship between society and nature, through a project of bringing planetary systems under human control and the 'making' of new climates").

60 For a discussion of the importance of categorization and naming to the formulation of climate engineering research programs, *see* Rose C. Cairns, *Climate*

Geoengineering: Issues of Path-Dependence and Socio-Technical Lock-In, 5 *WIREs Climate Change* 649, 656–57 (2014).

61 *See also* David W. Keith et al., *supra* note 5 (listing field experiments and identifying their relationships to various geoengineering approaches).

62 *See* Solar Radiation Management Governance Initiative, Solar Radiation Management, The Governance of Research 23 (2011) [hereinafter SRMGI], http://bit.ly/2axM2j7 ("The difference between a climate aerosol experiment and a [solar radiation management/geogengineering] experiment might only be the intentions of the researcher").

63 *See* Haida Salmon Restoration Corporation, *GeoEngineering*, http://bit.ly/22liMC1 ("By reading this discussion, you are Geoengineering. We all do. Nearly every activity we humans do, affects the environment in some way").

64 NRC, Reflecting Sunlight, *supra* note 1, at 10 ("Much of the required research on albedo modification overlaps considerably with basic scientific research that is needed to improve understanding of the climate system").

65 ARM Climate Research Facility, Campaign: ARM Cloud Aerosol Precipitation Experiment (ACAPEX) 2015.01.14 to 2015.02.12, http://1.usa.gov/1zyKhLs (describing an experiment which involved land-, ship-, and airplane-based monitoring of interaction of particulates with precipitation from concentrated channels of atmospheric water vapor – "atmospheric rivers" – over the Pacific and the coast of California).

66 *See, e.g.*, Government Response to the House of Commons Science and Technology Committee 5th Report of Session 2009–10: The Regulation of Geoengineering 2 (Sept. 2010).

67 For a discussion of this question in relation to SRM in particular, *see* Andy Parker, *Governing Solar Geoengineering Research as It Leaves the Laboratory*, 372 *Philosophical Transactions of the Royal Society* A 20140173 (2014).

68 Secretariat of the Convention on Biological Diversity, Geoengineering in Relation to the Convention on Biological Diversity: Technical and Regulatory Matters 141–42 (Sept. 2012), bit.ly/1meRtbd (discussing examples); *see also* Chiara Armeni & Catherine Redgwell, International Legal and Regulatory Issues of Climate Geoengineering Governance: Rethinking the Approach, CGG Working Paper No. 23, at 28–30 (2015), bit.ly/1SsWqLt (discussing the distinction and approach taken by international law). This distinction also lies at the crux of the ongoing dispute over Japanese whaling, which the Japanese insist occurs for research purposes. *See Whaling in the Antarctic (Austl. v. Japan: N.Z. Intervening)*, Judgment, 2014 I.C.J. 148, para. 127 (Mar. 31), https://perma.cc/252G-Y25Z ("the Court's examination of the evidence with respect to JARPA II will focus on whether the killing, taking and treating of whales in pursuance of JARPA II is *for purposes of* scientific research"; original emphasis).

69 *See* Resolution LC-LP.2 (2010) on the Assessment Framework for Scientific Research Involving Ocean Fertilization (adopted Oct. 14, 2010), bit.ly/1ndAQoO (The Framework itself is Annex 6 of the Resolution). Specifically, the Assessment Framework is "designed for Contracting Parties to evaluate proposed activities that fall within the scope of resolution LC-LP.1 (2008)," which declares "ocean fertilization" to be subject to the London Convention and states that "given the present state of knowledge, ocean fertilization activities other than legitimate

scientific research should not be allowed." Resolution LC-LP.1(2008) on the Regulation of Ocean Fertilization (adopted Oct. 31, 2008), bit.ly/1QoscOo.

70 *See, e.g.*, Parson & Keith, *supra* note 2, at 1278; Lynn M. Russell et al., *Ecosystem Impacts of Geoengineering: A Review for Developing a Science Plan*, 41 *Ambio* 350, 363 (2012) ("small-scale (i.e., on the order of 10 km in size) geoengineering field experiments").

71 NRC, Reflecting Sunlight, *supra* note 1, at 6–9.

72 Alan Robock, *20 Reasons Why Geoengineering Is a Bad Idea*, 64 *Bulletin of the Atomic Scientists* 14–18, May/June 2008 (suggesting that for some geoengineering technologies deployment is the only real means of testing available).

73 UBA, *supra* note 55, at 21.

74 *See* David Collinridge, The Social Control of Technology (St. Martin's Press, 1980).

75 Bracmort & Lattanzio, *supra* note 10, at 4 ("By the time a technology is widely deployed, it may be impossible to build desireable oversight and risk management provisions without major disruptions to established interests").

76 SRMGI, *supra* note 62.

77 *See, e.g.*, Armeni & Redgwell, *supra* note 6, at 36; Schäfer et al., *supra* note 1, at 115–21 (discussing "[t]echnique-specific policy considerations"); Royal Society, *supra* note 1, at 47, 51–52.

78 Royal Society, *supra* note 1, at ix.

79 Asilomar Scientific Organizing Committee, The Asilomar Conference Recommendations on Principles for Research into Climate Engineering Techniques 8 (Nov. 2010) [hereinafter Asilomar Recommendations].

80 *Id.*

81 *See* NRC, Carbon Removal, *supra* note 10, at 4 tbl.S.1 ("Overview of General Differences Between Carbon Dioxide Removal Proposals and Albedo Modification Methods"), 20 ("Box 1.1 Why There Are Two Separate Reports"). Somewhat confusingly, the NRC reports refer to both CDR and SRM as forms of "climate intervention." *See id.* at 2, box S.1. This Chapter adopts the Asilomar terminology and refers to SRM only as an "intervention."

82 NRC, Reflecting Sunlight, *supra* note 1, at 1.

83 *Id.* at vii.

84 NRC, Carbon Removal, *supra* note 10, at 104.

85 Cao Long et al., *supra* note 7.

86 Boucher et al., *supra* note 55, at 29–30.

87 *Id.* at 23.

88 Bellamy, *supra* note 9, at 12–13 (summarizing categorizations discussed in 12 leading publications).

89 Royal Society, *supra* note 1, at 38–39; Bracmort & Lattanzio, *supra* note 10, at 5 (same).

90 Boucher et al., *supra* note 55, at 28.

91 NRC, Reflecting Sunlight, *supra* note 1, at 152 (discussing the "double moral hazard" that attends geoengineering).

92 Albert C. Lin, *Does Geoengineering Present a Moral Hazard?*, 40 *Ecology Law Quarterly* 673, 689 (2013).

93 *Id.* at 707 (identifying "a considerable danger that geoengineering will undermine mitigation and adaptation efforts").

94 *See* Jane C. S. Long & Dane Scott, *Vested Interests and Geoengineering Research*, 29 *Issues in Science and Technology*, 45–52 Spring 2013 ("The chief executive

officer of Exxon Corporation, Rex Tillerson, articulated his opinion about climate change, glibly commenting: '... we'll adapt to that. It's an engineering problem and it has engineering solutions.' The opinion espoused by Tillerson reflects his company's vested interests").

95 *See* Clive Hamilton, "Ethical Anxieties about Geoengineering: Moral Hazard, Slippery Slope and Playing God," paper presented at Australian Acad. Sci. Conference: Geoengineering the Climate, Canberra, Sept. 27, 2011, at 11–12, http://bit.ly/2auDaeo ("'Easy' options – ones which are relatively inexpensive and do not require major economic and social upheavals – will be particularly attractive. The possible negative side-effects of geoengineering may seem less worrisome compared with the already-felt effects of global warming").

96 *See* SRMGI, *supra* note 62, at 20–22; *see also* John D. Graham & Jonathan B. Wiener, Risk vs. Risk: Tradeoffs in Protecting Health and the Environment 230 (Harvard University Press, 1997) (discussing the problem of an "omitted voice," i.e., the absence of affected parties from the decision process and the concomitant disproportionate influence of organized interests).

97 Ken Caldeira & David W. Keith, *The Need for Climate Engineering Research*, XXVII *Issues in Science and Technology* (Fall 2010), http://bit.ly/1mgtF79. *See also* V. P. Meleshko et al., *supra* note 32 (identifying this inclination in the Russian aerosol injection experiment described in section 6.1).

98 *See* Parson & Keith, *supra* note 2, at 1279 (discussing concern about forum-shopping by those seeking to conduct field tests without international sanction).

99 Kintisch, *supra* note 28 (quoting "Yuri A. Izrael, ... [s]aid to be a close confidant of Prime Minister Vladimir Putin, ... also a prominent member of the Russian Academy of Sciences," as saying: "We really will be able to control the climate").

100 *See, e.g.*, John Vidal, *Bill Gates Backs Climate Scientists Lobbying for Large-Scale Geoengineering, The Guardian*, Feb. 6, 2012, bit.ly/1bYJHKN.

101 *See* Grant Wilson, *supra* note 51.

102 *See* BBC News, *Why So Many Shipowners Find Panama's Flag Convenient*, Aug. 5, 2014, bbc.in/1oe8LUl ("Most merchant ships flying Panama's flag belong to foreign owners wishing to avoid the stricter marine regulations imposed by their own countries"); *see also* Tracy Hester, *Remaking the World to Save It: Applying U.S. Environmental Laws to Climate Engineering Projects*, in *Climate Change Geoengineering: Philosophical Perspectives, Legal Issues, and Governance Frameworks* 263, 298 (Wil C. G. Burns & Andrew L. Strauss, eds., Cambridge University Press 2013) (describing how Planktos, for-profit company planning to conduct an OIF field test, decided to avoid US EPA jurisdiction by switching to a non-US-flagged vessel).

103 Rickels et al., *supra* note 54, at 134.

104 *See* Robock, *supra* note 72, at 17–18; Mike Hulme, *Climate Change: Climate Engineering through Stratospheric Aerosol Injection*, 36 *Progress in Physical Geography*, 694 (Aug. 2012), at 694, 697.

105 *See* Bracmort & Lattanzio, *supra* note 10, at 8 ("Innovative and entrepreneurial organizations seldom mobilize themselves to put complex technologies 'on the shelf'").

106 Schäfer et al., *supra* note 1, at 107.

107 *See* Cairns, *supra* note 60, at 5.

108 A. Jones et al., *The Impact of Abrupt Suspension of Solar Radiation Management (Termination Effect) in Experiment G2 of the Geoengineering Model Intercomparison Project (GeoMIP)*, 118 *Journal of Geophysical Research Atmosphere* 9743 (2013).
109 Cairns, *supra* note 60, at 651.
110 *Id.*
111 Long & Scott, *supra* note 94.
112 Such "publication bias" occurs where "publication of study results is based on the direction or significance of the findings." Annie Franco et al., *Publication Bias in the Social Sciences: Unlocking the File Drawer*, 345 *Science* 6203 (Sept. 19, 2014), bit.ly/1J5HEaU.
113 *See, e.g.*, Kurt A. Spokas et al., *Biochar: A Synthesis of Its Agronomic Impact beyond Carbon Sequestration*, 41 *Journal of Environmental Quality* 973, 977 (2011), bit.ly/1RAqkoC ("Approximately 50% of the compiled studies observed short-term positive yield or growth impacts … However, due to potential publication biases, these percentages should only be taken as reflective of the studies presented here and not as evidence of an overall biochar likelihood of producing positive impacts").
114 *See* Dan M. Kahan, The Politically Motivated Reasoning Paradigm, *Emerging Trends in Social & Behavioral Sciences* (forthcoming), http://ssrn.com/abstract=2703011 ("individuals whose pro-market sensibilities predispose them toward rejection of climate change rate the strength of evidence that humans are responsible for global warming much more highly after being exposed to information on geoengineering – a technological "fix" that would obviate the need for limiting greenhouse gas emitting forms of commerce; likewise, individuals whose anti-market sensibilities predispose them to credit evidence on climate change in fact treat such evidence as *less convincing* after being told about research on geoengineering than when they are first briefed on the need to impose CO_2 emissions on industry"; original emphasis).
115 Mason Inman, *supra* note 22.
116 *See* Shobita Parthasarathy et al., A Public Good? Geoengineering and Intellectual Property, STTP Working Paper 10–1, at 4 (2010), https://perma.cc/AWH9-5B5Q ("Intellectual property (IP) is often unrecognized as a form of governance, but it shapes the development of technology in pivotal ways").
117 Anthony E. Chavez, *Exclusive Rights to Saving the Planet: The Patenting of Geoengineering Inventions*, 13 *Northwestern Journal of Technology and Intellectual Property* 1, 10, 13 (2015).
118 *Id.* at 13–17.
119 Memorandum submitted by Tim Kruger et al. to the House of Commons following oral testimony before Science and Technology Committee ¶5 (Feb. 2010), bit.ly/1PbVFQh.
120 IASS Code of Conduct, *supra* note 56, at 89 (identifying the "non-disclosure of commercial information" as a "barrier to transparency" and noting that environmentally protective national laws and international treaties often include exceptions to disclosure requirements for proprietary information); Parthasarathy et al., *supra* note 116, at 10–12 (highlighting several pitfalls in business-as-usual patenting, including the grant of overbroad patents that set the stage for future litigation, concentrated patent ownership, and ownership by "non-practicing entities").
121 Armeni & Redgwell, *supra* note 68, at 13 (noting potential conflict between environmental protections and intellectual property protections under the

TRIPS international agreement). Various other international agreements also contain provisions drafted to protect intellectual property rights. *See* Raymundo Valdés & Maegan McCann, *Intellectual Property Provisions in Regional Trade Agreements: Revision and Update*, WTO Staff Working Paper, No. ERSD-2014-14 (2014), bit.ly/1R87luE.

122 *See generally A Handbook on the WTO TRIPS Agreement* (Antony Taubman, Hannu Wager, Jayashree Watal, eds., Cambridge University Press 2012).

123 *See* Patricia L. Judd, *Retooling TRIPS*, 55 *Virginia Journal of International Law* 117, 137–138 (2014); *see also* Peter K. Yu, *TRIPs and Its Achilles' Heel*, 18 *Journal of Intellectual Property Law* 479 (2010–2011) (stating that, notwithstanding its enforcement provisions, it is unsurprising that TRIPs "fails to induce stronger global enforcement of intellectual property rights").

124 *See* Erica C. Smit, *Geoengineering: Issues of Accountability in International Law*, 15 *Nevada Law Journal* 1078–81 (2015) (summarizing TRIPs and describing its potential relevance to geoengineering governance); Rafael Leal-Arcas, *Climate Change and International Trade* 447–450 (Edward Elgar, 2013) (noting potential implications of international intellectual property protections for geoengineering technology development and exchange).

125 *See* Achim Maas & Irina Comardicea, *Climate Gambit: Engineering Climate Security Risks?*, in *Backdraft: The Conflict Potential of Climate Change Adaptation and Mitigation* 37, 43 (Geoffrey D. Dabelko et al., eds., Woodrow Wilson International Center for Scholars 2013) (envisioning an "arms race" for geoengineering capacity); P. Michael Link et al., *Possible Implications of Climate Engineering for Peace and Security*, 94 *Bulletin of the American Meteorological Society* ES13 (Feb. 2013), bit.ly/1niPBj3; Robert J. Lempert & Don Prosnitz, RAND, *Governing Geoengineering Research: A Political and Technical Vulnerability Analysis of Potential Near-Term Options* at xi (2011), bit.ly/1ZFdm6e (anticipating "significant challenges for risk management, national security, and international governance"); Clive Hamilton, *The Return of Dr Strangelove: The Politics of Climate Engineering as a Response to Global Warming* (June 2010), bit.ly/1SNchUu (characterizing international relations issues arising from geoengineering as analogous to those arising from nuclear proliferation).

126 Joshua B. Horton & Jesse L. Reynolds, *The International Politics of Climate Engineering: A Review and Prospectus for International Relations*, 18 *International Studies Review* 438 (2016) (reviewing literature that discusses potential direct and indirect adverse effects of geoengineering research and deployment on international relations).

127 *Id.* at 5–6, 10; *see also* Chad M. Briggs, *Is Geoengineering a National Security Risk?*, Geoengineering Our Climate? (Oct. 2013), bit.ly/1JaP5ho ("The security concern with SRM is therefore that related heating or cooling will result in non-linear stresses on environmental systems, where vulnerable parts of the system will be damaged or collapse, resulting in cascading impacts which may in turn destabilize vulnerable social, political, and economic systems which rely upon a steady-state environment").

128 *See, e.g.*, ETC Group, Climate & Geoengineering, bit.ly/1PXK4I2 ("ETC Group opposes geoengineering and other false solutions to climate change … and supports peasant-led agroecological responses to the climate crisis"); Anchorage

Declaration, Indigenous Peoples' Global Summit on Climate Change, Apr. 24, 2009, bit.ly/1Kdqh2K ("We challenge States to abandon false solutions to climate change that negatively impact Indigenous Peoples' rights, lands, air, oceans, forests, territories and waters. These include ... geo-engineering techniques"); Asia Indigenous Peoples' Declaration on the 21st Session of the UNFCCC-Conference of Parties (COP21), Sept. 18, 2015, https://perma.cc/NP7J-AZVJ (same).

129 *See* Daniel Bodansky, *May We Engineer the Climate?*, 33 *Climatic Change* 309 (1996) (discussing possibility); *see also* Joshua B. Horton, *Geoengineering and the Myth of Unilateralism: Pressures and Prospects for International Cooperation*, 4 *Stanford Journal of Law, Science, and Policy* 56 (2011).

130 UBA, *supra* note 55, at 124 (discussing these three categories); Bellamy, *supra* note 9 (anticipating third category and naming it "clumsy" governance); *see also* Lempert & Prosnitz, *supra* note 125, at xi ("if U.S. policymakers believe that some type of SRM technology is possible, they ought to prefer the Strong Norms policy to No Norms or Ban").

131 To oversimplify: Both principles impose a duty of care on states before they authorize or undertake a risky or potentially risky activity; whereas prevention relates to well-understood risks, precaution relates to uncertainties. "These principles require countries to take action where significant (and irreversible) environmental harm is foreseeable." International Law Association, Washington Conference: Legal Principles Relating to Climate Change 16 (2014), https://perma.cc/2F66-4RY7. For a summary description of both principles, their embodiment in international law, and an explanation of how they interact, *see* IASS Code of Conduct, *supra* note 56, at 41.

132 *See* Chapter 3 of this volume; *see also, e.g.*, United Nations Environment Programme, Declaration of the United Nations Conference on the Human Environment, UN Doc A/Conf.48/14/Rev.1, June 16, 1972, 11 ILM 1416, principles 6, 7, 15, 18 and 24 (calling for preventive action); Convention on Biological Diversity, June 5, 1992, 1760 UNTS 79, *reprinted in* 31 ILM 818 (1992), preambular recital 8 and art. 1 (similar); Rio Declaration on Environment and Development, June 13, 1992, A/CONF.151/5/Rev.1, *reprinted in* 31 ILM 874 (1992), principle 15 ("In order to protect the environment, the precautionary approach shall be widely applied by States according to their capabilities"); Donald R. Rothwell & Tim Stephens, *The International Law of the Sea*, ch. 19 (Hart Publishing, 2016) (discussing invocations of the precautionary principle in international disputes and recognition of the principle as an "approach" if not "customary international law").

133 *See* Chapter 4 of this volume; *see also, e.g.*, Ling Zhu & Yachao Zhao, *Polluter-pays Principle – Policy Implementation*, 45 *Environmental Policy and Law* 34 (Mar. 2015) (discussing the relationship of the polluter pays and prevention principles); Martha Cecilia Paz, *Precautionary Principle: Case Law in Colombia*, 3 *Civil Legal Science* 108 (2013); David W.-L. Wu, *Embedding Environmental Rights in Section 7 of the Canadian Charter: Resolving the Tension Between the Need for Precaution and the Need for Harm*, 33 *National Journal of Constitutional Law* 191 (2013).

134 *See, e.g.*, Robock et al., *supra* note 72, at 18.

135 Jesse L. Reynolds & Floor Fleurke, *Climate Engineering Research: A Precautionary Response to Climate Change?*, 7 *Carbon & Climate Law Review* 101 (2013); P. Toussaint, *Between a Rock and a Hard Place: Climate Impacts, Geoengineering*

and the Precautionary Principle, 24 *Environmental Law & Management* 235 (2012); Lauren Hartzell-Nichols, *Precaution and Solar Radiation Management*, 15 *Ethics, Policy & Environment* 158 (2012).

136 J. Virgoe, *International Governance of a Possible Geoengineering Intervention to Combat Climate Change*, 95 *Climatic Change* 103, 117 (2009).

137 *See* NRC, Reflecting Sunlight, *supra* note 1, at 12 ("Ultimately, the goal is to ensure that the benefits of the research are realized to inform civil society decision making, the associated challenges are well understood, and risks are kept small"); IASS Code of Conduct, *supra* note 56, at 56; SRMGI, *supra* note 62, at 20.

138 Bodansky, *supra* note 4, at 541.

139 SRMGI, *supra* note 62. SRMGI's founders explained the need for their initiative in this way: "Research into SRM … presents some special potential risks. Governance arrangements for managing these risks are mostly lacking and will need to be developed if research continues." *Id.* at 9.

140 NRC, Carbon Removal, *supra* note 10, at 12.

141 *See also* UBA, *supra* note 55, at 121.

142 Asilomar Recommendations, *supra* note 79, at 17.

143 *See* IASS Code of Conduct, *supra* note 56, at 6, 24; *see also* NRC, Reflecting Sunlight, *supra* note 1; UBA, *supra* note 55, at 150 ("as a starting point for governance they demonstrate that the science community is aware of the wider implications and of the need to act responsibly within a political context"); SRMGI, *supra* note 62; BPC Task Force, *supra* note 13.

144 Tim Kruger, A *Commentary on the Oxford Principles*, Geoengineering Our Climate?, May 14, 2013, https://perma.cc/B4A9-QKAK ("[The Oxford Principles] subsequently formed the basis of discussions [at the Asilomar Conference] and the Asilomar Principles for Responsible Conduct of Climate Engineering Research"); *see also* Oxford Principles: History, bit.ly/1Oeojhb. For a thorough commentary on the Oxford Principles by their authors, see Steve Rayner et al., *The Oxford Principles*, 121 *Climatic Change* 499 (2013).

145 Government Response to the House of Commons Science and Technology Committee 5th Report of Session 2009–10: The Regulation of Geoengineering 6 (Sept. 2010), bit.ly/1UPAhEW.

146 Oxford Principles, bit.ly/1QFGNQP.

147 *See* Don C. Smith & Jessica Marie Richards, *Social License to Operate: Hydraulic Fracturing-Related Challenges Facing the Oil & Gas Industry*, 1 *Oil and Gas, Natural Resources, and Energy Journal* 81, 89–95 (Apr. 2015) ("[social license] is an ongoing social contract with society that allows a project to both start and continue operating in a community").

148 Andy Parker, Oliver Morton & George Collins, Forum for Climate Engineering Assessment, *Reflecting on the "Berlin Declaration,"* July 16, 2015, https://perma.cc/8Z95-9FGL; Draft Proposed Berlin Declaration, https://perma.cc/Y9J2-HQRL.

149 *See* SRMGI, *supra* note 62, at 30.

150 *See* Alan Robock, *Is Geoengineering Research Ethical?*, 4 *Security & Peace* 226 (2012); Mark B. Brown & David H. Guston, *Science, Democracy, and the Right to Research*, 15 *Science and Engineering Ethics* 351 (2009); US National Bioethics Advisory Commission, Cloning Human Beings: Report and Recommendations of the National Bioethics Advisory Commission, vol. 2, at 6 (1997) ("Because science

is both a public and social enterprise and its application can have profound impact, society recognizes that the freedom of scientific inquiry is not an absolute right and scientists are expected to conduct their research according to widely held ethical principles").

151 *See* IASS Code of Conduct, *supra* note 56, at 12–16 (discussing scientific research as a subject of international law and in particular as among other protected human rights); California Constitution, art. 35, sec. 5, https://perma.cc/DD7S-57DB ("There is hereby established a right to conduct stem cell research which includes research involving adult stem cells, cord blood stem cells, pluripotent stem cells, and/or progenitor cells").

152 Jack Stilgoe, *Geoengineering as Collective Experimentation*, 22 *Sci. & Eng'g Ethics* (Apr. 2015, 851, 855 (listing technologies characterized by "the impossibility of control in a scientific sense").

153 *See* Joshua S. Gans & Fiona Murray, *Funding Scientific Knowledge: Selection, Disclosure and the Public-Private Portfolio*, in *The Rate and Direction of Inventive Activity Revisited* 51, 53 (Josh Lerner & Scott Stern, eds., University of Chicago Press 2012) (describing rubrics used by various private, public, and governmental entities for the selection and disclosure of funded research, and identifying implications of the differences between those rubrics).

154 *See* Jason J. Blackstock et al., Centre for International Governance Innovation, Workshop Report: Designing Procedural Mechanisms for the Governance of Solar Radiation Management Field Experiments, Feb. 23–24, 2015, bit.ly/1Jc1ZFX.

155 *See* John E. Barkstrom, *Recombinant DNA and the Regulation of Biotechnology: Reflections on the Asilomar Conference, Ten Years After*, 19 *Akron Law Review* 81, 87 (1985) ("The NIH Guidelines are virtually the only form of government control for recombinant DNA research and even that control focuses primarily on contracts involving government funding for research procured through the NIH").

156 Arnold S. Relman, *Dealing with Conflicts of Interest*, 310 *New England Journal of Medicine* 1182 (1984) ("The Journal has had no stated policy on [author conflicts of interest] until now, but my editorial associates and I think it is time we formulated one"); James B. Conant, *Education and the National Science Foundation*, 34 *American Scientist* 94, 98–99 (Jan. 1946) (anticipating potential conflicts of interest in implementation of a proposed scientific fellowship program).

157 *See, e.g.*, J. S. Ancker & A. Flanagin, A *Comparison of Conflict of Interest Policies at Peer-Reviewed Journals in Different Scientific Disciplines*, 13 *Science and Engineering Ethics* 147 (2007) (examining requirements of disclosures relating to financial and other relationships between authors, subjects, funders, and peer-reviewers).

158 5 U.S.C. app. 2 (2012); Federal Advisory Committee Management, 66 Fed. Reg. 37,728 (July 19, 2001) (final rule instructing federal agencies on implementation of FACA policies); *see also* Memorandum from Vanessa T. Vu, Director of EPA Science Advisory Board Staff Office, to Clean Air Scientific Advisory Committee (CASAC) regarding Federal Advisory Committee Act (FACA) Policies (May 3, 2005), bit.ly/234Foaj (describing how FACA requirements apply to CASAC activities).

159 *See, e.g.*, *R. J. Reynolds Tobacco Co. v. U.S. Food & Drug Admin.*, Case No. 14–5226 (D.C. Cir. Jan. 15, 2016) (overturning trial court decision holding that conflicts of

interest should have led the Food and Drug Administration, pursuant to federal ethics laws, to exclude three experts from panel that authored a report finding that menthol cigarettes have adverse health impacts comparable to tobacco cigarettes); Glenn W. Suter & Susan M. Cormier, *The Problem of Biased Data and Potential Solutions for Health and Environmental Assessments*, 21 *Human and Ecological Risk Assessment: An International Journal* 1736 (2014) ("despite peer review, much of the scientific literature is biased. Sources of bias include publication practices, research design and implementation, funding influences, investigator expectations, statistical methods, confounding, suppression, and fraud"); Institute of Medicine, Nat'l Acad. Sci., Conflict of Interest in Medical Research, Education, and Practice (Bernard Lo & Marilyn J. Field eds. 2009) (describing sources of conflicts).

160 *See, e.g.*, Franco et al., *supra* note 112, at 6203 ("There is a strong relationship between the results of a study and whether it was published, a pattern indicative of publication bias … [W]hat is perhaps most striking in [our data] is not that so few null results are published, but that so many of them are never even written up (65%)"); Kerry Dwan et al., *Systematic Review of the Empirical Evidence of Study Publication Bias and Outcome Reporting Bias – An Updated Review*, 8 *PLoS ONE* e66844 (July 5, 2013), bit.ly/202OWA4 ("There is strong evidence of an association between significant results and publication; studies that report positive or significant results are more likely to be published and outcomes that are statistically significant have higher odds of being fully reported").

161 *See* the discussion of SPICE in section 6.1; Erin Hale, *supra* note 36.

162 David E. Winickoff & Mark B. Brown, *Time for a Government Advisory Committee for Geoengineering Research*, 29 *Issues in Science and Technology* 79 (Summer 2013); BPC Task Force, *supra* note 13, at 24.

163 Long & Scott, *supra* note 94 ("Peer review will remain necessary – in part, to help balance the exuberance of individual scientists – but by itself will probably be inadequate … In this "red team/blue team" model, one team develops the research, and the other tries to ferret out all the problems"); Caldeira & Keith, *supra* note 97 ("a red team/blue team approach, wherein one team is tasked with showing how an approach can be made to work, and another team is tasked with showing why the approach cannot produce a system that can actually diminish environmental risk at an acceptable cost").

164 M. G. Morgan et al., *Needed: Research Guidelines for Solar Radiation Management*, 29 *Issues in Science and Technology* 37 (Spring 2013), http://bit.ly/2aL9jBp.

165 *Id.*

166 *See* Stefan Schäfer & Nigel Moore, Brief Summary of Workshop: Understanding Process Mechanisms for the Governance of SRM Field Experiments 7 (2014), bit.ly/206Qeub; Blackstock et al., *supra* note 154, at 6–7.

167 Blackstock et al., *supra* note 154, at 6–7.

168 *Id.*

169 *See* Morgan et al., *supra* note 164, at 41–44 (weighing both options and recommending the first).

170 42 U.S.C. § 4332(2)(C).

171 43 CFR § 46.305(a) (Public involvement in the environmental assessment process).

172 40 CFR § 1503 (Commenting), § 1505.1 (Agency decisionmaking procedures).

173 *See generally* Christopher Wood, *Environmental Impact Assessment: A Comparative Review* (2nd edn Routledge 2013) (discussing the USA, Canada, Britain, Australia, New Zealand, South Africa, Netherlands, and the European Union).

174 *Pulp Mills on the River Uruguay (Arg. v. Uru.)*, Judgment, at 83 (Apr. 20, 2010), https://perma.cc/SWH3-CBF5 ("it may now be considered a requirement under general international law to undertake an environmental impact assessment where there is a risk that the proposed industrial activity may have a significant adverse impact in a transboundary context, in particular, on a shared resource"); *see also* IASS Code of Conduct, *supra* note 56, at 71 ("As fundamentally a national instrument, EIA is entrenched in the domestic law of a large number of States").

175 *See* International Maritime Organization, *Assessment Framework*, bit.ly/239Z5gT (the Assessment Framework "does not contain a threshold below which experiments would be exempt from its assessment provisions. Every experiment, regardless of size or scale, should be assessed in accordance with the entire Assessment Framework"); Resolution LC-LP.2(2010) on the Assessment Framework for Scientific Research Involving Ocean Fertilization (adopted Oct. 14, 2010), bit.ly/1ndAQoO (the Framework itself is Annex 6 of the Resolution).

176 IASS Code of Conduct, *supra* note 56, at 73 ("This draft Code recommends that all scientific research involving geoengineering conducted in the open environment should undergo a prior assessment"); Nigel Moore et al., Inst. for Adv. Sustainability Studies, Workshop Report: Procedural Governance of Field Experiments in Solar Radiation Management 10 (Mar. 2015); *see also* Blackstock et al., *supra* note 154, at 4 ("EIAs would be particularly ineffective at the technology development and process study levels, because any environmental impacts arising from such small-scale experiments would likely be negligible and therefore fail to trigger full EIA processes; as such, there would be limited opportunities for public engagement").

177 Department of Energy, *Categorical Exclusion (CX) Determinations by CX*, 1.usa.gov/1L1eVPH (listing seven categories of research activity to which categorical exclusions apply); 45 C.F.R. § 640.3 (discussing actions by or permitted or funded by NSF requiring an environmental assessment and categorical exclusions).

178 Blackstock et al., *supra* note 154, at 4 ("policy makers should not assume a governmental action trigger, given both the potential for private funders to support experimentation and uncertainty regarding other regulatory triggers, such as permits for releases into air or water").

179 *See* Larry Mayer, *Arctic Marine Research: The Perspective of a US Practitioner*, in *Arctic Science, International Law and Climate Change* 83, 93 (S. Wasum-Rainer, I. Winkelmann, & K. Tiroch eds., Springer 2012) ("In some cases the constraints of the environmental permitting process have delayed or prevented critical scientific studies").

180 *See* Moore et al., *supra* note 176, at 13.

181 Pak-Hang Wong, *Consenting to Geoengineering*, 29 *Philosophy & Technology* (published online June 7, 2015), http://bit.ly/2aNAL2G.

182 *See* Jesse Reynolds, *The Regulation of Climate Engineering*, 3 *Law, Innovation and Technology* 113, 125 (2011) ("Given this complex political landscape, establishing legitimacy will be both crucial and difficult for any regulatory scheme"); Adam Corner & Nick Pidgeon, *Geoengineering the Climate: The Social and Ethical Implications*, 52 *Environment: Science and Policy for Sustainable Development* 24,

29 (2010) ("the prospect of [climate engineering] is something that all citizens could reasonably claim to have a legitimate stake in"); IASS Code of Conduct, *supra* note 56, at 81 (noting calls in Aarhaus Convention art. 1 and Rio Declaration principle 10 to states to guarantee access to information).

183 *See* Pidgeon et al., *supra* note 38 (describing public engagement as a process of framing issues by posing basic questions such as "What is a development for? What is the need? Who owns it? Who will be responsible if things go wrong?"); Wong, *supra* note 181 (discussing three models of public consent and two approaches to consent to climate engineering, and noting problems of resolving disagreements over issues underlying the choice of model while also seeking consent).

184 *See* Wong, *supra* note 181, at 174; UBA, *supra* note 55, at 131 ("A polarized debate ... would make it difficult for a state to adopt and implement any policy on geoengineering"); Bodansky, *supra* note 4, at 547 ("Ultimately, the[] success [of nonbinding norms developed by scientists, social scientists, philosophers and lawyers] in forestalling more drastic regulation hinges on their public credibility, rather than on their legal status or source"). For discussion of the need for and role of public engagement in the process of commencing research into another new and contentious technology, see Stilgoe et al., *supra* note 45.

185 Pidgeon et al., *supra* note 183, at 452–54 (describing both the general pattern and a particular example of public questions and doubts about SPICE); Jack Stilgoe et al., *supra* note 45 (describing how the actual experiment was less concerning to the public than what the experiment represented); *see also* W. Carr et al., *Swimming Upstream: Engaging the American Public Early on Climate Engineering*, 70 *Bull. of the Atomic Scientists* 38 (2014) ("such ideas are easier in theory than in practice. Meaningful public engagement can require more of scientists and regulatory agencies than many realize").

186 Bellamy, *supra* note 9, at 33; *see also* Macnaghten & Szerszynski, *supra* note 59, at 466 (challenging the assumption "that debates around [SRM] are debates about a unified, stable, technological object ... rather than a more complex conversation in which the very nature of geoengineering is put into question").

187 Blackstock et al., *supra* note 154, at 6 ("without a politically agreed program of SRM research, any decision made on a single experiment could, in effect, turn that experiment into a political referendum on all further SRM research").

188 *See* Carr et al., *supra* note 185, at 44 ("Upstream engagement does not necessarily guarantee better climate engineering science, technology, or regulation. The quality of the outcome depends on the effectiveness of the process itself"); Macnaghten & Szerszynski, *supra* note 59; A. Corner, *Perceptions of Geoengineering: Public Attitudes, Stakeholder Perspectives, and the Challenge of "Upstream" Engagement*, 3 *WIREs Climate Change* 451 (2012).

189 *See* Fund for Innovative Climate and Energy Research, *Answers to Frequently Asked Questions*, stanford.io/1McyLso ("**Q.** What is the source and size of the fund? Who administers the fund? **A.** Since its inception in 2007, FICER has given out grants to 13 research projects and various scientific meetings totaling $8.5 million. Internationally known climate scientists Dr. David Keith of Harvard University ... and Dr. Ken Caldeira of the Carnegie Institution for Science select projects that receive support from the fund. While Mr. Gates provides input from time

to time on the fund, Drs. Keith and Caldera make final decisions on projects"); GAO-10–903, *supra* note 6, at 18 ("We identified approximately $100.9 million in geoengineering-related funding across USGCRP [U.S. Global Change Research Program] agencies in fiscal years 2009 and 2010, with about $1.9 million of this amount related to research directly investigating a particular geoengineering approach").

190 *See* James Boyle & Jennifer Jenkins, *Intellectual Property: Law & The Information Society; Cases & Materials* 2–3 (Center for the Study of the Public Domain 2014) (summarizing the basic policy goals of IP law); Daniel Cressey, *Cancelled Project Spurs Debate over Geoengineering Patents, Nature: News*, May 23, 2012 (corrected May 24, 2012), http://bit.ly/1JbNSjX.

191 Rayner et al., supra note 144.

192 *See, e.g.*, Arunabha Ghosh, *Environmental Institutions, International Research Programmes, and Lessons for Geoengineering Research*, Geoengineering Our Climate?, Working Paper, at 12–13 (Feb. 2014), bit.ly/1mApChj ("Government-funded research should … remain in the public domain, while privately funded work should have limits on proprietary knowledge").

193 Chavez, *supra* note 117, at 32.

194 Aladdin Tingling Diakun, Clearing the Air on "Geoengineering" and Intellectual Property Rights; Towards a Framework Approach, Master's Research Paper presented to the University of Waterloo in fulfillment of the MRP requirement for the degree of Master of Arts in Global Governance 29 (2015) ("an effective analytical framework will ultimately need to evaluate the tradeoffs associated with IPRs [intellectual property rights] and [climate engineering] on a case-by-case basis. In particular, the scale and distribution of potential impacts … as well as their relative intensity, will to a large degree determine how concerned we should be about IPRs").

195 Parthasarathy et al., *supra* note 116, at 10–12 (discussing creation of the Atomic Energy Commission by the Atomic Energy Act of 1946 and the Commission's role in steering decisions about protection of intellectual property).

196 *Id.* at 12–13. Those recommendations are: (1) stop issuing broad patents; (2) create an interagency geoengineering patent task force; (3) add climate engineering to the existing sensitive application warning system within the PTO [Patent and Trademark Office] and require patent review by the Interagency Geoengineering Patent Task Force; and (4) offer non-patent-based innovation incentives.

197 Ghosh, *supra* note 192, at 12.

198 *See, e.g.*, David Reichwein et al., *State Responsibility for Environmental Harm from Climate Engineering*, 5 *Climate Law* 142 (2015); Juan Moreno-Cruz et al., *A Simple Model to Account for Regional Inequalities for Solar Radiation Management*, 110 *Climate Change Journal* 649 (2010).

199 *See, e.g.*, Joshua B. Horton et al., *Liability for Solar Geoengineering: Historical Precedents, Contemporary Innovations, and Governance Possibilities*, 22 *NYU Environmental Law Journal* 225 (2015); M. Bunzl, *Geoengineering Harms and Compensation*, 4 *Stanford Journal of Law, Science & Policy* 70 (2011).

200 Toby Svoboda & Peter Irvine, *Ethical and Technical Challenges in Compensating for Harm due to Solar Radiation Management Geoengineering*, 17 *Ethics, Policy & Environment* 157 (2014).

201 *Id.* at 170; Pak-Hang Wong et al., *Compensation for Geoengineering Harms and No-Fault Climate Change Compensation*, Geoegineering Our Climate Working Paper No. 8 (2014), bit.ly/1OzeoG4.

202 NRC, Carbon Removal, *supra* note 10, at 104.

203 *Id.* at 110 ("In some contexts, it might be useful to treat various CDR proposals and albedo modification proposals jointly. This is especially true of those CDR approaches that raise novel risks and governance issues (e.g., ocean fertilization, ocean alkalinization)").

204 *See, e.g.*, Douglas G. MacMartin et al., *On Solar Geoengineering and Climate Uncertainty*, 42 *Geophysical Research Letters* 7156 (2015) (identifying uncertain parameters by examining discrepancies among model outputs); Robert Wood & Thomas P. Ackerman, *Defining Success and Limits of Field Experiments to Test Geoengineering by Marine Cloud Brightening*, 121 *Climatic Change* 459 (2013) (acknowledging multiple scientific uncertainties); *see also* Ben Parkes et al., *Crop Failure Rates in a Geoengineered Climate: Impact of Climate Change and Marine Cloud Brightening*, 10 *Environmental Research Letters* 5 (Aug. 2015), http://bit.ly/1NOAztI (inferring from model simulations expected effects of the increased cloud droplet formation caused by marine cloud brightening on crop failure rates in West Africa and China).

205 David W. Keith et al., *supra* note 5, at 3 tbl.1; *see also* Blackstock et al., *supra* note 154, at 2–3 (describing similar stages of SRM research); SRMGI, *supra* note 62, at 45–54 (proposing similar "categories" of research); BPC Task Force, *supra* note 13, at 23 ("Any research should be conducted using a phased approach, starting with low-risk and low-cost exploratory research in the laboratory and only gradually considering larger-scale, higher-risk research (including field experiments) as more is understood about the technology").

206 Jason J. Blackstock, The International Politics of Geoengineering, Memorandum submitted to House of Commons Sci. & Tech. Sub-Comm. (March 18, 2010), bit.ly/1j7aLvE.

207 Keith et al., *supra* note 5, at 4.

208 *Id.*; Blackstock, *supra* note 206.

209 Parson & Keith, *supra* note 2 (proposing a threshold based on "the product of area, duration, and size of radiative forcing perturbation").

210 *Id.*

211 *But see* Integrated Assessment of Geoengineering Proposals, Briefing Paper No. 3: Decisionmaking for Geoengineering – Why Will It Be Challenging? 3 (2014) ("The ability to accurately detect and attribute changes in the climate should be a prerequisite for implementing geoengineering"); NRC, Reflecting Sunlight, *supra* note 1, at 184 ("Current observing systems are insufficient to quantify the effects of any intervention at present. If albedo modification at climate-altering scales were ever to occur, it should be accompanied by an observing system that is appropriate for assessing the impacts of the deployment and informing subsequent actions").

212 International Maritime Organization, *Assessment Framework*, bit.ly/239Z5gT (the Assessment Framework "does not contain a threshold below which experiments would be exempt from its assessment provisions. Every experiment, regardless of size or scale, should be assessed in accordance with the entire Assessment Framework"); Resolution LC-LP.2(2010) on the Assessment Framework for

Scientific Research Involving Ocean Fertilization (adopted October 14, 2010), bit.ly/1ndAQoO (the Framework itself is Annex 6 of the Resolution).

213 *See, e.g.*, Jason J. Blackstock & Jane C. S. Long, *The Politics of Geoengineering*, 327 *Science* 527 (2010) ("Emerging national research programs – and even individual scientists – must forswear climatic impacts testing and carefully restrict subscale field-testing until approved by a broad, legitimate, international process").

214 Parker, Morton, & Collins, *supra* note 148.

215 Draft Proposed Berlin Declaration, https://perma.cc/Y9J2-HQRL.

216 Merriam-Webster's Dictionary, bit.ly/1SrKdao ("(1)(a) a legally authorized period of delay in the performance of a legal obligation or the payment of a debt (b) a waiting period set by an authority; (2) a suspension of activity").

217 Attendees at the 2014 Climate Engineering Conference rejected the draft Berlin Declaration over several concerns, including that it could be read as requiring a halt to even modeling and laboratory research efforts. *See* Parker, Morton, & Collins, *supra* note 148.

218 NRC, Reflecting Sunlight, *supra* note 1, at (quoting Royal Society: "a moratorium … would make it almost impossible to accumulate the information necessary to make informed judgments" and "is likely to deter only those countries, firms and individuals who would be most likely to develop the technology in a responsible fashion, while failing to discourage potentially dangerous experimentation by less responsible parties"); Daniel Bodansky, *Governing Climate Engineering: Scenarios for Analysis*, Harvard Project on Climate Agreements Discussion Paper at 22, Nov. 1, 2011, bit.ly/1ToopTh ("A moratorium could thus have the perverse effect of leaving the field of geoengineering research to less responsible countries that ignore the moratorium and engage in riskier activities"); BPC Task Force, *supra* note 13, at 29 (same).

219 *See, e.g.*, NRC, Reflecting Sunlight, *supra* note 1, at 133, 148; Draft Proposed Berlin Declaration, https://perma.cc/Y9J2-HQRL; Parson & Keith, *supra* note 2.

220 UBA, *supra* note 55, at 21.

221 Parson & Keith, *supra* note 2.

222 NRC, Reflecting Sunlight, *supra* note 1, at 184 ("If research and development on albedo modification were to be done at climate-altering scales, it should be carried out only as part of coordinated national or international planning, proceeding from smaller, less risky to larger, more risky projects; more risky projects should be undertaken only as information is collected to quantify the risks at each stage"); Draft Proposed Berlin Declaration, https://perma.cc/Y9J2-HQRL.

223 *Geoengineering III: Domestic and International Research Governance: Hearing Before the H. Comm. on Sci. & Tech.*, 111th Cong. 273–77 (statement of Prof. M. Granger Morgan, Carnegie Mellon University); *see also* SRMGI, *supra* note 62, at 55 ("take a 'hands-off' approach early in the research program and to gradually increase the extent of governance arrangements as research becomes increasingly risky, where risks are defined in terms of physical harms that may be caused by research and testing").

224 SRMGI, *supra* note 62, at 49; *see also* Morgan & Ricke, *supra* note 14, at 18 ("An 'allowed zone' might be defined in terms of a number of different variables or factors, such as the amount of radiative forcing, the duration of the forcing, and the impact that the experiment might have on ozone destruction").

225 Morgan & Ricke, *supra* note 14, at 18 (research in this zone should be allowed to proceed "without formal international approval, subject only to the requirements that their studies are publicly announced and all results are made public").

226 *See* Morgan testimony, *supra* note 223, at 273 (proposing a zone that would not impose any *new* checks in addition to existing environmental and other laws).

227 Armeni & Redgwell, *supra* note 68, at 29–30; *see also* Caldeira & Keith, *supra* note 97 (suggesting similar).

228 Keith et al., *supra* note 5, at 5 ("A research programme should be sequential and iterative … in the sense that one would not proceed to the next phase without a positive outcome from the prior phase. Determining that it is more difficult than expected to achieve desired outcomes, or that there are larger than expected undesired consequences would result in at least reconsideration and potentially a termination of any particular line of research").

229 Armeni & Redgwell, *supra* note 68, at 36; SRMGI, *supra* note 62, at 25; *see also* Asilomar Recommendations, *supra* note 79, at 19 ("One potential simplifying mechanism would be to subdivide the approaches into categories with common potential risks (and so likely similar issues of governance)").

230 Bellamy, *supra* note 9, at 33; *see also* BPC Task Force, *supra* note 13, at 31 (calling for "an incremental but proactive approach to international engagement").

231 *Id.*; *see also* Asilomar Recommendations, *supra* note 79, at 18 ("Recognizing the potential breadth of situations, a coordinated mix of types of oversight and governance likely has the potential to be most effective and accessible").

232 IASS Code of Conduct, *supra* note 56, at 7.

233 *Id.* at 9–11.

234 *See* Asilomar Recommendations, *supra* note 79, at 20 ("Developing an efficient and well-functioning system of review will probably require a mechanism for determining the specific institutional framework appropriate to evaluating activities of different scales and aggregate impacts").

235 Peter Healey & Steve Rayner, Key Findings from the Climate Geoengineering Governance Project, CGG Working Paper No. 25, at 16 (Mar. 2015), http://bit.ly/1lMVWTi; *see also* NRC, Reflecting Sunlight, *supra* note 1, at 12 ("[SRM] research is not specifically addressed by any federal laws or regulations").

236 NRC, Carbon Removal, *supra* note 10, at 107–110 (recommending the US Global Change Research Program take responsibility for coordinating various CDR research efforts and concluding that "many proposed CDR approaches do not pose novel risks or governance issues (e.g., land management, bio-energy carbon capture and storage)").

237 Notably, some CDR methods, deployed on a large scale, could have significant environmental and transboundary impacts – for instance, large monoculture forests could threaten biodiversity; large-scale management of liquid or sequestered CO_2 could pollute adjacent surface or groundwaters; so too can waste products from amine-capture in industrial CDR.

238 *See Pulp Mills*, *supra* note 174, at 203–219 (discussing when the obligation to conduct environmental impact assessments obtains).

239 The 2010 Joint Statement of the US Congress and the House of Commons on geoengineering announced parallel hearings but nothing more. H.R. Comm. on Sci. & Tech., House of Commons Select Comm. on Sci. & Tech.: Collaboration and Coordination on Geoengineering, Joint Statement, Mar. 18, 2010. Since then, only a handful of US government agencies have promulgated rules that can be read as contemplating geoengineering activities, and those examples do not mention geoengineering *research*. *See* Federal Requirements under the Underground Injection Control (UIC) Program for Carbon Dioxide (CO_2), 75 Fed. Reg. 77229 (Dec. 10, 2010) ("This final rule applies to owners or operators of wells that will be used to inject CO_2 into the subsurface for the purpose of long-term storage"). The German government, for its part, "does not intend to take an active role in supporting and funding research programmes in [climate engineering]." Chiara Armeni & Catherine Redgwell, Geoengineering Under National Law: A Case Study of Germany, CGG Working Paper No. 024, at 26 (Mar. 2015). The British Government has said that it is "too early to be able to establish appropriate regulatory frameworks for Geoengineering research or deployment," because it remains unclear "what needs to be regulated and why." UK Government Response to House of Commons Sci. & Tech. Comm. 5th Rep. of Sess. 2009–10: The Regulation of Geoengineering, at 1–2.

240 *See* UBA, *supra* note 55, at 16 ("[beyond national waters], it is arguable but not clear that cloud brightening would fall under United Nations Convention on the Law of the Sea (UNCLOS) provisions against marine pollution. The LP does not prohibit cloud brightening as long as sea water vapour is used and does not constitute dumping").

241 Catherine Brahic, *Company plans 'eco' iron dump off Galapagos*, New Scientist, June 22, 2007, bit.ly/1NhieSv.

242 LC-LP.1/Circ.14, Statement of concern regarding iron fertilization of the oceans to sequester CO_2, July 13, 2007, bit.ly/1V3qwD8.

243 Resolution LC-LP.1 (2008) on the Regulation of Ocean Fertilization (adopted Oct. 31, 2008), bit.ly/1QoscOo.

244 Resolution LC-LP.2 (2010) on the Assessment Framework for Scientific Research Involving Ocean Fertilization (adopted Oct. 14, 2010), bit.ly/1ndAQoO (Framework itself is Annex 6 of the Resolution).

245 CBD COP decision IX/16 C; X/29 para 13(e) and 57–62; X/33 para 8(w)-(x).

246 *See* UBA, *supra* note 55, at 157.

247 *See* UNESCO Doc. IOC-XXV/3 ¶142; UNESCO Intergovernmental Oceanographic Commission (IOC), Report of the Forty-third Session of the Executive Council, Paris, June 8–16, 2010; UNESCO Doc. IOC/EC-XLIII/3 ¶155 and Annex IX at 19 (2010); *see also* D. Wallace, Ocean Fertilization: A Scientific Summary for Policy Makers, IOC/BRO/2010/2 (2010), bit.ly/1ZB2BBF.

248 Martin Lukacs, Canadian Government 'Knew of Plans to Dump Iron into the Pacific', *The Guardian*, Oct. 17, 2012, bit.ly/1nup8yT.

249 *See* Bodansky, *supra* note 218, at 17 (noting persistent need for action on the part of national governments with respect to OIF, even after 2008 adoption of resolution under London Convention).

250 *See* Ghosh, *supra* note 192 (tabulating tasks involved in international governance of climate engineering research).
251 UBA, *supra* note 55, at 19.
252 Armeni & Redgwell, *supra* note 68, at 31–37.
253 *See, e.g.*, Ghosh, *supra* note 192, at 13 ("although several multilateral environmental treaties might have some relevance to geoengineering, there is a governance gap when it comes to research. Depending on the scale and scope of research, field testing and deployment, there are several aspects that could benefit from internationally coordinated efforts."); UBA, *supra* note 55, at 22–23 (examining and rejecting as candidates all existing international agreements relating to protection of the atmosphere).
254 Bodansky, *supra* note 218, at 18 ("Since research to date suggests that sulfur aerosols are likely to modify the ozone layer, they fall within the ambit of the ozone regime and could potentially be regulated by the Montreal Protocol.").
255 *See* SRMGI, *supra* note 62, Appendix at 9–10.
256 Convention on Long-Range Transboundary Air Pollution art. 2, bit.ly/1Aj8Ibr.
257 Bodansky, *supra* note 218, at 18. The volume of sulfate aerosols required for effective SRM deployment has been estimated to be roughly 3.5 kilotons per year, depending on the particulates used. *See* David W. Keith & Douglas G. MacMartin, *A Temporary, Moderate and Responsive Scenario for Solar Geoengineering, Nature Climate Change* (Feb. 2015), doi: 10.1038/nclimate2493, http://bit.ly/29yOeLa. Compare that to the 2010 emissions limitations for parties to the Convention on Long-Range Transboundary Air Pollution: the United Kingdom was limited to 980 kilotons per year, Russia to 4,297. *See* Protocol to the 1979 Convention on Long-Range Transboundary Air Pollution on Further Reduction of Sulphur Emissions, Annex II (1994), bit.ly/1VhvsGX.
258 UBA, *supra*, note 55, at 23 ("the institutional logic of the UNFCCC is directed at combating climate change … As a result, it might be intrinsically difficult … to pursue a precautionary approach that is *restrictive* to geoengineering. In addition, geoengineering does not fit easily with the overall approach of the UNFCCC aimed at mitigating greenhouse gas emissions and adapting to the impacts of climate change. The UNFCCC may thus best be considered a complementary forum that may be suitable for incentivising any 'encapsulated' geoengineering activities.").
259 SRMGI, *supra* note 62, at Appendix 6 ("The Achilles' heel of UNFCCC governance is its slow decisionmaking process and overburdened agenda. The Conference of the Parties (COP) has adopted a consensus-based (or at least consensus minus one) decision-making approach, which means that there is considerable diplomatic leeway for a small number of states to block decisions approved by the majority, and – crucially for SRM – to cause gridlock on all novel issues.").
260 Treaty on Principles Governing the Activities of States in the Exploration and Use of Outer Space, Including the Moon and other Celestial Bodies 9 ILM (1967); *see also* Armeni & Redgwell, *supra* note 68, at 18, 26.
261 SRMGI, *supra* note 62, at 10.

262 *See* Ghosh, *supra* note 192, at 7–8 (describing European Organization for Nuclear Research and International Thermonuclear Experimental Reactor as partial models climate engineering research governance).

263 World Nuclear Association, International Nuclear Waste Disposal Concepts, bit.ly/1ZK18cu (updated May 2016) (discussing EU resolution to establish European Repository Development Organisation); K. Smith et al., *BIOPROTA: International Collaboration in Biosphere Research for Radioactive Waste Disposal*, 76 *Mineralogical Magazine* 3233 (2012), bit.ly/1lr2Psc ("In 2012 there are 20 funding organizations in 14 countries from East Asia, Europe and North America, with additional participation from associated technical support organizations and research institutions.").

264 John Komen, *The Emerging International Regulatory Framework for Biotechnology*, 3 *GM Crops & Food: Biotechnology in Agriculture and the Food Chain* 78 (2012) (focusing discussion on Convention on Biological Diversity and Cartagena Protocol on Biosafety).

265 *See, e.g.*, NRC, Reflecting Sunlight, *supra* note 1, at 154; Jane C. S. Long et al., *Start Research on Climate Engineering*, 518 *Nature* 29 (Feb. 5, 2015) ("governance and experimentation must co-evolve."); Parson & Keith, *supra* note 2, at 1278.

7

Conclusions and Recommendations

Michael B. Gerrard and Tracy Hester

7.1. INTRODUCTION

As prior chapters have shown, some form of carbon dioxide removal (CDR) is essential if the world is going to stay within 2°C of warming.[1] Some form of solar radiation management (SRM) may well be attempted in the years to come. There is no regulatory process in place or under active development, either domestically in the United States or globally, for either CDR or SRM. This situation needs to change.

The governance and legal challenges facing CDR and those facing SRM are profoundly different. Most forms of CDR have only local impacts, suggesting that international governance is not required. CDR needs to be affirmatively encouraged, and incentives need to be put into place to accelerate research and deployment. On the other hand, SRM, if deployed at a scale that would make a real difference, poses very substantial risks around the world, and therefore requires international governance.

Because the issues concerning CDR and SRM are so different, this concluding chapter will for the most part treat them separately. We are going to reshuffle the categories in one respect, however. Ocean iron fertilization is technically a form of CDR, but it differs so much from all the others, and may have the potential for widespread impacts if done on a large scale, that it should be governed more like SRM and will not be further discussed in the CDR section that follows.

Two points should be made at the outset that apply to both CDM and SRM – the importance of considering (a) indirect effects, and (b) scale.

An extremely important indirect effect of SRM is that reliance on it can allow CO_2 levels to continue to rise in the atmosphere, leading to more severe ocean acidification, which can devastate marine ecosystems without increasing temperatures. Looking only at temperatures – the principal target of SRM – misses this central issue in evaluating SRM.

With respect to scale, a major attraction of SRM is that it could be achieved by an entirely manageable number of aircraft – perhaps in the dozens. On the other hand, bioenergy with carbon capture and storage (BECCS), a leading candidate for air capture, would need vast amounts of land, as discussed below. Direct air capture could require the manufacture and deployment of millions or tens of millions of devices – something of roughly the scale of the passenger vehicle industry – and solutions to the problem of what to do with the vast amounts of CO_2 that would be captured. Thus a device that works in a laboratory may be wholly unsuitable for mass deployment.

It is possible that both SRM and CDR will be required. Professor Wally Broecker has suggested that, in the event the climate became intolerable, fast-acting SRM could be temporarily deployed while CDR was doing its slow work of lowering atmospheric CO_2 levels; after some years, after the CO_2 levels had lowered sufficiently, the SRM could be phased out.[2] Even with that approach, major efforts to reduce greenhouse gas (GHG) emissions will still be necessary because no combination of SRM and CDR can suffice if GHGs continue to be released in significant quantities on a net basis.[3]

7.2. CARBON DIOXIDE REMOVAL

Chapter 1 showed that all the Intergovernmental Panel on Climate Change (IPCC) scenarios for holding global temperature increases under 2°C involved a substantial program of CDR. Even with the most rapid shift away from fossil fuels imaginable, global emissions will still be well above the levels that would stabilize temperatures at a tolerable level. (Even 2°C is intolerable to many, as agreed at Paris, but it will be so difficult to stay below that level that we will take it as a baseline goal here.)[4] Thus we need to be moving into the unexplored realm of net negative emissions.[5]

No CDR technology is operating commercially today that could attain anywhere near the scale needed to meet the need, and new technologies must be developed.[6] Several laboratories around the world are working on direct air capture technologies, with financial assistance from a few extremely wealthy donors (most prominently Bill Gates) and from some government and university grants. However, the funding for this R&D is far lower than what would be justified by the need it seeks to meet. Congressional and, more recently, presidential hostility to climate action restrains the availability of federal assistance in the United States.

Moreover, there is little financial incentive to develop CDR technology because most places allow carbon dioxide to be emitted into the atmosphere for free. A carbon tax would have a transformative effect here; it could create a market for CDR services and quickly make them attractive businesses. However, political resistance has so far blocked serious consideration of a carbon tax in the United States.

The IPCC placed primary reliance on BECCS as the means of CDR. Not only is there only a modest amount of research occurring now on BECCS, but what we know so far suggests that it would require massive amounts of land – so much that it could displace much agricultural activity and drive up food prices. It would also consume a great deal of water.[7]

Under an estimate from Phil Williamson of the Natural Environment Research Council in the UK, the use of BECCS to remove 600 gigatonnes of CO_2 over the twenty-first century (the medium estimate of what is needed) would probably require crops to be planted solely for the purpose of CO_2 removal on between 430 million and 580 million hectares of land – about one-third of the current total arable land on the planet, or about half the land area of the United States.[8] Dr. Williamson also observed:

> Unless there are remarkable increases in agricultural productivity, greatly exceeding the needs of a growing global population, the land requirements to make BECCS work would vastly accelerate the loss of primary forest and natural grassland. Thus, such dependence on BECCS could cause a loss of terrestrial species at the end of the century perhaps worse than the losses resulting from a temperature increase of about 2.8°C above pre-industrial levels.[9]

Aside from the serious environmental issues this raises, acquiring the extraordinary amount of land needed for BECCS on this scale will raise its own legal challenges. Some of the land that might be suitable is no doubt publicly owned; governments will need to decide whether to divert it from its current use to BECCS. Some of the land is now in private ownership, and difficult questions of valuation and the use of eminent domain to acquire it will certainly arise.

The daunting amount of land that would be required for BECCS leads to serious doubts about its practical viability. In the words of Sean Low of the Institute for Advanced Sustainability Studies in Potsdam, Germany:

> [t]he necessary scale and speed of implementation proposed by these modeling scenarios is belied by the fact that BECCS – like most other approaches currently grouped under the term "climate engineering" – faces huge unknowns in material and political support, and in human and ecological impact. The incorporation of an imaginary technological base to meet a politically negotiated goal that would otherwise be unfeasible has been interrogated as "magical thinking" and "policy-based evidence-making," and

brought under renewed questioning the processes by which the demands of politics can drive scientific expertise toward speculative conclusions.[10]

A study by Pete Smith of the Institute of Biological and Environmental Sciences at the University of Aberdeen estimated that BECCS would require roughly $29 billion to $36 billion per year in 2030, and $123 billion to $138 billion per year in 2050, representing around 5 percent of the projected total global energy system investments required by 2050.[11] Use of BECCS would also have serious implications for international trade in fossil fuels, agriculture products, and biomass.[12] A study by Lena R. Boysen and colleagues of the Max Planck Institute for Meteorology in Hamburg went so far as to declare that, unless global emissions are severely reduced, BECCS could not be accomplished at the necessary scale "without eliminating virtually all natural ecosystems" and consuming such massive amounts of fertilizer and water as to have other severe ecological impacts.[13]

Direct air capture units require much less land than BECCS, and they can be put on relatively small parcels, so the land acquisition would not be as challenging. However, unless the captured CO_2 can be put to useful purposes or disposed of in alternate ways, very large amounts of subsurface land will be required for injecting the CO_2 for permanent storage, raising additional property rights and public acceptance issues. A pipeline infrastructure would be needed to carry the CO_2 from the capture units to the storage reservoirs.

All of this highlights the urgency for intensified research to find alternatives, and for the development of some mechanism to create incentives to employ CDR, such as a carbon tax.

7.3. SOLAR RADIATION MANAGEMENT

It appears increasingly unlikely that temperatures will stay below a 2°C increase, and the Paris Agreement objective of trying to stay as close as possible to 1.5°C seems unreachable in the absence of some technological revolution coupled with a persistent and strong political will that we have not seen. As previously noted, even if all the non-binding Paris Agreement pledges are met, temperatures are likely to rise well above 2°C – perhaps to 3.5°C, according to one study.[14] Before the November 2016 US election, it was going to be challenging for the USA to meet its target; new policies would be needed on top of the two bedrocks of the policies then in place – the Clean Power Plan and the emissions standards for light-duty motor vehicles.[15] That election brought to power a president who repudiated the first and may repudiate the second, and who will certainly not introduce any new climate-friendly programs, thus making it extremely difficult for the USA to continue the downward emissions path needed for a 2°C trajectory. The other major emitting countries so far are not backing away from their Paris Agreement pledges, but the example set by the United States has shown how vulnerable climate action is to political vagaries.

Even if temperatures are kept within 2°C in the twenty-first century, there will be a significant (though not readily quantifiable) chance that climate change will cause or worsen a catastrophe somewhere in the world. For example, it is imaginable that a city could be devastated by a storm, with others thought to be in danger, or that a region would be subjected to such high and prolonged heat that mass deaths are occurring. In such an event, considerable pressure could build to deploy airplanes or other methods in a hasty program of SRM. Other countries might strongly object, and the stage could be set for a major conflict.

Even without such a sudden crisis, the ongoing degradation of the climate will lead to more widespread consideration of SRM. As discussed in Chapter 6, the conduct of and even proposals for small-scale climate engineering experiments have already led to rancorous disagreements. On a larger scale, this book has shown that neither international nor domestic laws offer a governance framework which explicitly addresses the large-scale policy and legal questions posed by SRM regulation. These include (among many others) whether to act, and, if so, the appropriate manner and scale of action, who has power to authorize action, what to do if SRM is undertaken without the required authorization, the need to continue certain SRM measures indefinitely, how liability would be ascertained, who will pay, and how any victims would be identified and compensated.

For all these reasons, it is extremely important to establish an international governance mechanism for the deployment of SRM. We can begin to identify core principles that should govern SRM efforts going forward, either as an interim strategy or as an ultimate governance framework. These basic principles should be endorsed and followed prior to any substantive deployment of SRM:

(a) **Tailor the governance to the technology.** Not all climate engineering technologies pose the same challenges, opportunities, or risks. Large-scale release of aerosols into the upper atmosphere, ocean iron fertilization, and cirrus cloud stripping, for example, all pose different hazards and uncertainties.

(b) **Distinguish governance of research and small-scale field tests from management of larger deployments.** Governance and controls for actual deployment of SRM technologies should not apply wholly to research efforts at a much smaller scale.[16] Several sets of voluntary principles have been formulated to govern research,[17] but research should not await adoption of a comprehensive international agreement on climate engineering. Much more research is needed before sound decisions can be made on deployment, and there is no time to waste. Moreover, a well-designed research program should not have significant border or transboundary impacts.

(c) **Craft governance that can also respond on regional and local scales as well as globally.** SRM and some other climate engineering technologies (such as marine cloud brightening or ocean iron fertilization) may have differing effects on a regional or local scale that may require localized governance or liability approaches.

(d) **Governance should assure that climate engineering remains reversible and additive.** SRM governance should assure that research or deployment is reversible, monitored, and consistent with other climate change responses such as emissions mitigation, adaptation, and resilience requirements. SRM should not substitute or replace these other strategies, especially solely on grounds of costs, unless it offers equivalent or superior benefits and protections, and satisfies the other principles described above.

Governance will likely lag SRM efforts in the absence of a moratorium or full ban. An evolutionary approach keyed to existing domestic and international authorities that focuses on local and regional scales might offer the best initial strategy to protect public and ecological interests. Implementing this evolutionary approach should follow the guidelines set out below:

(a) **Integrate existing international and domestic legal authorities into a consistent and logical governance framework for SRM.** For example, research in the United States could proceed under a framework guidance document from relevant federal agencies that explains how existing federal and state environmental laws set out minimal requirements for development and deployment of climate technologies. One precedent for such an approach is the Coordinated Framework for the Regulation of Biotechnology, issued by the White House Office of Science and Technology Policy in 1986 and updated in 1992 and 2017. Such a framework could serve as a model that can guide independent state or local efforts so that they remain consistent, address the most relevant concerns and risks, and not unduly suppress appropriate research.

(b) **Assure each nation's governance framework approach coordinates in a consistent and coherent fashion with each other.** SRM technology almost inherently will have cross-border effects, and attempts to extend regulatory jurisdiction across national lines will create confusion and loopholes, as well as discourage responsible research. A model framework could minimize these disruptions if it allows similar approaches within comparable nations.

(c) **Encourage the development of parallel technologies needed to assure compliance.** Some SRM techniques might be carried out on a stealth basis, creating challenges for international governance. To the extent that the detection and attribution of SRM efforts may require new modeling and sensing technologies, governance frameworks and requirements should promote or require the development of these technologies in parallel with SRM itself.

(d) **Remain open to bottom-up, emergent governance of SRM.** Given the past difficulties in negotiating global multinational agreements on climate change responses, it will likely take years (if not decades) to reach agreement on a global governance framework for SRM and other climate engineering technologies. If nations with the capacity to take early action or conduct substantive research can develop regional or bilateral agreements within a larger informal framework, this approach may allow faster action and more flexibility and capacity for growth and adaptation. This strategy would also provide a more logical framework for SRM efforts designed to operate on a regional or local scale (e.g., Arctic ice restoration, local heat wave mitigation).

7.4. POTENTIAL VEHICLES FOR MANAGEMENT AND GOVERNANCE

If a climate engineering project or test affects only a small area wholly within the boundaries of one nation, that nation's domestic laws and institutions should govern it. Most CDR would fall within this category. If it affects a region with more than one nation, a regional agreement and governance body would be the appropriate level to oversee the project. If a project or demonstration may have impacts across regions or globally (e.g., SRM and perhaps ocean iron fertilization), then an international decision-making process for the approval of climate engineering deployment would be appropriate.

The United Nations Framework Convention on Climate Change (UNFCCC) could serve as a useful existing platform to conduct discussions and hopefully reach international agreement. Whether the language of the UNFCCC is broad enough to encompass the governance of climate engineering has been subject to some dispute,[18] but in any event a supplemental agreement or protocol would probably need to be adopted in order to establish an effective governance regime. The UNFCCC is an established body with participation by essentially every country in the world. The flip side of its comprehensive membership is that its operations can be unwieldy. The Paris Climate Agreement – the first international agreement in which virtually every country undertook to act against climate change – was adopted in 2015, 23 years after the 1992 adoption of the UNFCCC convention, which foreshadowed the need for such a comprehensive agreement. It can be read to implicitly call for carbon dioxide removal, but it did not address solar radiation management at all.[19] A global agreement on climate engineering might be somewhat less complex and would build on the body of knowledge and practice created under the UNFCCC, but it would still take years and multiple conferences to achieve.

Thus smaller and perhaps unofficial groups should begin the process of drafting an agreement, so that that a fully formed draft (perhaps with options) can be presented to the UNFCCC or other international or multilateral body. One effort potentially moving in that direction is the Carnegie Climate Geoengineering Governance Initiative, launched in 2016 by the Carnegie Council for Ethics in International Affairs. It is directed by Janos Pasztor, formerly senior climate advisor to UN Secretary General Ban Ki-Moon. As discussed in Chapter 6, there have been several unofficial conferences and committees that have already produced voluntary guidelines for the conduct of research in (as opposed to deployment of) climate engineering technologies.

7.5. ESSENTIAL ELEMENTS OF A GOVERNANCE REGIME

However it is negotiated, an international agreement on climate engineering should have the following elements:

Definition of large-scale projects – A careful definition will be needed for which projects require international authorization. Such authorization should be required for projects that could have impacts across national borders (unless sufficiently covered by bilateral or multilateral agreements covering the full region affected) or globally. On the other hand, no international authorization should be required for CDR (except ocean iron fertilization or other techniques that could have large-scale impacts) or for experimentation that cannot have large-scale impacts. (The regulation of climate engineering research, as opposed to deployment, is discussed in Chapter 6.)[20] One hypothesized variety of climate engineering that is on the borderline would be efforts aimed at limited geographic areas – e.g., attempting to lower temperatures in a limited area, such as a city that is experiencing a dangerous heat wave, or in areas of the Arctic that are rapidly melting. Criteria would need to be established for assessing such efforts.[21]

Authorization required for large-scale projects – No covered project may be undertaken without this authorization. Any state seeking authorization would need to consent to the jurisdiction of the international tribunal designated to handle resulting disputes. A state receiving authorization should be able to seek reimbursement of the costs of the project from the Climate Engineering Fund described below.

Authorization process – Requiring approval from the full UNFCCC for a particular project would be too time-consuming or cumbersome. Instead, a subsidiary body or the like should be established with the authority to grant or deny approvals and impose conditions. The process used should include environmental impact assessment and an opportunity for public review and input. Expedited procedures should be devised that could be employed in the event of an imminent threat of a climate catastrophe.

Authorization principles – Principles for granting or withholding approval should be established. Anthony E. Chavez has set forth one useful set of proposed principles, including cost–benefit analysis; consideration of alternatives; intergenerational equity; regional equity; reversibility of consequences; and containment of effects.[22] As Professor Wil Burns has argued, human rights principles (which overlap with some of Professor Chavez's principles) should also be considered, especially since certain climate engineering actions could benefit certain parts of the world but risk harm to others.[23]

Eligible applicants – Only states may obtain authorization to conduct large-scale climate engineering projects. States may contract with private entities to carry out the projects, but the states must retain supervision, control, and liability.

Monitoring, reporting, and verification – The deployment of any SRM project would involve a myriad of technical details.[24] A system will need to be established to monitor the conduct and impacts of any SRM projects; report the results of the monitoring; verify the results; and take action if needed. This approach will require the designation of an entity to oversee SRM projects and ensure that they meet the conditions under which they were approved. This entity should have the power to act quickly in the event of violations or unanticipated adverse impacts, including to issue and enforce orders to stop or modify a project.

Transparency – Data from any CDR and SRM projects should be made publicly available so that the world scientific community can have the benefit of the information developed. While in a few instances private parties may want to protect information they develop, confidential business information should be narrowly construed in this context.

Public deliberation – SRM and large-scale CDR are too important to be decided upon in secret. Any governance regime should include provisions for broad dissemination of information about the deliberations, and for the press, nongovernmental organizations, and business associations to observe the official meetings. The ensuing public debates will inform the governmental negotiators.

Impact assessment – Sound governmental decision-making and informed public deliberation about climate engineering depend on the development and publication of the relevant scientific information. Environmental impact assessment has emerged as a major method for the gathering, analysis, and public release of information relevant to governmental decisions. As discussed in Chapters 3 and 4, the obligation for performing an environmental impact assessment is now solidly enshrined in international and US domestic law, and it should be included in any climate engineering governance scheme. Such assessments should include independent review by outside parties.

Liability – Projects whose activities and impacts are limited to one state should be subject to the liability rules and institutions of that state. Projects with transboundary impacts should be subject to an international liability scheme. That scheme will need to establish the necessary standards of conduct, and the burdens of proof of causation and injury.

Compensation – Any effective liability regime should include a compensation mechanism to provide injured parties with payment for injuries without the need for additional admissions of liability by the responsible state. While it may not be possible to define in advance the nature of the injuries, criteria should be established concerning the general sorts of injuries that should fall within the ambit of a compensation scheme, and the types of parties that should be eligible for compensation.

Climate Engineering Fund – A climate engineering project that goes seriously awry could cause massive damage. Paying for those damages could be beyond the capacity of all but the largest states; and states (even the largest), if found liable, may refuse to pay, in part as a result of their domestic political situations. Thus a fund should be established in advance to assure payment of damages. This fund could also cover the costs of the climate engineering itself, if the state performing the work requests this funding. (A state may not want international funding if that might impair its sole control over the activity.) Climate engineering, if needed, will have been necessitated by the emissions of the world's GHG emitters, and not just those of the state undertaking the project. For that reason, the financial burdens should be shared equitably by the states that have contributed the most to the load of GHGs in the atmosphere. An extra portion should be borne by the state that has undertaken the work, to help give it added incentive to be careful. But it would be inequitable to require that state alone to shoulder the entire burden of a climate engineering program that has been authorized by, and conducted in accordance with the requirements imposed by, an international body. Moreover, such a sole liability burden could deter a state from undertaking a climate engineering program that the international community has determined to be globally necessary.

An ideal mechanism to pay for the Climate Engineering Fund might be a carbon tax, which could generate considerable revenues that could be used not only for climate engineering projects and compensation, but also for adaptation and mitigation efforts.[25] Such a tax would also send price signals that could discourage the generation of GHGs and encourage the development of alternative technologies, especially in the energy sector. However, the negotiation of such an international tax would be extraordinarily difficult and may not be realistic. Special difficulties would arise with the rapidly developing states, such as China and India, whose share of current emissions considerably exceeds their share of historic emissions, and whose economies (especially in the case of India) have a long way to catch up with the standards of the developed countries. In the absence of something resembling a carbon tax, it may be necessary to seek voluntary contributions from states. That is the method that has been used in attempting to pay for the Green Climate Fund, but very limited success has been achieved, and contributions have been subject to the vicissitudes of domestic politics.

Penalties – Substantial penalties should be imposed on any state that conducts climate engineering (or that tolerates such conduct by a private party within its borders or under its control) without authorization or in excess of any authorization it has received for such projects.

Dispute resolution – A tribunal will need to be designated to resolve disputes that arise in these processes. The most obvious kind of dispute will involve a liability claim by a party that claims it has suffered injury as a result of a climate engineering project. This tribunal could also resolve disputes concerning the necessity for or scope of authorization for particular projects.

7.6. CONCLUSION

Economic and emissions trends, and the persistent failure of the international community to develop and implement an effective way to control greenhouse gas emissions, mean that a massive global program of carbon dioxide removal is almost certainly necessary if we are to avoid the worst impacts of climate change. Fortunately the adverse impacts of most kinds of CDR are almost entirely local, and thus an international governance regime is not needed to regulate them. However, international governance – perhaps in the form of economic incentives such as a carbon tax – is necessary in order to induce the nations of the world to undertake CDR at the needed scale.

On the other hand, solar radiation management could have massive extraterritorial impacts. The odds of catastrophic climate impacts have become high enough that there is a significant chance that someone will attempt to deploy SRM, even before we have any clear idea of its positive and negative impacts. Therefore, a research program is necessary to reduce the uncertainties, and there is no time to waste in developing an international governance regime to decide whether and how to deploy SRM and to deal with its consequences.

NOTES

1 *See also* Glen P. Peters, The 'Best Available Science' to Inform 1.5°C Policy Choices, 6 *Nature Climate Change* 646 (2016).
2 Wally Broecker, Does Air Capture Constitute a Viable Backstop against a Bad CO2 Trip? 1 *Elementa: Science of the Anthropocene* 1 (2013).
3 *See* David W. Keith, Toward a Responsible Solar Geoengineering Research Program, 33 *Issues in Science & Technology* 71–77 (Spring 2017).
4 *See* Todd Sanford et al., The Climate Policy Narrative for a Dangerously Warming World, 4 *Nature Climate Change* 164 (March 2014).
5 *See* Kevin Anderson & Glen Peters, The Trouble with Negative Emissions, 354 *Science* 182 (2016).
6 CO_2 is used in various industrial and commercial processes; it is captured from industrial operations that produce large quantities as an unwanted byproduct, but not at close to the scale needed to seriously address the climate problem.

7 Robert Amos, Bioenergy Carbon Capture and Storage in Global Climate Policy: Examining the Issues, 4 CLIMATE CHANGE LAW REVIEW 187 (2016).
8 Phil Williamson, Scrutinize CO2 Removal Methods, 530 *Nature* 153, 154 (2016).
9 *Id.*
10 Sean Low, The Futures of Climate Engineering, 5 *Earth's Future* (Jan. 17, 2017), doi: 10.1002/2016EF000442.
11 Pete Smith et al., Biophysical and Economic Limits to Negative CO2 Emissions, 6 *Nature Climate Change* 142 (2016), doi: 10.1038/NCLIMATE2870.
12 Matteo Muratori et al., Global Economic Consequences of Deploying Bioenergy with Carbon Capture and Storage (BECCS), 11 *Environmental Research Letters* 095004 (Aug. 31, 2016), doi: 10.1088/1748–9326/11/9/095004.
13 Lena R. Boysen et al., The Limits to Global-Warming Mitigation by Terrestrial Carbon Removal, 5 *GeoHealth* 463 (May 17, 2017), doi: 10.1002/2016EF000469.f; Lydia J. Smith & Margaret S. Torn, Ecological Limits to Terrestrial Biological Carbon Dioxide Removal, 118 *Climatic Change* 89 (2013), doi: 10.1007/s10584-012-0682-3. *See also* Sivan Kartha & Kate Dooley, *The Risks of Relying on Tomorrow's "Negative Emissions" to Guide Today's Mitigation Action*, Stockholm Environment Institute, Working Paper 2016-08 (2016); Raphael Slade et al., Global Bioenergy Resources, 4 *Nature Climate Change* 99 (2014), doi: 10.1038/nclimate2097; Holly Jean Buck, Rapid Scale-up of Negative Emissions Technologies: Social Barriers and Social Implications, 139 *Climatic Change* 155 (2016), doi: 10.1007/s10584-016-1770-6.
14 Andrew Jones et al., Deeper, Earlier Emissions Cuts Needed to Reach Paris Goals, *Climate Interactive* (May 18, 2016); Joeri Rogelj et al., Paris Agreement Climate Proposals Need a Boost to Keep Warming Well Below 2°C, 534 *Nature* 631, 634 (2016).
15 John Larsen et al., Taking Stock: Progress Toward Meeting US Climate Goals (Rhodium Group, January 28, 2016).
16 Chapter 6 discusses ways in which the line might be drawn between research and deployment.
17 These are described in National Research Council, *Climate Intervention: Reflecting Sunlight to Cool Earth* (National Academies Press 2015), pp. 149–75.
18 *See* Wil Burns & Simon Nicholson, Governing Climate Engineering, in *New Earth Politics: Essays from the Anthropocene* (Simon Nicholson & Sikina Jinnah, eds., MIT Press 2016) pp. 343–66 (arguing that the UNFCCC can be read to encompass climate engineering); John Virgoe, International Governance of a Possible Geoengineering Intervention to Combat Climate Change, 95 *Climatic Change* 103 (July 2009) (arguing the opposite).
19 *See* Joshua B. Horton, David W. Keith, & Matthias Honegger, Implications of the Paris Agreement for Carbon Dioxide Removal and Solar Geoengineering, Harvard Project on Climate Agreements (July 2016).
20 *See also* Anna-Maria Hubert & David Reichwein, *An Exploration of a Code of Conduct for Responsible Scientific Research Involving Geoengineering* (Institute for Advanced Sustainability Studies, May 2015).
21 *See* Tracy Hester, *A Matter of Scale: Regional Climate Engineering and the Shortfalls of Multinational Governance*, 3 *Carbon and Climate Law Review* 168 (2013).
22 Anthony E. Chavez, *Using Legal Principles to Guide Geoengineering Deployment*, 24 *NYU Environmental Law Journal* 59 (2016).

23 *See* William C. G. Burns, Human Rights Dimensions of Bioenergy With Carbon Capture and Storage: A Framework for Climate Justice in the Realm of Climate Geoengineering, in *Climate Justice: Case Studies in Global and Regional Governance Challenges* (Randall Abate, ed., Environmental Law Institute 2016), pp. 149–70; William C. G. Burns *The Paris Agreement and Climate Geoengineering Governance: The Need for a Human Rights-Based Component* (Center for International Governance Innovation 2016).

24 David W. Keith & Douglas G. MacMartin, A Temporary, Moderate and Responsive Scenario for Solar Geoengineering, 5 *Nature Climate Change* 201 (2015), doi: 10.1038/nclimate2493.

25 Taxes would be imposed at the domestic level, even if organized within a multinational framework. Imposition and collection of taxes by an international organization would cause multiple problems.

Index